WIRELESS NETWORK
COEXISTENCE

Wireless Network Coexistence

Robert Morrow

McGraw-Hill

New York   Chicago   San Francisco   Lisbon
London   Madrid   Mexico City   Milan   New Delhi
San Juan   Seoul   Singapore   Sydney   Toronto

The McGraw·Hill Companies

Library of Congress Cataloging-in-Publication Data

Morrow, Robert (Robert K.)
  Wireless network coexistence / Robert Morrow.
     p.   cm.
  Includes index.
  ISBN 0-07-139915-1
    1. Wireless LANs. 2. Internetworking (Telecommunication) I. Title.

TK5105.78.M67   2004
004.6'8—dc22

2004051962

1 2 3 4 5 6 7 8 9 0 DOC/DOC 0 9 8 7 6 5 4

ISBN 0-07-139915-1

The sponsoring editor for this book was Judy Bass and the production supervisor was Sherri Souffrance. It was set in Vendome by MacAllister Publishing Services. The art director for the cover was Handel Low.

Printed and bound by RR Donnelley.

 This book was printed on recycled, acid-free paper containing a minimum of 50% recycled, de-inked fiber.

McGraw-Hill books are available at special quantity discounts to use as premiums and sales promotions, or for use in corporate training programs. For more information, please write to the Director of Special Sales, Professional Publishing, McGraw-Hill, Two Penn Plaza, New York, NY 10121-2298. Or contact your local bookstore.

To my parents

without whom none of this would have been possible

CONTENTS

Contents

Contents

PREFACE

As wireless devices proliferate, they will be required to operate in an environment filled with interference from other coexisting devices. For example, both 802.11* *Wireless Fidelity* (Wi-Fi) and 802.15.1 Bluetooth have revolutionized the way computers and other digital devices communicate. Wi-Fi is the most successful implementation of the *wireless local area network* (WLAN), and Bluetooth is a *wireless personal area network* (WPAN) used primarily for replacing cables with a virtual link. Each of these functions is extremely useful, and as a result they can often exist side by side in many applications. Unfortunately, the most common implementation of Wi-Fi uses the same 2.4 GHz frequency band as Bluetooth, so interference between the two is inevitable in many situations. In fact, some Wi-Fi systems engineers have even placed signs prohibiting Bluetooth operation in the same vicinity as their network.

Two other wireless networks are entering the fray with their own specialized target markets. For high-speed multimedia applications, 802.15.3 has been developed. Using the trade name WiMedia, it can operate in the 2.4 GHz band or via *ultra-wideband* (UWB). In either case, coexisting with other users is an important consideration. The other network emerging from the development stage is 802.15.4, commonly known as ZigBee. This is a low-speed network used for instrumentation and control. It can operate in the 2.4 GHz band as well, complicating coexistence even further.

Some have used the terms "interoperability" and "coexistence" interchangeably, but they are most definitely not the same thing. *Interoperability* can be defined in two slightly different ways. First, it refers to the capability of devices from different manufacturers to communicate on a particular network. For example, if Bluetooth modules from manufacturer *A* and manufacturer *B* can connect and communicate at all levels for which they are certified, then they are interoperable. Interoperability can also refer to the ease with which a single client or server can support different WLAN/WPAN specifications. If a single module supports both Bluetooth and Wi-Fi operations, then the two protocols might share various components such as radio and clock circuitry. If this is done seamlessly, then these implementations are interoperable.

So what do we mean by *coexistence*? This term refers to the capability of a wireless network to operate properly in an environment in which noise and interference are present. For reliable operation, a node in a wireless network should be able to

- Coexist with thermal noise
- Coexist with incoming multipath components

---

*All of the 802 designations in this book are trademarked by the *Institute of Electrical and Electronics Engineers* (IEEE).

- Coexist with other nodes within its own network
- Coexist with independent wireless networks using the same protocols
- Coexist with independent wireless networks using different protocols
- Coexist with other services

All the previous criteria will be addressed in this book, in which we focus primarily on Wi-Fi, Bluetooth, WiMedia, and ZigBee wireless networks. For example, a Wi-Fi node should be able to operate over a communication channel that contains thermal noise and multipath components. It should be able to properly coordinate its transmit and receive times with other nodes within its own network so efficient communication can take place. It should be able to operate in the presence of other nearby independent Wi-Fi networks and Bluetooth piconets. Finally, it should be able to accept interference from a microwave oven but avoid producing harmful interference to a nearby *global positioning system* (GPS) receiver.

Determining each system's performance in an interference environment is a difficult task indeed, requiring daunting mathematics and/or complex laboratory procedures. Research in this area has been intense and productive. Much of the literature focuses on modeling a particular aspect of coexistence with great precision, but other aspects are often either approximated or ignored. Consequently, it could be argued that a simpler model will still yield sufficient accuracy. Considerable effort has been dedicated in this book toward using simplified mathematical models whenever possible, while maintaining sight of the goal of obtaining accurate practical results.

One of the challenges when writing about different WLAN and WPAN systems is trying to integrate terms that can be different in each WLAN/WPAN rendition. For example, the terms *packet* and *frame* both refer to a collection of data bits, but they can be used in different ways. The 802.11 standard uses the term *frame* except for 802.11g, which sometimes uses the term *packet*. The 802.15.1 standard always uses *packet*, and 802.15.3 always uses *frame*. The 802.15.4 standard uses *packet* when referring to a collection of bits to be transmitted, but uses *frame* for a collection of bits at a higher protocol layer. We try to adhere to these terms when referring to a particular protocol. However, in general we refer to a collection of data bits as a *packet*.

The journey begins in Chapter 1, "Introduction," where we show how the various unlicensed bands used by WLAN and WPAN systems are regulated, how a typical network is structured, and operational highlights of Wi-Fi, Bluetooth, WiMedia, and ZigBee. Next, propagation in the 900 MHz, 2.4 GHz, and 5.x GHz bands are studied in Chapter 2, "Indoor RF Propagation and Diversity Techniques," along with the effect that multipath and antenna directivity have on range and interference. Chapter 3, "Basic Modulation and Coding," covers basic methods for placing data on a carrier, and the various advanced techniques for data transmission are examined in Chapter 4, "Advanced Modulation and Coding." The various transmitter and receiver specifications are detailed in Chapter 5, "Radio Performance."

Channel access methods are studied in Chapter 6, "Medium Access Control," beginning with a general analysis and progressing to the MAC used by WLAN and WPAN. The focus here is primarily on operation within a single network. Chapter 7, "Passive Coexistence," broadens the outlook to examining the ability of a network to coexist with other similar and dissimilar networks while passively accepting their interference. In Chapter 8, "Active Coexistence," we look at how performance can be improved when a network takes active measures to improve its capability to coexist. Finally, Chapter 9, "Coexistence with Other Wireless Services," examines how WLAN and WPAN devices can coexist with other wireless systems both within and outside the 2.4 GHz band.

This book is directed at WLAN and WPAN design and deployment engineers so they can better understand the factors associated with the reliable operation of these networks, especially when multiple networks are physically close to each other and have the potential to adversely affect performance from their interference. Comments or questions are welcome.

Robert Morrow, Ph.D.
Centerville, Indiana, USA
rkmorrow@ieee.org

ACKNOWLEDGMENTS

Much of the information in this book originated in my short courses on wireless networking, which were sponsored by the fine people at Besser Associates. Thanks are due to Les Besser, Jeff Lange, and others for providing the opportunity to present this material to clients throughout the world.

Thanks also to Professor Jim Lehnert at Purdue University who set me on the path to a Ph.D. and taught me how to think efficiently. A fellow amateur radio enthusiast, Ken Wyatt of Agilent, provided insight into how systems are tested for conformance to rules set by the various regulatory domains. June Namgoong created a routine to plot the output of switched diversity antennas to combat Rayleigh fading.

Thanks so much to those who kindly provided material for some of the figures in the book: Eric Meihofer, Matthew Shoemake, Shawn Rogers, Tom Stansell, and Per Enge. Richard Langley of the University of New Brunswick assisted with information on *ultra-wideband* (UWB) and *global positioning system* (GPS) coexistence. Also, thanks go to the IEEE for their generous sharing of published research written by so many dedicated people intent on pushing the state of the art to new heights. Excerpts from IEEE Standards are reprinted with permission, and are Copyright © 2003 by IEEE. The IEEE disclaims any responsibility or liability resulting from the placement and use in the described manner.

Finally, my heartfelt appreciation belongs to my editor Judy Bass at McGraw-Hill for her patience and confidence in this project.

BIOGRAPHY

 Robert Morrow received a Bachelor of Science degree from the United States Air Force Academy, a Master of Science degree from Stanford University, and a Ph.D. degree from Purdue University, all in electrical engineering. Bob taught undergraduate engineering and was the Director of Research at the U.S. Air Force Academy, where he also piloted the aircraft used by the Wings of Blue parachute team. A few years earlier he jumped out of a perfectly good airplane several times so the parachute team wouldn't accuse him of being chicken. As Deputy Head of the Electrical and Computer Engineering Department at the Air Force Institute of Technology, he flew a desk and taught graduate-level engineering courses. At various times during his Air Force career, Bob instructed in turbojet, turboprop, and reciprocating military aircraft. He is a senior member of the *Institute of Electrical and Electronics Engineers* (IEEE), has published over two dozen technical articles, holds a U.S. patent, and is an Advanced Class Amateur Radio operator. He authored the book *Bluetooth Operation and Use*, published by McGraw-Hill in 2002. Bob presently serves as a consultant and short course instructor in short-range wireless networks.

CHAPTER 1

Introduction

Everyone who has used wireless devices has probably experienced interference in one form or another. Cordless phones interfere with one another and with the baby monitor. Cell phones are disrupted and calls are dropped for a variety of reasons. A *Wireless Fidelity* (Wi-Fi) node's throughput might slow to a crawl when Bluetooth users are nearby. Such interference is the price we pay for the incredible convenience of tetherless communications. Interference can be one-way, as when the microwave oven creates annoying audio distortion in the 2.4 GHz cordless phone, but the phone has absolutely no effect on the oven. Of far greater significance is the mutual interference among independent wireless devices that is becoming more and more common as these devices proliferate into every aspect of our lives. This type of interference is more serious because it impacts the operation and reliability of all affected communication links.

Perhaps no coexistence issue has drawn more attention than that between Wi-Fi (IEEE 802.11b) and Bluetooth (IEEE 802.15.1). This is due to two major factors. First, Wi-Fi is a *wireless local area network* (WLAN) to enable a user to conveniently access client and server devices, and Bluetooth is a *wireless personal area network* (WPAN) used primarily as a cable replacement technology between a client and a peripheral. As such, both are likely to be found in the same general area and even be installed in the same host computer. Second, both Wi-Fi and Bluetooth operate within the same frequency band. Although modulation and coding techniques differ significantly between them, overlap between *radio frequency* (RF) carriers may cause so much signal distortion that both systems perform poorly when they are located close to each other.

Although Wi-Fi and Bluetooth are mature WLAN and WPAN specifications, several other wireless networks have been introduced and will soon be competing for scarce electromagnetic spectrum. The IEEE 802.15.3 *high-rate* (HR) WPAN has been developed to provide raw data rates up to 55 Mb/s, and the IEEE 802.15.4 *low-rate* (LR) WPAN is intended to be an ultra-low-cost network providing data rates up to 250 kb/s. Furthermore, plenty of other wireless systems are available, many of which also reside in the 2.4 GHz *industrial, scientific, and medical* (ISM) band, which could cause significant interference to WLAN and WPAN operations.

This chapter begins with a look at reasons for choosing wireless over wired, followed by a description of some of the government regulations for operating in the unlicensed bands used by WLAN and WPAN devices. Next, the model used to build a network hierarchy is introduced and we show which parts of this model apply to coexistence. The basic operation of each WLAN and WPAN is summarized. We conclude with a description of how coexistence issues can be quantified, along with an outline of efforts by the *Institute of Electrical and Electronics Engineers* (IEEE) to resolve coexistence problems.

Why Wireless?

Both WLAN and WPAN systems aim at replacing a wired network with a wireless equivalent. For example, wired Ethernet, also known as IEEE 802.3, can connect computers, peripherals, and gateways with cable so that cross-platform communications can occur. Different cables, connectors, protocols, and data rates have been used as Ethernet has matured, and many desktop and laptop computers come equipped with built-in Ethernet capabilities. IEEE 802.11 and its variants are an attempt to replace the Ethernet cable with a wireless link instead. Similarly, Bluetooth, also called IEEE 802.15.1, is primarily aimed at replacing the cable between a host and a peripheral to gain versatility while costing little more than the cable itself. Other WPAN protocols have also been introduced, each of which has its own intended market emphasis.

Characteristics of Wired and Wireless Networks

Several obvious and not-so-obvious characteristics differentiate the wired and wireless network. Wired network endpoints must be relatively static, because it's difficult to move devices around without disconnecting and reconnecting the various network cables. Eliminating this constraint is perhaps the main motivator for going wireless. Not only can an endpoint be easily moved to a new location, but it could perhaps also be operated while in motion (roaming).

The hardware and lower-level software are usually simpler in a wired system, so the engineering and bill of materials are often less costly as well. Of course, lots of cable (surprise!) must be purchased, deployed, and connected. This cable can be expensive and time consuming to install, maintain, and change. Placing cable into the walls of a building is not only a large project, but it also requires committing to a technology that will probably become obsolete long before the life of the building has ended. Choosing a wireless network eliminates most of these wiring headaches.

On the other hand, cable provides a dedicated channel that will have few interferers. The channel tends to be quiet for two reasons. First, noise is usually lower within a shielded or twisted pair cable than over a wireless link, and second, signal attenuation is much lower over the cable. Both of these factors result in significantly higher signal-to-noise ratios than can be expected over a wireless link. High signal attenuation between the wireless transmitter and receiver means that bit errors are much more common. File transfers over the network must usually take place error-free, so significant overhead is often added to the information for error control, reducing efficiency even further.

Wireless network reliability can be improved by selecting a data transmission speed such that sufficient bit energy is available to partially offset the high channel attenuation. As a rule of thumb, a wired network can send data about 10 times faster than a wireless network of the same vintage. For example, the wired *universal serial bus* (USB) 1.x operates at speeds of about 12 Mb/s, whereas Bluetooth sends data at a raw rate of 1 Mb/s, about 10 times slower. 100 Base-T Ethernet operates at 100 Mb/s, whereas IEEE 802.11b exchanges data at up to 11 Mb/s. IEEE 1394 (FireWire) and USB 2.0 operate at speeds as high as 480 Mb/s, while their contemporary wireless counterparts, IEEE 802.11a and 802.11g, communicate at up to 54 Mb/s, again slower by about a factor of 10. (It's interesting to note that the data rate of a typical fiber-optic channel exceeds its contemporary wired channel also by a factor of 10.) Both modulation and error control are discussed further in Chapter 3, "Basic Modulation and Coding" and Chapter 4, "Advanced Modulation and Coding."

The higher signal attenuation over the wireless link does have an advantage. If two networks operating on the same carrier frequency are sufficiently separated from each other in distance, their mutual interference will be negligible. This is called *frequency reuse,* and it forms the basis by which the cellular network can process several million simultaneous phone calls over a few hundred carrier frequencies. Likewise, low-power wireless networks such as those examined in this book can operate at full speed as long as other networks sharing the same channel are far enough away. This issue is discussed in Chapter 5, "Radio Performance."

Due to low cable attenuation for runs of a few tens of meters or less, all signals on the cable have essentially the same amplitude regardless of where the signal originated, so wired network nodes are physically able to transmit and receive at the same time. This means that a node can listen to its own transmission to make sure it isn't being corrupted by another node transmitting simultaneously. If such a collision occurs, all affected nodes can stop sending immediately and enter a random backoff routine to prevent another jamming situation. Conversely, high signal attenuation between the transmitter and receiver in a wireless link usually means that activating a wireless node's transmitter disables the receiver. Thus, the occurrence of a collision cannot be determined in real time. This can cause significant delays and a loss of communication efficiency. We quantify this loss of efficiency in Chapter 6, "Medium Access Control."

From a security point of view, wireless signals are easier to intercept, disrupt, and jam, either intentionally or unintentionally, because the signals aren't confined to a cable. Examining intentional disruption and taking appropriate countermeasures is the task of the security engineer. Modeling the effect of unintentional interference and working out appropriate countermeasures are what coexistence is all about.

Categories of User Information

User information carried over a wired or wireless network can be roughly divided into three categories: *asynchronous*, *isochronous*, and *synchronous*. Asynchronous data mostly consists of user-initiated file transfers or network control operations. Isochronous data is unidirectional audio or video, sometimes called *streaming data*, and synchronous data primarily consists of real-time, two-way voice. Each of these has its own transfer requirements, as depicted in Table 1-1.

These characteristics dictate how each type of data should be transferred, especially over the error-prone wireless channel. For example, asynchronous data must be highly accurate, because a single bit error can be disastrous in a file. Therefore, high latency (delay) is acceptable if a frame needs to be sent several times before it is received error free. On the other hand, real-time, two-way voice requires low latency to ensure a minimal delay between the time words are spoken and the time the words are heard at the other end of the link. As a result, little if any time is available to resend a bad frame, so bit errors must be tolerated.

It's obvious, then, that interference will affect these different types of data in different ways. If the channel is error-prone, multiple transmissions of asynchronous data frames will be required, which in turn reduces network throughput for all users. Fortunately, the bursty channel loading of such transfers means that other users may not notice the network slowdown if they're mostly idle during this time. Synchronous frames that contain bit errors will exhibit varying degrees of audio distortion, and their continuous channel-loading characteristic could limit the coexistence capability between several similar streams. Isochronous data resides in the middle ground, where interference may increase network loading from frame retransmissions, but audio or video distortion may also occur from frames that cannot be received error-free before they must be discarded to make way for the following frames.

Table 1-1

Transfer requirements for asynchronous, isochronous, and synchronous data

Data type	Accuracy	Latency	Bit rate	Channel balance	Channel loading
Asynchronous	High (required)	High (acceptable)	Medium/high	Asymmetric	Bursty
Isochronous	Medium	Medium	Medium/high	Asymmetric	Continuous
Synchronous	Low (acceptable)	Low (required)	Low	Symmetric	Continuous

To dB or Not to dB

Transmit (TX) power levels in most indoor communication systems are typically a few hundred milliwatts or less. By the time this signal reaches its destination receiver, power has often dropped by four to six orders of magnitude. Keeping track of numbers with such a wide range is difficult enough, not to mention the errors that can occur when performing estimations involving multiplication and division.

In light of these issues, communication engineers (and lots of other engineers, for that matter) like to think in terms of *decibels* (dB) instead of linear numbers for three reasons:

- RF propagation exhibits logarithmic characteristics in several respects.
- Large signal strength variations can be expressed within a relatively small range of numbers.
- Mathematical multiplication becomes addition, and division becomes subtraction, so mental calculations are easier to do.

It's important to remember that dB is a dimensionless quantity, so it is always calculated as a ratio between two values having the same units. In terms of power, dB is calculated using the formula $P_{(dB)} = 10\log(P_2/P_1)$, where P_2 and P_1 are the two powers being compared.

Power in WLAN and WPAN systems is most conveniently represented in terms of dBm, which is a power in dB relative to 1 *milliwatt* (mW), that is,

$$P_{(dBm)} = 10 \log\left(\frac{P}{1 \text{ mW}}\right) = 10 \log P \tag{1.1}$$

when P is expressed in mW. Now the numbers are much easier to work with. For example, a 100 mW transmitter produces a power of +20 dBm, and if the resulting signal at a receiver is 0.01 *microwatts* (μW), then it can be written as −50 dBm. The ratio of these two powers is 70 dB, which represents the signal loss from transmitter to receiver. This ratio is found by simply taking the difference between the TX and *receive* (RX) powers that are given in dBm. That operation is much easier to accomplish than trying to mentally divide 100 mW by 0.01 μW without losing track of the decimal place.

A couple of things are worth pointing out in the previous example. First, it's sometimes helpful when expressing powers in dBm to put a plus sign (+) in front of positive values to emphasize that the power listed is higher than 1 mW; similarly, the minus sign (−) designates a power below 1 mW. Second, when expressing the ratio of two powers that are given in dBm, the result is always expressed in dB, not dBm. That's because the ratio is the same regardless of how the two powers are expressed, whether in W, mW, μW, or other appropriate units.

Table 1-2

Power levels and their dBm equivalents

Power level	Equivalent dBm
1 W	+30 dBm
100 mW	+20 dBm
10 mW	+10 dBm
5 mW	+7 dBm
3 mW	+5 dBm
2 mW	+3 dBm
1 mW	0 dBm
1 μW	−30 dBm
1 *nanowatt* (nW)	−60 dBm
1 *picowatt* (pW)	−90 dBm

Table 1-3

Common dB representations in communications engineering

dB representation	Meaning
dBm	Relative to 1 mW
dBr	Relative to the highest power in a spectral plot
dBc	Relative to the RF carrier power
dBi	Relative to power radiated from an isotropic source
dBd	Relative to power radiated from a half-wave dipole antenna

Some commonly used power levels and their corresponding dBm values for short-range wireless transmitters and receivers are given in Table 1-2. The logarithmic nature of dB units is clearly shown here. Multiplying or dividing a power by a factor of 10 is equivalent to adding or subtracting 10 dB, respectively. Multiplying or dividing by a factor of 3 is equivalent to adding or subtracting about 5 dB, and multiplying or dividing by 2 is equivalent to adding or subtracting 3 dB. By remembering these simple rules, you should be able to convert any power into its dBm equivalent with an accuracy of about 1 or 2 dB.

Aside from powers expressed relative to 1 mW, comparisons can be made to other power levels and their dB representations given the proper suffix. Table 1-3 shows some of these as they apply to different aspects of communication system engineering.

Regulation of Unlicensed Bands

The United States allows intentional radiators (transmitters) the use of several ISM and *Unlicensed National Information Infrastructure* (U-NII) bands for a variety of purposes without requiring a license to transmit. The three most popular ISM bands are depicted in Figure 1-1. We will refer to these as the 900 MHz, 2.4 GHz, and 5.7 GHz bands.* The 900 MHz band is where some IEEE 802.15.4 networks reside. IEEE 802.11b, 802.11g, 802.15.1, 802.15.3, and 802.15.4 operate within the 2.4 GHz ISM band, and IEEE 802.11a operates in the 5.7 GHz band.

The U-NII bands (see Figure 1-2) in the United States were created to provide greater flexibility in the way data is modulated onto the carrier. To add to the confusion, one of these bands overlaps most of the 5.7 GHz ISM band. IEEE 802.11a (and HiperLAN/2 in Europe) use carrier frequencies in the 5 GHz region.

Contrary to popular belief, "unlicensed" doesn't mean "unregulated," and indeed most countries strictly regulate the use of unlicensed frequencies. Frequency administration falls under the *Federal Communications Commission* (FCC) in the United States. Other governments often have rules that are quite different from, and sometimes incompatible with, FCC regulations. Naturally, this situation has the potential to cause a regulatory nightmare for the various

Figure 1-1
The 900 MHz, 2.4 GHz, and 5.7 GHz ISM bands in the United States

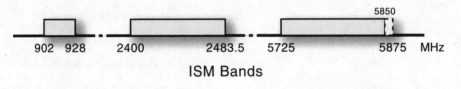

ISM Bands

Figure 1-2
The U-NII bands

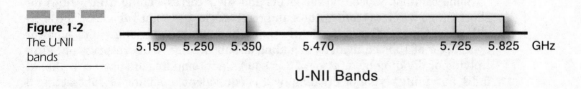

U-NII Bands

*The 900 MHz ISM band is sometimes called the 902 MHz band or the 915 MHz band, the 2.4 GHz ISM band may be listed as the 2.45 GHz band, and the 5.7 GHz ISM band can be called the 5.8 GHz band. The terminology may be slightly different, but the band limits are the same.

WLAN and WPAN specifications in their quest to become worldwide standards for short-range wireless. Also, manufacturers would like to avoid having to design and assemble a different module for each regulatory domain whenever possible.

In 1992 the *International Telecommunications Union* (ITU), which is part of the United Nations, formed its *Radiocommunication Sector* (ITU-R) in an attempt to ensure rational, efficient, and economical use of the radio frequency spectrum. Every few years the ITU conducts a *World Radiocommunications Conference* (WRC), where member nations agree on how the radio spectrum is allocated and used. Furthermore, groups associated with marketing wireless networks are working with various governments in attempts to bring their regulations into alignment with the WLAN and WPAN specifications.

In order to keep this section to a reasonable size, we will concentrate on FCC rules as they apply in the United States, and the rules for operating WLAN and WPAN devices in Canada are similar. For operating rules in other countries, check with the appropriate regulatory agency. Also, we cover only the highlights of these rules. Check the corresponding regulatory documents for the complete set.

FCC Part 15 Requirements

FCC rules are divided into several parts, with Part 15 (created in 1975 and revised several times since) being devoted to intentional, unintentional, or incidental radiators that are allowed to operate without an individual license. *Intentional radiators*, which include WLAN and WPAN transmitters, deliberately produce and radiate RF within their assigned frequency bands. *Unintentional radiators* are circuits such as wireless receivers and computer clocks that deliberately create and use signals at RF within the confines of the device itself, but some of this energy invariably radiates. *Incidental radiators* are those devices that generate RF energy as a result of normal operations, but don't use this energy in any appreciable way. Examples of incidental radiators include motors and generators. Designers of wireless communication systems are most interested in the rules pertaining to receivers and digital circuits as unintentional radiators, and transmitters as intentional radiators. Of course, intentional radiators play the predominant role in coexistence, so that is what we will emphasize.

In general, FCC Part 15 devices operating as intentional radiators

- cannot cause interference to licensed users.
- have no regulatory protection against interference from other users, licensed or unlicensed.
- must have a permanently mounted antenna (preferred) or one that uses a unique connector.
- require government certification prior to marketing.

When examining these factors, it becomes obvious that the FCC is most inter-ested in a device's transmission of electromagnetic energy as a potential source of interference to other users. In other words, FCC rules are meant to enhance coexistence among intentional radiators. Aside from placing limits on peak TX power and average power density within the band of use, the FCC also places limits on how much radiation is allowed to occur outside the band of operation.

The FCC rulebook reads like a legal document (which it is) and can be some-what difficult to interpret by those without a legal background. Fortunately, sev-eral Web sites offer white papers that can help translate these documents into engineer's talk, but care must be taken to insure that the papers contain the lat-est FCC information. Several frequency bands are covered by Part 15 regula-tions, and each band has its associated power limitations and modulation requirements. We'll concentrate on the ISM and U-NII bands because these are of primary interest to wireless network engineers.

Wireless technology is progressing rapidly, and the FCC is scrambling to keep up by issuing rules that not only accommodate new technology, but also enhance the efficient use of the limited radio spectrum. Amendments to Part 15 rules are being published at a rapid pace, so it's important to check for the latest of these at www.fcc.gov before committing to a design that may no longer be compliant or following constraints that no longer exist.

The FCC also places limits on the RF exposure that people are allowed when either using wireless devices or when they are in the vicinity of their use. These limits are covered in Parts 1 and 2 of the FCC regulations, along with various international publications. Generally, devices conforming to Part 15 TX power limitations also conform to *specific absorption rate* (SAR) limits for the human body, but a designer should check the applicable regulations to make sure.

General Rules for ISM Communication

Many of the FCC emission limits, especially for devices operating using the low-power rules in Section 15.249, are given in terms of electric field intensity rather than TX *equivalent isotropic radiated power* (EIRP). The formula relating the two quantities is

$$P = \frac{4\pi R^2 E^2 G}{120\pi} \tag{1.2}$$

where P is in watts, R is the distance from the antenna having gain G, and E is the electric field strength in volts/meter. For G = 1 and D = 3 m, the formula sim-plifies to

$$P = 0.3E^2 \tag{1.3}$$

For communication in the 900 MHz, 2.4 GHz, and 5.7 GHz ISM bands using a wide variety of modulation methods, both analog and digital, the average TX power output under Section 15.249 is limited to 50 *millivolts* (mV) per meter measured at a distance of 3 meters, which corresponds to an EIRP of about 0.75 mW (−1.25 dBm). The measurement resolution is 100 kHz for frequencies up to 1,000 MHz and 1 MHz for carrier frequencies above 1,000 MHz. Signals occupying a bandwidth greater than the resolution bandwidth can have proportionally higher TX power. Many manufacturers design these low-power transmitters to produce 1 mW of output power, with antenna and coupling inefficiencies dropping the field strength below the legal limit. Examples of low-power ISM products are baby monitors, some walkie-talkie toys, and analog cordless telephones.

Curiously, the 5.7 GHz ISM upper band limit shown in Figure 1-1 is extended to 5,875 MHz for such low-power use. Peak TX power can be as much as 20 dB above the average as long as the averaging time doesn't exceed 100 ms. For increased range, FCC Section 15.247 allows higher TX power if either digital modulation, *frequency hopping spread spectrum* (FHSS), or a hybrid of the two is used.

For TX antenna gain greater than 6 dBi in portable devices using digital modulation or FHSS, the allowable peak TX power must be reduced by 1 dB for each dB that the antenna gain exceeds 6 dBi. The FCC allows higher TX powers for certain operations using a fixed point-to-point link. For improved reliability over long distances, these links often use highly directional antennas with considerable gain. Indeed, the FCC often mandates the use of directional antennas to prevent interference to other users. Although fixed point-to-point links are growing in popularity, their typical locations on antenna towers, hilltops, or on building rooftops usually mean that their interference to other "regular" WLAN/WPAN users is minimal.

The FCC sets limits for out-of-band radiation for all ISM transmissions. Maximum out-of-band TX power in any 100 kHz bandwidth must be at least 20 dB below the in-band 100 kHz segment containing the highest TX power level, with additional limitations in certain restricted bands. Spurious emissions above 1 GHz cannot exceed −41.3 dBm/MHz, again with lower limits in restricted bands and below 1 GHz.

All the WLAN and WPAN networks presented in this book operate on carrier frequencies that are part of the FCC unlicensed band structure in the United States with one exception. IEEE 802.15.4 specifies a single channel at 868.3 MHz, which is within the 868 to 868.6 MHz license-free band in parts of Europe, but not in the United States at the power levels allowed by the specification. Coexistence issues on this channel are essentially identical to the co-channel operation of multiple networks in the 900 MHz ISM band.

The 900 MHz ISM Band

The 900 MHz ISM band was the first of the three ISM bands to host wireless communications in consumer products. These were primarily cordless telephones, baby monitors, extension speakers, and custom wireless networks. This band is available in the United States but not in many other countries, and few standardized operating procedures exist. Consequently, most of the products operating in this band are custom designs that are not compatible across manufacturers; that is, they are not interoperable.

Digital Modulation in the 900 MHz ISM Band   The FCC requires that such digitally modulated signals using the 900 MHz ISM band conform to the following rules:

- TX peak output power not more than 1 W (+30 dBm)
- Minimum TX signal bandwidth of 500 kHz at the −6 dB points
- Peak TX *power spectral density* (PSD) not more than 8 dBm in any 3 kHz bandwidth segment

These rules allow for a wide variety of modulation techniques and transmitted data rates, giving designers significant flexibility in creating new standards and custom wireless systems. Some cordless telephones and custom WLAN systems in the United States, along with 802.15.4 devices in this band, use digital modulation and conform to these rules.

FHSS in the 900 MHz ISM Band   FHSS transmissions in the 900 MHz band can be either full channel (at least 50 hop frequencies) or reduced channel (25 to 49 hop frequencies). Full-channel FHSS must conform to the following:

- At least 50 nonoverlapping hop channels
- Peak TX power not more than 1 W (+30 dBm)
- Hop channel bandwidth less than 250 kHz
- Hop channel separation at least the −20 dB hop bandwidth, but not less than 25 kHz
- Hop channels selected pseudorandomly with equal dwell time in each channel on average
- Cumulative dwell time in each channel not more than 0.4 seconds during each 20-second time period

Reduced-channel FHSS in the 900 MHz ISM band has the following differences from full-channel FHSS:

- At least 25 nonoverlapping hop channels
- Peak TX power not more than 250 mW (+24 dBm)
- Hop channel bandwidth at least 250 kHz, but not more than 500 kHz
- Cumulative dwell time in each channel not more than 0.4 seconds during each 10-second time period

Some cordless telephones in the United States use FHSS in the 900 MHz ISM band.

The 2.4 GHz ISM Band

Most wireless networking devices available at the beginning of the twenty-first century operate at 2.4 GHz because that is the only practical unlicensed band that is (mostly) allocated worldwide. That's good news, of course, but there's a reason the band is available throughout the world, and that reason is microwave ovens. These ovens became popular long before any other general use of these high frequencies was envisioned, and the nominal microwave oven frequency of 2.45 GHz was chosen because water molecules readily absorb RF energy at this frequency and convert it to heat. These ovens operate at several hundred watts of power, and as we will discover in a later chapter, these can be a significant source of interference to wireless users in the 2.4 GHz band.

Low-Power Operation in the 2.4 GHz ISM Band   For operation above 1,000 MHz, FCC rules require a bandwidth resolution of 1 MHz on the averaging detector used for measuring TX field strength at not more than 50 *mV per meter* (mV/m) measured at 3 m from the antenna. This corresponds to a TX output power of approximately −1.25 dBm/MHz. The 802.15.3 WiMedia specification assumes that its transmitters will be certified under FCC Section 15.259. For the typical 15 MHz TX bandwidth of such signals, maximum TX power is −1.25 + 10 log (15) = +10.5 dBm for a uniform PSD. Nominal TX power for 802.15.3 transmitters is given in the specification as +8 dBm to ensure compliance for its nonuniform PSD. An advantage of certifying the transmitter under Part 15.259 is that TX signal harmonics, which are further spread in proportion to the harmonic being considered, are also measured at 1 MHz resolution. Wider bandwidth harmonics are allowed to have proportionally higher total power, reducing transmitter design effort and cost.

Digital Modulation in the 2.4 GHz ISM Band   All 802.11 and many other communication signals contain information that is digitally modulated onto a (usually) fixed RF carrier. The FCC requires that such digitally modulated signals conform to the following rules:

- TX peak output power not more than 1 W (+30 dBm)
- Minimum TX signal bandwidth of 500 kHz at the −6 dB points
- Peak TX PSD not more than 8 dBm in any 3 kHz bandwidth segment

Prior to mid-2002, the FCC required digital modulated transmissions to include a processing gain of 10 dB or greater, corresponding to at least 10 chips per data bit, in a *direct sequence spread spectrum* (DSSS) signal for TX output powers higher than about 0.75 mW (−1.25 dBm). (See Chapter 4 for a description of DSSS operation.) This processing gain provided some interference immunity for improved coexistence. The DSSS requirement was eliminated in May 2002 because the FCC decided that manufacturers were already sufficiently motivated to build interference immunity into their products, and it was felt that compelling manufacturers to meet a minimum DSSS processing gain placed additional unwarranted restrictions on system design.

The original 802.11 standard, released in 1997, conformed to FCC Part 15 regulations existing at that time, so it meets the above rules as well as having a DSSS processing gain of just over 10 dB. On the other hand, 802.11b uses *complementary code keying* (CCK) to obtain its higher data rates, which isn't traditional DSSS at all, and 802.11g employs *orthogonal frequency division multiplexing* (OFDM) instead of DSSS. Although the FCC issued a special ruling to allow CCK operation under the old DSSS rules, both CCK and OFDM conform to the later FCC digital modulation regulations for use in the 2.4 GHz band.

FHSS in the 2.4 GHz ISM Band   The original 802.11 specification had provisions for FHSS, but most chip manufacturers abandoned FHSS for the faster data rates provided by CCK in the 802.11b standard when it was released. Unlike 802.11, 802.15.1 Bluetooth devices use only FHSS in the 2.4 GHz ISM band, but like 802.11, Bluetooth was developed when the FCC rules were more stringent than they are now. In May 2002, the FCC loosened slightly the rules for FHSS systems using at least 75 hop channels (full-channel FHSS), and they also formalized rules for devices with fewer than 75 hop channels (reduced-channel FHSS).

For full-channel FHSS, transmitted signals must conform to the following:

- At least 75 nonoverlapping hop channels
- Peak TX power not more than 1 W (+30 dBm)
- Hop channel separation at least the −20 dB hop bandwidth, but not less than 25 kHz
- Hop channels selected pseudorandomly with equal dwell time in each channel on average
- Cumulative dwell time in each channel not more than 0.4 seconds during each 30-second time period

The Bluetooth specification limits maximum TX power to 100 mW (+20 dBm) to conform to European FHSS regulations. The previous FCC rules also imply that each hop channel is allowed a maximum bandwidth of about 1 MHz so that all hop channels fit within the 2.4 GHz ISM band limits. (When Bluetooth was first developed, FCC rules dictated a maximum hop bandwidth of 1 MHz.) The minimum hop rate is 2.5 hops/second (75/30) to conform to the cumulative dwell time requirements, assuming that a transmission occupies nearly the entire dwell time during each hop. Bluetooth uses 79 hop channels, each with a maximum −20 dB bandwidth of 1 MHz, so Bluetooth occupies 79 MHz of the available 83.5 MHz bandwidth in the 2.4 GHz ISM band. The hop rate for the Bluetooth radio is nominally 1,600 hops/second.

Reduced-channel FHSS devices must conform to the following changes from full-channel FHSS operation:

- At least 15 nonoverlapping hop channels
- Peak TX power not more than 125 mW (+21 dBm)
- Intelligent hopping techniques may be used to avoid interference to other transmissions
- Average occupancy time on any channel not greater than 0.4 seconds during any 0.4 × (number of hop channels)-second time period

In other words, by restricting its maximum TX power to +21 dBm or lower, a FHSS device is permitted to reduce the minimum number of hop channels from 75 to as few as 15. As a consequence, the −20 dB maximum TX bandwidth within a hop channel can now be as high as about 5 MHz when 15 hop channels are used without exceeding the 2.4 GHz band limits. Several other countries have permitted FHSS transmitters to operate with as few as 15 or 20 hop channels, so the FCC adoption of reduced-channel FHSS brings the United States into closer alignment with other parts of the world. The average occupancy time of 0.4 seconds during any 0.4 × (number of hop channels)-second time period once again implies a minimum hop rate of 2.5 hops/second.

Because the Bluetooth specification limits TX power to +20 dBm maximum, these signals technically fall under the reduced-channel FHSS rules. However, early Bluetooth specifications 1.0A, 1.0B, and 1.1 conformed to full-channel FHSS requirements. Reduced-channel FHSS rules enable a system to avoid using hop channels containing interfering signals. Hop channel adaptation through intelligent hop techniques must be done "independently and individually," which means that no method of coordinating with other FHSS systems may be employed. This so-called *adaptive frequency hopping* (AFH) can greatly improve reliability, and Bluetooth specification 1.2 includes AFH. Furthermore, because reduced-channel FHSS enables a transmitter to use a wider bandwidth in each hop channel, this could possibly be exploited in a future specification to increase Bluetooth data rates.

The 5.7 GHz ISM Band

One way to escape the pandemonium at 2.4 GHz is to move operations to an unlicensed band in the 5.x GHz range. Fabrication processes have recently been improved to the point that RF designs at these carrier frequencies are relatively efficient and low in cost. The result is an increased interest in 802.11a implementations, which also have built-in error correction and higher data rates than 802.11b in the 2.4 GHz band.

Digital Modulation in the 5.7 GHz ISM Band   When TX powers above −1.25 dBm are used with digital modulation, the band limits become 5,725 to 5,850 MHz (refer to Figure 1-1), and FCC regulations require

- TX peak output power not more than 1 W (+30 dBm).
- Minimum TX signal bandwidth of 500 kHz at the −6 dB points.
- Peak TX PSD not more than 8 dBm in any 3 kHz bandwidth segment.

802.11a conforms to these regulations, and the European HiperLAN/2 also conforms to these regulations provided its operating frequencies remain within the 5.7 GHz ISM band or the 5.x GHz U-NII bands discussed later.

FHSS in the 5.7 GHz ISM Band   Unlike its 2.4 GHz ISM counterpart, only full-channel FHSS is allowed in the 5.7 GHz ISM band. These systems must conform to the following provisions:

- At least 75 nonoverlapping hop channels
- Peak TX power not more than 1 W (+30 dBm)
- Maximum TX signal bandwidth on any hop channel of 1 MHz at the −20 dB points
- Hop channel separation at least at the −20 dB hop bandwidth but not less than 25 kHz
- Hop channels selected pseudorandomly with equal dwell time in each channel on average
- Cumulative dwell time in each channel not more than 0.4 seconds during each 30-second time period

As of this writing (mid-2004), no major wireless networks use FHSS in the 5.7 GHz ISM band.

The 5.x GHz U-NII Bands

In January 1997, the FCC opened three 100 MHz-wide segments located between 5.15 and 5.825 GHz to encourage the development of short-range, high-data-rate wireless networks. The FCC added a fourth 255 MHz-wide segment in November 2003 to better align the U.S. bands to those available in Europe and other regulatory domains. These U-NII bands are depicted in Figure 1-2. Although the various segments have different TX power limitations, great flexibility exists in the type and bandwidth of signals from intentional radiators that can operate in the U-NII bands.* We will refer to the two disjoint U-NII segments, along with the partially overlapping 5.7 GHz ISM frequencies, as the 5.x GHz band. This is where IEEE 802.11a networks reside.

Like the ISM bands, the U-NII bands have their own antenna rules and general radiation limitations. A gain antenna of up to 6 dBi can be used without any TX power reduction. For TX antenna gains greater than 6 dBi, the allowable peak TX power and PSD must be reduced by 1 dB for each dB that the antenna gain exceeds 6 dBi. Maximum out-of-band TX power must not exceed an EIRP of −27 dBm/MHz for the lower three bands. For the upper band, emissions from the band edge to 10 MHz above or below the band edge must not exceed −17 dBm/MHz EIRP, and for frequencies 10 MHz or greater above or below the band edge, emissions must not exceed −27 dBm/MHz EIRP. The ratio of peak excursion of the modulated envelope to the peak TX power must not exceed +13 dB across any 1 MHz bandwidth, or the emission bandwidth, whichever is less. Finally, devices operating in the U-NII bands must automatically discontinue transmission in the absence of information to send or if operational failure occurs.

Some of these U-NII bands require unlicensed equipment to implement *dynamic frequency selection* (DFS) and *transmit power control* (TPC) to reduce any potential interference to licensed users that share these bands. U-NII devices implementing DFS are required to monitor a channel for 60 seconds to ensure that it is not occupied by a radar system before commencing operation. The minimum detection threshold is −62 dBm if the device's EIRP is less than 200 mW; otherwise, the detection threshold is −64 dBm. The devices also are required to monitor the channel during operation, and if a radar appears at a later time then the network must cease normal traffic within 200 *milliseconds* (ms). The controller then coordinates migration of the network to a new unoccupied channel within 10 seconds, and the network must remain away from the former channel for at least 30 minutes. If no unoccupied channel is found, then the network must shut down. Of course, if DFS is expanded to include sensing the presence of other WLAN users, then coexistence is improved accordingly.

For operation in some of the U-NII bands, TPC is required for devices having TX EIRP of 500 mW (+27 dBm) or higher, and the TPC-equipped device must have the capability to operate at least 6 dB below the +30 dBm maximum EIRP. Generally, a WLAN controller using TPC will be programmed with regulatory and local maximum TX power levels for the current channel being used. Actual TX power when communicating with a particular node may be determined by

---

*The FCC rules listed in this section are current as of November 2003 and were changed from previous editions as a result of measures adopted at WRC-03. Additional changes will occur as new international regulations are implemented.

path loss and link margin estimates, up to the maximum allowed. TPC improves coexistence with other users by reducing the range of potential interference to other networks. Both DFS and TPC are covered in greater detail in Chapter 8, "Active Coexistence."

Operation in the 5.15 to 5.25 GHz U-NII Band Intentional radiators within the 5.15 to 5.25 GHz U-NII band must conform to the following regulations:

- Operation is restricted to indoor use only.
- Peak TX power must be the lesser of either 50 mW (+17 dBm) or 4 dBm + 10 log B, where B is the −26 dB TX bandwidth in MHz.
- Peak TX PSD cannot exceed +4 dBm in any 1 MHz band segment.
- A TX antenna must be integral to the device.

Devices operating within this U-NII band are restricted to indoor use to prevent interference with *mobile satellite service* (MSS) equipment, such as Iridium and Odyssey, which provide satellite telephone and data services. The maximum allowable peak TX power of +17 dBm is reached for signal bandwidths B of 20 MHz or greater; otherwise, power is restricted to 4 dBm + 10 log B. The latter formula corresponds to a power density of 2.5 mW/MHz. Figure 1-3 plots maximum TX power as a function of bandwidth for the three U-NII band segments. TX EIRP levels can be up to 6 dB higher using a gain antenna.

Operation in the 5.25 to 5.35 GHz and 5.470 to 5.725 GHz U-NII Bands Intentional radiators within the 5.25 to 5.35 and 5.470 to 5.725 GHz U-NII bands must conform to the following:

- Peak TX power must be the lesser of either 250 mW (+24 dBm) or 11 dBm + 10 log B, where B is the −26 dB TX bandwidth in MHz.
- Peak TX PSD cannot exceed +11 dBm in any 1 MHz band segment.
- TPC is required if TX EIRP is +27 dBm or higher.
- DFS is required.

It can be inferred from these rules that, in these U-NII band segments, the TX power density is limited to 12.5 mW/MHz, and the maximum allowable TX power of +24 dBm is reached for signal bandwidths of 20 MHz or more (see Figure 1-3). TX EIRP levels can be up to 6 dB higher using a gain antenna.

Operation in the 5.725 to 5.825 GHz U-NII Band Intentional radiators within the 5.725 to 5.825 GHz U-NII band must conform to the following:

- Peak TX power must be the lesser of either 1 W (+30 dBm) or 17 dBm + 10 log B, where B is the −26 dB TX bandwidth in MHz.
- Peak TX PSD cannot exceed +17 dBm in any 1 MHz band segment.

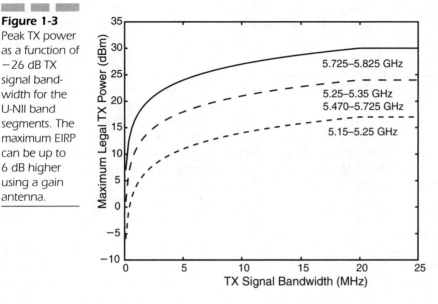

Figure 1-3
Peak TX power as a function of −26 dB TX signal bandwidth for the U-NII band segments. The maximum EIRP can be up to 6 dB higher using a gain antenna.

In this U-NII band segment, TX power density is limited to 50 mW/MHz, and the maximum allowable TX power of +30 dBm is reached for signal bandwidths of 20 MHz or more (see Figure 1-3). TX EIRP levels can be up to 6 dB higher using a gain antenna.

This particular band segment is better suited than the others for high-power point-to-point communications across distances of about 1 km or greater. Not only is the allowable TX power significantly higher, but directional antennas with gains up to +23 dBi can be used to concentrate the signal toward its destination without requiring a corresponding reduction in TX power. (Antenna gain and directivity will be examined further in Chapter 2, "Indoor RF Propagation and Diversity Techniques.") For example, a signal of at least 20 MHz bandwidth can legally send +20 dBm of TX power into an antenna with a +23 dBm gain, for an EIRP of +43 dBm, equivalent to 20 W.

5.7 GHz ISM Versus 5.7 GHz U-NII

The FCC has designated three different but overlapping unlicensed bands that start at 5.725 GHz and have a different upper limit. The widest is the 5.725 to 5.875 GHz ISM band for general use with TX power limited to about −1.25 dBm. Next comes 5.725 to 5.850 GHz ISM for use with digital modulation or FHSS, with a maximum TX power of +30 dBm. Finally, the 5.725 to 5.825 GHz U-NII band is available for TX powers up to +30 dBm and modulation methods limited only by their PSD and the ratio of peak envelope excursion to peak TX power.

Within the two segments in which TX power can be as high as +30 dBm, the FCC requires a signal PSD not more than +8 dBm in any 3 kHz ISM band segment, whereas the U-NII band can contain a signal PSD not more than +17 dBm in any 1 MHz band segment. Which is more restrictive? The answer can easily become an "apples versus oranges" issue due to the different bandwidths in question, but the U-NII rules allow for signal PSD to be less homogeneous than the ISM rules demand. Furthermore, a fixed-carrier (non-FHSS) digital signal with a bandwidth less than 500 kHz cannot be used with TX powers above −1.75 dBm in the 5.7 GHz ISM band, but such a signal conforms to U-NII rules for maximum power levels of +14 dBm (which is 17 dBm + 10 log 0.5). If the signal's bandwidth is 500 kHz, then ISM rules permit TX powers up to +30 dBm compared to just above +14 dBm under U-NII regulations. (Note, though, that signal bandwidths are measured at the −6 dB points for ISM, but at the −26 dB points for U-NII.) Once the TX signal bandwidth reaches 20 MHz, then U-NII rules allow TX powers up to +30 dBm as well.

From a coexistence point of view, low-power ISM communications could be placed in the 25 MHz-wide segment from 5.850 to 5.875 GHz to avoid high power ISM and U-NII transmissions. High-power ISM devices could avoid high-power U-NII transmissions by using the 25 MHz-wide segment from 5.825 to 5.850 GHz.

Ultra-Wideband Operations

Ultra-wideband (UWB) transmitters operate with bandwidths that are several GHz wide, but the total energy per MHz is so low that most ordinary receivers are, theoretically at least, unaffected by the presence of UWB signals. Although originally developed for ranging and imaging applications, a great deal of effort has been devoted toward using UWB for communications. One of the main advantages of UWB is its relatively flat PSD over a wide bandwidth for the more efficient use of the RF spectrum. Rather than forcing multiple users into a narrow bandwidth (such as placing several Wi-Fi and Bluetooth signals within only 83.5 MHz that makes up the 2.4 GHz ISM band), the same number of UWB users would coexist over several GHz instead. In this way, mutual interference may be significantly reduced. It's possible, then, that UWB could solve many of the coexistence problems that plague traditional wireless networks.

The FCC defines a UWB signal as one that has a −10 dB fractional bandwidth greater than 0.20, or a −10 dB bandwidth of at least 500 MHz regardless of fractional bandwidth. The FCC Part 15 divides the UWB rules into several categories. The category of interest here covers handheld systems for communication between a transmitter and receiver. The latter must conform to the following regulations:

■ The transmitter must operate only when sending information to a receiver, and transmissions must cease unless an *acknowledgment* (ACK) is returned by the receiver at least every 10 seconds.

- The antenna must be mounted on the device itself, using either a permanent attachment or a unique connector.
- Operation indoors or outdoors is permitted, but operation on aircraft, ships, or satellites is prohibited.
- Using UWB for applications involving toys is prohibited.
- TX bandwidth must be contained between 3,100 and 10,600 MHz, with very low emissions allowed outside this range.
- Maximum EIRP radiated emissions must conform to the values listed in Table 1-4.
- Peak EIRP within a 50 MHz bandwidth centered on the TX center frequency must not exceed 0 dBm.

These TX power levels appear to be extremely low, mainly because the FCC is taking a cautious approach to UWB until it can be shown that it can indeed coexist with other services. Indeed, the −41.3 dBm/MHz limit for intentional radiation within the UWB band is the general spurious emission limit above 1,000 MHz for non-UWB Part 15 radios. The 3,100 to 10,600 MHz UWB operating region is 7,500 MHz wide. By expressing this number as 38.8 dB, we can calculate the maximum average TX power allowed by the FCC to be −2.5 dBm (or 0.56 mW), derived from −41.3 + 38.8, assuming that the transmitted PSD is uniform within the operating frequency range. The peak TX power within 25 MHz on either side of the frequency on which the highest radiation occurs is limited to 0 dBm. The FCC also removed its general Part 15 prohibition against damped wave transmissions when UWB devices conformed to these rules.

Table 1-4

Average UWB power limits for handheld communication devices

Frequency (MHz)	Resolution bandwidth	Maximum average EIRP
960–1,610	1 MHz	−75.3 dBm
1,164–1,240	1 kHz	−85.3 dBm
1,559–1,610	1 kHz	−85.3 dBm
1,610–1,990	1 MHz	−63.3 dBm
1,990–3,100	1 MHz	−61.3 dBm
3,100–10,600	1 MHz	−41.3 dBm
Above 10,600	1 MHz	−61.3 dBm

Some of the restrictions on out-of-band emissions are relaxed if the UWB device is used indoors only; that is, they are made for fixed operation from the AC power mains. See www.fcc.gov for more details. Restrictions in the 1,164 to 1,240 MHz and 1,559 to 1,610 MHz bands are placed by the FCC in response to concerns that UWB could interfere with GPS navigation. UWB emissions below 960 MHz must conform to FCC Part 15.209. For example, from 216 to 960 MHz, emissions are limited to −49.2 dBm/100 kHz.

FCC Product Certification

FCC certification is required before any product can be sold that has the potential to produce RF interference. This includes products containing computer circuitry, receivers, and transmitters. Product development can take place without a special license, and certification is actually one of the last steps to be accomplished before the wireless product is brought to the marketplace.

Certification is done on a production-ready prototype. After the prototype is built, it should be tested for FCC compliance, either in-house or via one of the many test facilities available for such purposes. Next, the device is submitted to an authorized FCC testing facility called a *Telecommunications Certification Body* (TCB). The TCB performs the required tests and, if all is well, issues an FCC ID number that must be affixed to each product sold. The FCC has authorized several TCBs, so Part 15 certification consumes only a week or two of time for a properly designed product.

Because many wireless devices are supposed to be available for worldwide sale, the eventual goal of the certification process is to allow one nation's certification to be valid worldwide. A two-phase *Mutual Recognition Agreement* (MRA) has been signed between the United States and the *European Union* (EU), the *Asian-Pacific Economic Conference* (APEC) countries, and other countries in the western hemisphere (*Inter-American Telecommunications Commission* [CITEL]) in an effort to achieve this goal. Phase 1 allows the countries to accept each other's laboratory test data for checking against their own certification rules, and Phase 2 allows for full acceptance of each other's certification from authorized testing bodies [Cok01].

Interference Temperature

In November 2003, the FCC released a document soliciting comments on the feasibility of establishing an interference temperature metric to quantify and measure the level of interference in an attempt to expand available unlicensed operation into some of the existing fixed, mobile, and satellite frequency bands. Interference temperature is defined as the measure of the temperature equivalent in *Kelvin* (K) of RF power per unit bandwidth generated by undesired emitters plus noise sources (I + N) present at a receiver, not including its own internal noise. The concept of interference temperature is analogous to antenna temperature, which is widely used as a component of the total noise temperature of a receiver system.

This concept represents a major shift in the way the FCC manages spectrum, which in the past has been based upon controlling TX emissions and occupied frequencies. The change in philosophy was motivated by increased spectrum demand, coupled with rapid technical advances in radio systems, including interference-mitigating mechanisms such as spread spectrum. Interference is increasingly being characterized by protocols using different signaling waveforms that are deployed in relatively high local densities with short-range links and low TX power. The FCC is thus motivated to assess interference based upon real-time measurements of actual spectrum use that takes into account all undesired RF energy available at a particular receiver's antenna, rather than simply placing global limits on EIRP.

For the interference temperature limit to function properly, the affected receiver, usually part of a licensed service, needs to measure interference temperature in its band of operation. The node then communicates that information to unlicensed devices subject to the limit, either directly or through a monitoring station. A response would be needed that restricts the operation of these devices to maintain an interference temperature below the required limit. Furthermore, if multiple unlicensed devices are operating, a means by which they reduce their collective interference would need to be negotiated. These are daunting challenges, especially because this requires communication links to be established between devices that operate with vastly different *physical layer* (PHY) and *medium access control* (MAC) protocols.

The OSI Model

In order to reduce the complexity of designing a digital communication system, the *International Standards Organization* (ISO) established a subcommittee to research the need to develop a standardized, layered approach to general computer communications. In 1982, the *Open Systems Interconnection* (OSI) reference model, shown in Figure 1-4, was completed. By working through the layers of the model a designer can create the communication system in an orderly manner, and a particular layer can be changed without affecting the other layers [Sta88]. The disadvantage to using this method is that redundancy, along with its resulting inefficiencies, is inevitable.

The PHY contains the actual physical interface and the rules for its use. In wireless communications, the PHY is usually RF and the modulation and detection processes are listed in the appropriate specification. The physical layer is made reliable, and the data link layer provides link connection and detachment rules. This layer contains the MAC, which is a set of rules that determine, the structure of basic data packets and channel access procedures, as well as the *logical link control* (LLC), which provides the protocol for link establishment and detachment. The *network layer* provides a transparent transfer of data between transport entities on each end of the communication link. A properly implemented network layer relieves the higher layers from requiring any knowledge of the method by which data moves from source to destination. In other words,

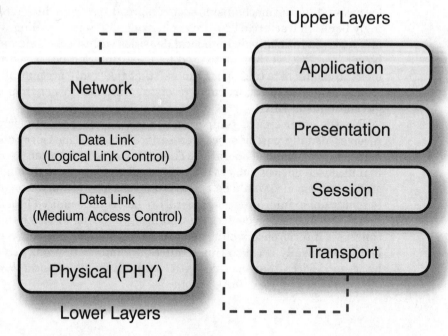

Bluetooth can appear to be a serial cable to higher layers, and Wi-Fi can appear to be an ordinary wired Ethernet connection.

The functions of the remaining higher layers are less well defined. The *transport layer* includes optimization routines and other *quality of service* (QoS) methods for efficient data exchange, and the *session layer* contains the method for controlling dialog between applications on either end of the link. Finally, the *presentation layer* resolves differences between format and data representations between entities, and the *application layer* provides the means by which applications can access the OSI environment. As we move up the OSI layers, their implementation gradually changes from hardware, through firmware, and finally into software.

It is most useful to concentrate on the PHY and MAC layers of the OSI model when analyzing the ability of networks to coexist. The modulation method and how it is implemented in the radio determine how interference affects the reliability of detecting desired incoming transmissions. The MAC layer sets the rules for when a transmission can occur across the shared RF spectrum, and how this is done determines throughput for the network as a whole, and for individual nodes. Solutions to various coexistence challenges, however, often employ all layers of the OSI model, especially when two or more independent wireless networks are collocated on the same host.

The IEEE 802 Standards and the OSI Model

The 802.11 and 802.15 standards, along with several others in the IEEE 802 group, apply only to the PHY and MAC layers of the OSI model. The data link layer above MAC is addressed by 802.1 and 802.2, and 802.10 covers security at several OSI layers. Layers above data link are not part of the 802 standard, but instead are the domain of other protocols such as the *Internet protocol* (IP), and these include additional processing at each OSI layer, depending upon the network being considered. Figure 1-5 shows the relationship of these entities relative to the OSI model.

A complete 802.11-based WLAN also requires 802.1, 802.2, and 802.10 just to address the physical and data link layers of the OSI model. New entities, such as IEEE 802.1x for enhanced security, are added as the need arises. As pointed out earlier, a detailed examination of coexistence requires concentrating only on the PHY and MAC layers, both of which are included in 802.11 and 802.15, so that is where we will focus our attention. Because of this, we often treat the IEEE designation (for example, 802.11b) and the trade name (Wi-Fi) as synonymous. In general, though, the IEEE designation encompasses only PHY and MAC, and the trade name refers to the entire set of OSI layers, or to the layers above PHY and MAC.

Figure 1-5 also shows some of the other PHY and MAC standards in addition to 802.11 and 802.15. To satisfy your burning curiosity, here they are:

- **802.3** *Carrier sense multiple access with collision detection* (CSMA/CD) access
- **802.4** Token bus access method
- **802.5** Token ring access method
- **802.12** Demand priority access method

Figure 1-5
Relationship of the various IEEE 802 standards to lower layers of the OSI model

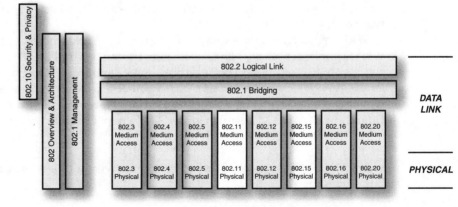

- **802.16** Fixed broadband wireless access
- **802.20** Mobile broadband wireless access

This is not a complete list, and some of those standards not shown have become essentially obsolete. Others will no doubt be added to this list as time goes on, but they all cover only the PHY and MAC and require the other 802 and higher layers for a complete system design.

IEEE 802.11a/b/g WLAN (Wi-Fi)

The IEEE adopted 802.11 as the first international WLAN standard in 1997. The goals of the standard were to describe a WLAN that delivers services equivalent to a wired network with high throughput, reliable data delivery, and continuous network connections. The 802.11 network is further enhanced by allowing for mobility and power savings that are transparent to the user. The general philosophy was to distribute decision making to the mobile stations to eliminate bottlenecks and provide fault tolerance. The architecture is flexible, supporting small or large networks that can be either semipermanent or permanent [Oha99].

Market Applications

As of this writing (mid-2004), the great majority of WLAN deployments are 802.11b Wi-Fi networks, although significant inroads are being made by both 802.11a and 802.11g. The most common market application for such a network is access to Internet services by portable laptop computers and mobile *personal digital assistant* (PDA) devices. With raw data rates up to 11 Mb/s, and actual throughput at the application layer of about 5 Mb/s, 802.11b provides fast enough rates for email, Web surfing, and even small-screen, real-time video transfers.

Wi-Fi wireless networks have an important market advantage over custom WLAN implementations: They could substitute for wired 802.3 (Ethernet) applications that have already been established. Standard IPs such as the *file transfer protocol* (FTP) and *hypertext transfer protocol* (HTTP) were enhanced through wireless connectivity. The major operating systems (Microsoft Windows XP, Linux, and Mac OS X) have integrated Wi-Fi networks into their software in such a way that users can seamlessly connect to any Wi-Fi-equipped node. In its ideal configuration, users can access authorized services anywhere, with Wi-Fi providing the last few meters of wireless connectivity into a network of arbitrarily large size.

Summary of Wi-Fi Operation

The 802.11 standard supports several different network topologies, but the most common is the *basic service set* (BSS), consisting of a number of *stations* (STA) communicating with an *access point* (AP) in a star configuration (see Figure 1-6). Often the STA takes the form of a *network interface card* (NIC), but the increasing trend is for the WLAN to be integrated into a desktop or laptop PC. The AP can be a separate entity or an NIC-PC combination programmed to act as an AP. Access to a server, or even to other STA nodes, is accomplished through the AP. See Figure 1-7 for a picture of a Wi-Fi NIC and AP.

Wi-Fi was developed with data transfer as its primary use, so the aim was high accuracy and relatively high transfer speeds. Depending upon TX power, RX sensitivity, path clutter, interference levels, and so on, the typical maximum range between STA and AP is between 10 and 100 meters, with even greater ranges possible using fixed point-to-point systems with directional antennas.

The standard was modified over the years to increase its speed and flexibility as technology development and customer demand warranted. Each modification resulted in a new letter suffix to the 802.11 name to reflect specific capabilities and compatibility, or lack thereof, with earlier standards.

802.11 The original 802.11 standard defined MAC layer operation and three PHY layers: DSSS and FHSS in the 2.4 GHz band, and *diffuse infrared* (DFIR)

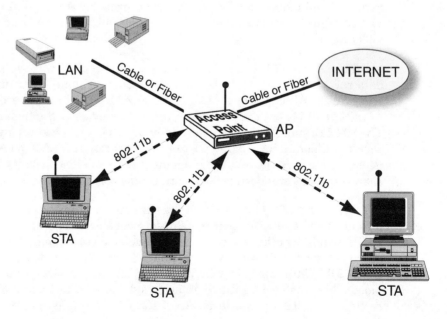

Figure 1-6
Most Wi-Fi networks are configured as a BSS, where STA communicate with an AP that in turn is connected to a backbone network.

Figure 1-7
A typical Wi-Fi
NIC plugs into
the Personal
Computer
Memory Card
International
Association
(PCMCIA) slot
of a computer.
The AP can be
a separate unit,
as shown, or a
software-
configured PC
equipped with
Wi-Fi.

Network Interface Card

Access Point

as an optical alternative. For FHSS, the hop rates aren't specified, but most 802.11 equipment hops between 10 and 50 channels per second, depending upon the manufacturer. The 75 hop frequencies are identical to those used by 802.15.1 Bluetooth. The DSSS implementation in the United States specifies 11 fixed channels, starting at 2,412 MHz and spaced every 5 MHz. All three implementations provide either 1 or 2 Mb/s raw data rates over distances up to about 100 m for RF and room-wide for IR. TX power levels for DSSS and FHSS can be up to +30 dBm and still conform to FCC regulations, but are more typically +15 to +20 dBm.

802.11b   The 802.11b standard includes the use of CCK modulation, enabling data rates of 5.5 and 11 Mb/s while producing a signal spectrum that appears identical to ordinary 1 or 2 Mb/s DSSS 802.11. The channels are also identical to DSSS 802.11, but the maximum range is shorter at the higher data speeds. The 802.11b standard is extremely popular, having reached deployment levels of several million units. A large part of the research in WLAN-WPAN coexistence is directed toward the 802.11b implementation and its capability to coexist with Bluetooth. This standard is backward compatible with 802.11.

802.11a   The 802.11a standard has become commercially viable with the advent of relatively inexpensive chipsets capable of operating in the 5.x GHz U-NII bands. The higher carrier frequencies dictate smaller antennas and a shorter range for a given TX power and RX sensitivity compared to 802.11b, but the 54 Mb/s data rate is fast enough to convey *high definition television* (HDTV) video. These faster data rates are achieved using OFDM and built-in error correction. Typical TX power levels are +10 to +17 dBm in the three U-NII bands.

Equipment conforming only to this standard is *not* compatible with 802.11, 802.11b, or 802.11g.

802.11g The 802.11g standard, ratified in June 2003, uses the same OFDM modulation method and signal structure found in 802.11a, but it operates in the 2.4 GHz ISM band instead of the 5.x GHz U-NII band. An FCC rule change was required for the modulation technique used in 802.11g to be legal in the United States, and this occurred in May 2002. Typical TX power levels are +15 to +20 dBm. This standard is backward compatible with 802.11b.

The Wi-Fi Alliance

During deployments of early 802.11b networks, it was soon discovered that equipment from different manufacturers wouldn't function together. To enhance 802.11b interoperability, the *Wireless Ethernet Compatibility Alliance* (WECA) was formed by Aironet, Intersil (formerly Harris), Agere (formerly part of Lucent), Nokia, 3Com, and Symbol Technologies. An interoperability test plan was released in April 2000 for manufacturers of 802.11b devices to use for testing product operation against other 802.11b units. WECA changed its name to the *Wi-Fi Alliance* in October 2002.

The Wi-Fi Alliance coined the commercial name *Wi-Fi*™ to indicate product interoperability. In a manner similar to Bluetooth certification, a manufacturer submits its 802.11b product to a third-party laboratory for testing and, if successful, the Wi-Fi logo can be affixed to the product. The majority of 802.11b products carry the Wi-Fi label.

With the proliferation of 802.11a and 802.11g, interoperability testing has expanded into these WLAN standards. The Wi-Fi Alliance has decided that the Wi-Fi name should represent all 802.11-based WLAN systems. As pointed out earlier, some of the later 802.11 versions are backward compatible with older versions and some aren't, so it's important that users select the proper equipment when building a new WLAN or adding AP or STA units to an existing network. The Wi-Fi Alliance Web site is located at www.wi-fi.org.

IEEE 802.15.1 WPAN (Bluetooth)

The concept behind Bluetooth had its origins in 1994 when Ericsson began researching the idea of replacing cables connecting accessories to mobile phones and computers with wireless links. As technical details began to emerge, Ericsson quickly realized that the potential market for Bluetooth products was huge, but cooperation throughout the world would be needed for the products to succeed. The first Bluetooth technical specification was released in 1999.

But why call it "Bluetooth"? There's no hint within the name itself that it represents a wireless communication system. Harald Bluetooth was a tenth-century Viking monarch who managed to unite Denmark and Norway, and because formalization of the concept of wireless cable replacement began in Scandinavia, it made some sense to identify its Viking origin. Furthermore, Harald's unifying approach to conquest meshed nicely with the goal of uniting computer and peripheral through a specification that would hopefully achieve worldwide acceptance.

Unlike the 802.11 standard, the Bluetooth specification addresses all layers of the OSI model and even extends beyond the model in directing how various definitions and instructions are to be conveyed to the end users. As a result, the specification is much longer (well over 2,500 pages, not including test documentation) than the 802.11 specification (about 770 pages for all renditions). Also, unlike 802.11, interoperability is part of the Bluetooth specification, because it was realized early in its development that customer satisfaction depended greatly on the capability of devices from different manufacturers to communicate reliably.

Market Applications

As the Bluetooth concept began taking shape in the mid-1990s, various usage models were created as possible applications of a short-range digital wireless system with data rates at the OSI application layer between about 100 and 500 kb/s. The function that a Bluetooth-enabled device is supposed to perform is based on its usage model, which is a real-world model of what a customer would expect from this device. Usage models were originally conceived for the following categories:

- **Three-in-one phone**   A handset that can operate with the (non-Bluetooth) cellular network, as a Bluetooth link to a phone base station, and as a walkie-talkie to another three-in-one phone.

- **Ultimate headset**   A headset that connects to a cell phone or any other device where headset operation is convenient.

- **Internet bridge**   A replacement of the cable between the computer and cell phone for connecting to the Internet through the cellular network.

- **Data access point**   A connection to a LAN service in a manner similar to Wi-Fi.

- **Object push**   The transfer or exchange of simple entities such as phone numbers or business cards.

- **File transfer**   Browsing, uploading, or downloading files between peers or between client and server.

- **Automatic synchronization**   Updating files to the latest version among copies on two or more storage devices.

Other usage models have been added as the capabilities of Bluetooth are matched to other areas where a wireless link would be a logical means for communication. Among these are the following:

- **Human interface device**   Provides a wireless connection of the mouse, keyboard, joystick, and so on to the computer or processing unit.

- **Audio/video distribution**   Transmits audio or video information to a display or transducer.

- **Audio/video remote control**   Controls multimedia devices in a manner similar to infrared but without the line-of-sight requirement.

- **Basic printing**   Enables mobile devices to print text messages such as email and simple formatted documents.

- **Basic imaging**   Provides the capability to exchange images of various sizes and formats.

- **Hardcopy cable replacement**   Enables wireless link substitution for the cable between the printer or scanner and its host computer.

- **Personal area network**   Integrates Bluetooth protocols into traditional Ethernet or IP-based communications for wireless access.

- **Operating a phone via an in-car device**   Places Bluetooth capability into a car so that it automatically finds and connects with any nearby authorized cellular telephone for hands-free use.

You'll notice that as the usage models proliferate, they are branching out into several diverse areas, even those that have in the past been predominately serviced by other wireless methods such as infrared.

The Bluetooth Protocol Stack

The various layers of the Bluetooth protocol stack are shown in Figure 1-8, and these layers can be compared to the OSI model in Figure 1-4. It's at once apparent that the protocol stack doesn't conform to the OSI model exactly, but the layers are still there and gradually transition from implementation in hardware and firmware (lower layers) to software (higher layers). If each of these groups of layers are separate entities, such as a PCMCIA card containing the lower layers and a laptop computer hosting the upper layers, they can communicate with each other through the *host controller interface* (HCI). It provides paths for data, audio, and control signals between the Bluetooth module and host.

The radio completes the physical layer by providing a transmitter and receiver for two-way communication. Data packets are assembled and fed to the radio by the baseband state machine. The link controller provides more complex state operations such as the standby, connect, and low-power modes. The baseband and link controller functions are combined into one layer in Figure 1-8 to

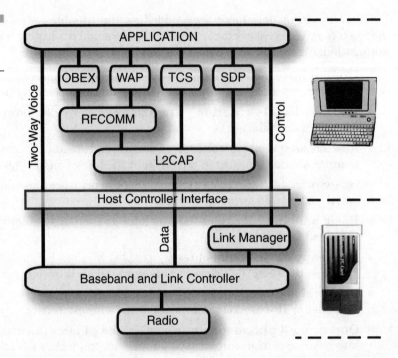

Figure 1-8
The Bluetooth
protocol stack

be consistent with their treatment in the Bluetooth specification. The link manager provides link control and configuration through the *link manager protocol* (LMP).

The *logical link control and adaptation protocol* (L2CAP) establishes virtual channels between hosts that can keep track of several simultaneous sessions such as multiple file transfers. L2CAP also takes application data, breaks it into Bluetooth-size morsels for transmission, and reverses the process for received data. *Radio frequency communication* (RFCOMM) is the Bluetooth serial port emulator, and its main purpose is to make an application think that a wired serial port exists instead of an RF link. Finally, the various software programs that are needed for the different Bluetooth usage models enable a familiar application to use Bluetooth. These include the *service discovery protocol* (SDP), *object exchange* (OBEX), *telephony control protocol specification* (TCS), and *wireless application protocol* (WAP). The L2CAP protocol layer and above, along with HCI, essentially have no 802.11 equivalent, because the 802.11 standard only reaches as high as the MAC layer of the OSI model.

Aside from data communications, Bluetooth has a special provision for synchronous, real-time, two-way, digitized voice as well. Once these voice packets are created by an application, they bypass most of the data protocol stack and are handled directly by the baseband layer. This reduces latency between the time the packets are created and the time they arrive at their destination. Control of the Bluetooth module usually proceeds from the application through HCI to the module and also bypasses the protocol layers used for handling the data communication process itself.

Summary of Bluetooth Operation

Although most Wi-Fi networks are of an *infrastructure* topology, where the AP is at a fixed location, Bluetooth predominately operates in an *ad hoc* fashion. Thus, links can be established between any set of in-range devices and participants can come and go as they please. The Bluetooth network is termed a *piconet*, meaning "small network." The device initiating communication is called the *master*, and the other participants are called *slaves*. When that link is later broken, the master/slave designations no longer apply. In fact, every Bluetooth device has both master and slave hardware.

When only one slave is used, then the link is called *point-to-point*. A master can control up to seven active slaves in a *point-to-multipoint* configuration. Slaves communicate only with the master, never with each other directly. Timing is such that two or more members of the same piconet cannot transmit simultaneously, so the participants won't collide with each other. Finally, communication across piconets can be realized if a Bluetooth device can be a slave in two piconets, or a master in one and a slave in another. Piconets configured in this manner are called *scatternets*. These various arrangements are depicted in Figure 1-9.

Bluetooth reduces the possibility of multiple piconets constantly interfering with each other by using FHSS, where the 2.4 GHz band is segmented into 79 channels, each 1 MHz wide, and each piconet under control of its master hops from channel to channel in what appears to be a random pattern. In this way, cross-piconet interference occurs only occasionally, and an error correction code and acknowledgment process provide error recovery. FHSS also acts as an effective antijam method against other WLAN/WPAN users in the 2.4 GHz band. If a wideband signal such as Wi-Fi is present, most of the band will still be open to Bluetooth devices. The issue becomes more troublesome for both Bluetooth and Wi-Fi if several Wi-Fi networks occupy most of the 2.4 GHz band, as we'll discover in later chapters.

Bluetooth-equipped devices find and connect to each other though processes called *inquiry* and *paging*, where a special abbreviated hop code is used to speed up the establishment of hop synchronization. During inquiry and paging, the hop channel set is reduced from 79 to 32, but even the abbreviated channel set is spread out over most of the 2.4 GHz band.

Figure 1-9
Point-to-point,
point-to-
multipoint, and
scatternet
topologies

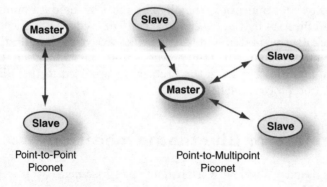

Point-to-Point
Piconet

Point-to-Multipoint
Piconet

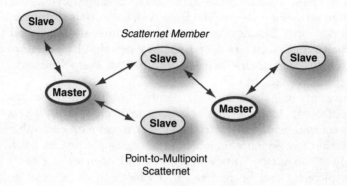

Point-to-Multipoint
Scatternet

Bluetooth Special Interest Group

Founded by Ericsson, Nokia, IBM, Intel, and Toshiba, the *Bluetooth Special Interest Group* (BSIG) was formed in February 1998. Even during its infancy, Bluetooth was clearly envisioned as a worldwide communication system as evidenced by Ericsson and Nokia representing Europe, IBM and Intel representing the Americas, and Toshiba representing Asia. The SIG founders were joined in December 1999 by Microsoft, 3Com, Lucent, and Motorola, and these nine entities are now called *promoters*. Promoters are responsible for upper-level SIG administration, and for administering the legal, marketing, and qualification processes [Bar00].

Some of the major functions of the SIG include

- Petitioning various government agencies to allow Bluetooth to operate in their countries without special requirements or restrictions.
- Handling legal issues related to SIG membership, intellectual property, and use of the trademark.

- Managing the process that tests devices to ensure compliance to the Bluetooth specification.
- Managing the interoperability test process to ensure Bluetooth devices from different manufacturers can communicate with each other.
- Managing technical working groups.
- Creating and publishing the Bluetooth specification.

BSIG membership is required before the Bluetooth specification can be used in designs, and members are granted access to the Bluetooth intellectual property and logo without paying royalties. The BSIG became a nonprofit corporation in early 2001, and its Web site can be accessed at www.bluetooth.org.

IEEE 802.15.3 High-Rate WPAN (WiMedia)

A significant group of applications is served by WPAN implementation, but at higher data rates than those supported by 802.15.1 Bluetooth. These include the transfer of isochronous data and large files, which IEEE 802.15.3 accomplishes without sacrificing the benefits of low complexity, cost, and power consumption. Several data rates are supported for different applications, with data rate scalability being part of the specification. A draft of the 802.15.3 specification was released in February 2003.

Market Applications

The focus of 802.15.3 is on multimedia streaming audio and video at data rates between 11 and 55 Mb/s. These rates support fast file transfers and permit multimedia exchanges ranging from high-fidelity audio to HDTV. A QoS capability is implemented to insure that these multimedia applications are served with as little distortion or interruption as technically feasible. Specific target markets include the following [Rob03]:

- Multimedia mass storage and transmission
- Digital camcorder and still camera transfers
- Advanced printer applications
- High-fidelity audio and video
- Other high-rate communication applications in computers, cell phones, and PDAs

Table 1-5

Technical
requirements
for consumer
A/V appli-
cations

Service	Payload rate	Transport type	Latency	Maximum bit error rate (BER)
Streaming video	100 kb/s–20 Mb/s	Isochronous	100 ms	1e-4 to 1e-8
Streaming audio	64 kb/s–1.5 Mb/s	Isochronous	100–1000 ms	1e-6
High-speed data	1–10 Mb/s	Asynchronous	Timeout	1e-6 with FEC/CRC
Interactive graphics	2–4 Mb/s	Asynchronous	5 ms	1e-5 to 1e-6
Two-way conferencing	16 kb/s–2 Mb/s	Synchronous	10 ms	1e-3 to 1e-6

Table 1-5 lists some of the requirements for consumer audio/visual applications that can be supported by WiMedia [Xtr01] with the author's modifications.

Summary of Operations

IEEE 802.15.3 HR-WPAN has been developed to have the following features:

- Raw data rates of 11, 22, 33, 44, and 55 Mb/s
- Versatile piconet topologies
- Fast connection time
- Efficient data transfer
- QoS support
- Built-in security
- Low complexity
- Low cost
- Low power consumption

A node can have two possible levels of capability. A *device* (DEV) provides basic data-exchange functionality within the 802.15.3 piconet. The *piconet controller* (PNC) is a DEV that provides piconet timing signals by transmitting periodic beacons and manages QoS, power-save modes, and piconet access. Not all DEVs need to have PNC capabilities.

A typical piconet is ctructured as shown in Figure 1-10. One DEV with the PNC capability serves as the piconet controller, and other DEVs in the piconet use the PNC beacons for timing information to prevent collisions. Unlike

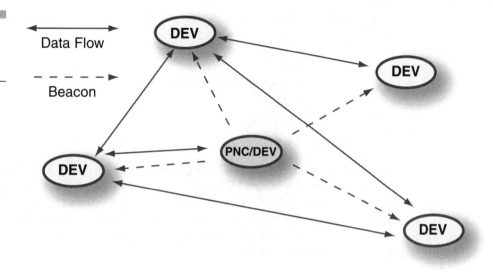

Figure 1-10
802.15.3
piconet
topology

802.15.1, though, a DEV in the 802.15.3 piconet can communicate directly with another DEV as long as the RF link is adequate.

Five channels are defined within the 2.4 GHz band, which the 802.15.3 piconets can operate on, but this channel set is reduced to three channels if 802.11 networks are present for improved WLAN coexistence. Typical TX power is +8 dBm. A DEV with PNC capability that wants to start a piconet scans the available channels, selects the one with the least interference, and then begins transmitting beacons. This announces to other DEVs that a PNC is available, and they can join the piconet at will, communicating with each other and/or with the PNC.

It's also possible for a DEV to form a subsidiary piconet, in which the original piconet allocates time for the subsidiary to communicate without interference. In this way, a good neighbor policy can be implemented across several 802.15.3 piconets. Although overall throughput is divided between these independent networks, thus slowing them all down to varying degrees, overall efficiency is improved by reducing the collision potential.

802.15.3a Alt-PHY

Because high data rates are a definitive goal of HR-PAN, there has been great interest in using UWB to increase data rates by an order of magnitude or more while providing seamless coexistence with other conventional RF systems. As a result, specification 802.15.3a is being developed to present an alternate physical layer (alt-PHY) to the somewhat conventional modulation methods used by 802.15.3 in the 2.4 GHz band.

Five criteria have been developed for the implementation of the alt-PHY [ROB02]:

- Broad market potential
- Compatibility with existing IEEE standards
- Distinct identity
- Technical feasibility
- Economic feasibility

In other words, alt-PHY isn't being developed merely as a technical showcase, but as a means by which HR-PAN can be given capabilities that cannot be met using the 2.4 GHz ISM band. The proposed data rates are 110, 200, and (optionally) 480 Mb/s to support video conferencing, home theater, interactive networking, and extremely high-demand file downloads. The target range at the highest data rate is about 1 meter.

WiMedia Alliance

The WiMedia Alliance was formed in September 2002 by Appairent Technologies, Eastman Kodak, Hewlett-Packard, Motorola, Philips, Samsung, Sharp Laboratories, STMicroelectronics, Time Domain, and XtremeSpectrum. WiMedia's goals consist of the following:

- Promote wireless multimedia connectivity
- Leverage standards-based technologies
- Establish a certification and logo program
- Support the development of higher-layer protocols and software specifications
- Liaison with key organizations as needed

WiMedia's initial focus is to take the 802.15.3 MAC specification with an appropriate PHY and develop a complete protocol stack, similar to what the Wi-Fi Alliance has done with 802.11 and its variants, in order to ensure interoperability between various manufacturers. WiMedia is directing its initial attention at the 802.15.3a PHY, and it may adopt other evolving multimedia technologies as they emerge. In a way, then, our referring to 802.15.3 and WiMedia as synonymous at the PHY and MAC layers is premature, since it's possible that WiMedia could expand beyond 802.15.3. However, because the other specifications (802.11 Wi-Fi, 802.15.1 Bluetooth, and 802.15.4 ZigBee) have trade names associated with them, we want to give 802.15.3 a name as well. So Wi-Media it is.

MultiBand OFDM Alliance   In October 2002, General Atomics, Intel, Staccato Communications, and Time Domain Corporation formed the *Multiband Coalition*, based upon their belief that the 802.15.3a UWB standard should include multiband technology. Several other companies joined the coalition, and in June 2003 its name was changed to the *MultiBand OFDM Alliance* (MBOA). Their technical proposal will be examined in Chapter 4.

IEEE 802.15.4 Low-Rate WPAN (ZigBee)

The 802.15.4 LR-WPAN supports low-data-rate wireless connectivity between fixed, portable, or mobile devices with very low battery consumption requirements while operating within a *personal operating space* (POS) of 10 meters or less. Certain applications may dictate a longer range at a lower data rate so that low battery consumption is maintained.

Market Applications

This technology is aimed at products with simple communication needs such as interactive toys, sensors, and automation. The purpose of 802.15.4 isn't to compete directly with 802.15.1 Bluetooth or 802.15.3 WiMedia, but instead it attempts to provide communication at lower data rates and the lowest possible cost [Bla01]. Lower data rates imply that networks using ZigBee should consist of nodes using a relatively low duty cycle to prevent frame collisions or long backlog delays. Most sensor applications readily meet this characteristic because the time between measurements is usually several orders of magnitude longer than the time needed to transmit a measurement. ZigBee market applications include [Zig03]

- Wireless home security
- Remote thermostats
- Remote lighting
- Call buttons for elderly or disabled
- Universal remote control
- Wireless keyboard, mouse, and games
- Wireless smoke and carbon monoxide detectors
- Industrial and building automation and control

Questions have been raised as to whether ZigBee and Bluetooth are aimed at the same market. Although certainly several areas of market overlap, the two systems have several important differences [Bah02]. Bluetooth is more suited for ad hoc networks, where users come and go at will, whereas ZigBee operates better with nodes that are reasonably static. A Bluetooth piconet is usually somewhat short-lived, is limited to eight active devices, and is able to transfer asynchronous, isochronous, and synchronous data with reasonable efficiency. A ZigBee network can contain up to 255 devices, but the network itself is most efficient when duty cycles are low and data frames are small. As such, isochronous and synchronous data links aren't supported. The ZigBee protocol stack firmware is only about 28 kB, compared to 250 kB for the Bluetooth stack [Zig03]. Finally, battery-powered, Bluetooth-equipped devices are expected to be periodically recharged, whereas ZigBee-equipped devices are expected to run for months or years on a primary (nonrechargeable) battery.

Summary of Operations

The 802.15.4 ZigBee WPAN has the following characteristics:

- Raw data rates of 250 kb/s, 40 kb/s, and 20 kb/s, depending on the band of operation
- Star or peer-to-peer network topologies
- Variable-length device addressing for improved efficiency
- Guaranteed time slot allocation for contention-free frame exchange
- Contention-based allocation for bursty data transfers
- Fully acknowledged frames for reliability
- Ultra-low power consumption
- Carrier energy detection to reduce collision potential
- Link-quality indication for coexistence assessment

Two different types of devices can participate in 802.15.4 communications. The *full function device* (FFD) can operate as a *PAN coordinator*, which is the principal PAN controller in a piconet; as a *coordinator*, which serves a secondary role in providing synchronization beacons to other piconet members; or as a *device*, which is the basic communication element. The *reduced function device* (RFD) provides only the simplest functionality and cannot serve as any type of coordinator. The FFD can communicate with any FFD or RFD within its piconet, but the RFD can communicate only with a single FFD. The RFD is useful in extremely simple applications in which cost and power consumption must be kept to a minimum, such as in a light switch or as a basic sensor.

At least one FFD serves as the PAN coordinator, and other FFDs or RFDs associate with it in a piconet, as depicted in Figure 1-11. In the star topology, each

Figure 1-11
802.15.4
piconet
topologies can
be set up as a
star or peer
to peer.

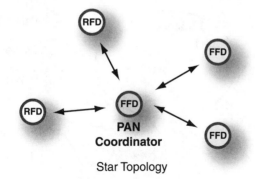

Figure 1-11
802.15.4
piconet
topologies can
be set up as a
star or peer
to peer.

Star Topology

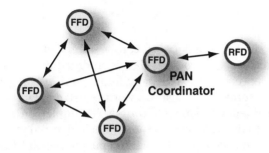

Peer-to-Peer Topology

piconet member communicates only with the PAN coordinator in a manner similar to 802.15.1 Bluetooth. This configuration may be best for applications involving home automation, PC peripherals, toys and games, and personal health care. Versatility can be enhanced through the peer-to-peer network, in which all FFDs within range can communicate directly with each other, while the RFD still communicates only with the PAN coordinator. This topology may be better suited to industrial control and monitoring, security, and surveillance.

One of the strengths of peer-to-peer structures is that they can be extended into the *cluster tree* topology (see Figure 1-12), where nodes are scattered throughout a relatively large area such that direct contact between all devices isn't possible. Instead, the PAN coordinator controls the establishment of various clusters, each with its own *cluster head* (CLH). Each cluster is established with peer-to-peer capabilities, along with a link to adjacent clusters. In this way, full communication between nodes can be obtained through a *store-and-forward* mechanism, where a message is passed from node to node until it reaches its destination. Cluster tree implementation details such as cluster setup, tear down, and traffic routing are not part of the 802.15.4 specification.

Figure 1-12
Cluster tree
network

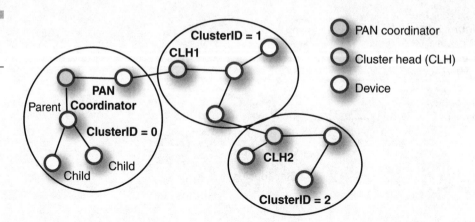

A peer-to-peer or cluster tree network can be extended into a *mesh network*, which provides automatic data routing that is updated as nodes enter the network. A self-healing feature also exists to maintain routing integrity when nodes disappear without warning. The mesh network is extremely useful for military and industrial sensor applications. Of course, its implementation can be extremely complex due to the high level of intelligence required in the routing algorithms in each node, and such algorithms aren't part of the 802.15.4 specification.

The 802.15.4 specification allocates channels in several license-free bands. One channel is at 868.3 MHz for operation in Europe, 10 channels exist in the 900 MHz ISM band for North America, and 16 channels are designated in the 2.4 GHz band for most of the world. TX power levels range from −10 to +10 dBm, with a typical power of 0 dBm, and multiyear battery life is an important goal for many applications.

802.15.4a Alt-PHY

In July 2003, a study group called IEEE 802.15.4a was established to examine the possibility of using UWB as an alternate physical layer for LR-WPAN communications. The two applications envisioned for UWB are precision location and networks requiring a high aggregate capacity [Lip03]. As we will discover in Chapter 6, network performance can come to a standstill if attempts are made to push its throughput above capacity. The store-and-forward requirements of the cluster tree network could experience this overload behavior, especially if several hops are required to move a frame to its ultimate destination. Another possible application for the Alt-PHY is a *radio frequency identification* (RFID) replacement technology that could read several hundred or thousand product tags in a short time when all are located in a small physical area. The study group is assigned to determine if applications exist that cannot be easily realized using 802.15.4 RF PHY.

ZigBee Alliance

The ZigBee Alliance was formed in October 2002 by Invensys, Mitsubishi Electric, Motorola, and Philips to focus on a connectivity standard for low-cost, low-power wireless technology. Membership is open to any interested party, from large corporations to end users. The ZigBee Alliance is developing and revising an interoperable protocol stack built on the PHY and MAC layers laid out in the 802.15.4 specification. This includes creating network, security, and application profiles.

Sources of Interference in the Unlicensed Bands

As more and more wireless devices appear in the 900 MHz, 2.4 GHz, and 5.x GHz bands, interference is expected to play a major role in determining the limits of system performance. Indeed, the cellular telephone network is so saturated with users in many urban sites that performance is limited from the *carrier-to-interference* ratio (C/I) measured at the antenna, or the *signal-to-interference ratio* (SIR) measured at the receiver's detector input, more so than from the *signal-to-noise* (S/N) ratio, or *SNR* (also measured at the detector input). Because noise and interference are both detrimental to reliable network operation, these factors are sometimes combined into a single term called the *signal-to-interference-and-noise ratio* (SINR).

Interference acts differently from noise in a number of ways. It usually varies in intensity and frequency, and it isn't caused by thermal effects. Interference can be especially disruptive if it has relatively high power compared to that from thermal noise. Finally, it can even take the form of deliberate jamming, although in most practical cases interference is accidental. As we discovered earlier, FCC rules require Part 15 devices to cope with interference on their own because they have no regulatory protection against it. Interference protection methods are also called antijam, and the implication is that the "jamming" is accidental. In consumer applications regulatory protection has been established against deliberate jamming.

Within the unlicensed bands discussed so far, the segment between 902 and 928 MHz is the most populated because inexpensive RF circuitry has been available for this band segment for a relatively long time. Many consumer products operate in this segment, mostly using TX powers of less than 0 dBm with narrowband analog or simple digital modulation. Although the sheer numbers of products operating in this band are high, the low TX power and emphasis on consumer applications mean that 900 MHz 802.15.4 operations in the home may have some coexistence problems, but the situation may be better in the office environment.

The greatest potential for interference exists in the 2.4 GHz ISM band, and that is where we focus most of our attention. Many different services operate in this band, including

- Microwave ovens
- 802.15.1 Bluetooth piconets
- 802.11b/g Wi-Fi networks
- 802.15.3 WiMedia networks
- 802.15.4 ZigBee networks
- Cordless telephones
- Custom devices
- Microwave lighting
- Licensed users such as amateur radio operators

Indeed, in several well-documented cases, Wi-Fi networks have become useless when a 2.4 GHz cordless telephone is activated within the same operating area. Also, some early Wi-Fi "hotspots" had signs asking that Bluetooth operations not be used in the vicinity. Is it really necessary to keep Wi-Fi and Bluetooth separated? We'll soon find out.

Interference is more sparse in the 5.x GHz band for two reasons. First, the U-NII band in the United States is 555 MHz wide, so independent networks in the same general area can easily find nonoverlapping channels in which to operate. Second, 5.x GHz radio technology is relatively new, and practical fabrication techniques have only recently been developed. As 5.x GHz radios become cheaper and begin to proliferate, interference will no doubt become an issue here as well.

Interference can appear in different ways, with characteristics that can be modeled in both time and frequency. In the time domain, interference can be either continuous or intermittent, and in the frequency domain it can be either narrowband or broadband. We examine specifics from these sources in other chapters, but it's clear that wireless devices need to incorporate some means of operating reliably in an interference-prone environment. Within the context of modulation and coding, for example, interference can be mitigated by

- DSSS and its variants (802.11b, 802.15.3, 802.15.4)
- FHSS (802.15.1)
- OFDM (802.11a and 802.11g)
- Error-detection codes (all 802.11 and 802.15 renditions)

- Error-correction codes (802.11a, 802.11g, and 802.15.1)
- Trellis-coded modulation (802.15.3)

Each of these methods will be discussed in detail throughout the book. Before we examine the effects of interference, though, it's important to understand how each WLAN/WPAN system operates. The technical details of their operation will be covered in subsequent chapters.

Wireless Network Coexistence Background

With the recent development of inexpensive RF fabrication technology that operates at high carrier frequencies, the 2.4 GHz ISM band has become the focus of attention for the deployment of a wide variety of communication systems. The low-cost aspect of these systems also means that a large number of them already exist, and these numbers will increase explosively as WLAN and WPAN products continue to flood the marketplace.

Throughout this book we will take a close look at the interference that various WLAN and WPAN devices must contend with, along with some of the approaches for solving the more critical situations. Because 802.11b Wi-Fi and 802.15.1 Bluetooth are the most common communication systems in this band, we will devote a lot of attention to the coexistence between these two systems. This includes interference between multiple deployments of the same system (Wi-Fi–on–Wi-Fi and Bluetooth-on-Bluetooth), as well as cross-network interference (Wi-Fi–on–Bluetooth and Bluetooth–on–Wi-Fi). Now that 802.11g systems are a reality, it's also important to analyze the interaction of Bluetooth FHSS and 802.11b CCK against 802.11g OFDM, and vice versa. Also, if 802.15.3 WiMedia and 802.15.4 ZigBee become successful, their effect on Bluetooth and 802.11b/g should be addressed. Finally, it's also necessary to consider coexistence with other RF emitters in the 2.4 GHz ISM band.

Good Neighbor Policy

The most straightforward solution to the interference problem is to simply increase the TX power so that the receiver at the other end has a better chance of extracting the desired information. If the solution were that easy, however, then this book would be extremely short. Squashing interference with higher TX power is not an acceptable approach because of the increased interference caused to other users in the band, not to mention decreased battery life and higher component cost. Furthermore, if every network node increased its TX power accordingly, then interference becomes even worse for everyone. Instead, our approaches will attempt to enhance reliable communication for all participants, not just for a single WLAN or WPAN user or a group of users. This is called the *good neighbor policy*.

Implementing such a policy is straightforward in most cases as long as the goal is enhancing the overall performance of all coexisting networks instead of simply maximizing throughput over a single link or within one network. Using just enough TX power for a link to be reliable, for example, will minimize the range at which a transmitting device interferes with other users. If all participating users implement TPC, then interference will be reduced for everyone. Conversely, if only a single node cranks up its TX power, the capability of other nearby links to communicate could be destroyed. Selecting the operating frequency through DFS distributes multiple networks throughout a given regulatory band, reducing mutual interference accordingly. Throughout this book we will invoke the good neighbor policy consistently when offering coexistence solutions and recommendations.

Noncollaborative Nodes

Different networks can either work with each other (within regulatory guidelines) or as individual entities when addressing coexistence. The *noncollaborative* approach assumes that the interfering systems have no means of communicating among them to negotiate access to the medium. This is almost always the case with wireless devices that have different hosts and are operating in different networks in a *separated topology*. Nodes operating in different networks can also be located within the same host in a *collocated topology*, which is defined by the IEEE as nodes that are physically separated by less than 0.5 m.

Each noncollaborative network works independently in an attempt to improve its own throughput, hopefully without adversely affecting other links. For example, a device can monitor activity on the communication channel and transmit only when it senses that no signal is present. This is called *carrier sense multiple access* (CSMA) and is used in both wired Ethernet (802.3) and several WLAN and WPAN implementations. Various types of spread spectrum can be employed to reduce catastrophic interference to a milder form. Most WLAN and WPAN networks operate using spread spectrum transmissions or a variation thereof.

Collaborative Nodes

The other approach to improving coexistence is through *collaborative* methods, which generally apply to collocated nodes. Nodes collaborate by communicating with each other, usually via a wired medium, and working out access to the wireless medium in such a way that mutual interference is reduced. The most basic form of collaboration is for a user to turn all devices off except for the one being used at the moment, but that isn't very efficient. Better methods are available to enhance throughput by rapidly switching transmitters and receivers on and off in such a way that interference is reduced, or through an interference cancellation process to remove unwanted artifacts from a receiver in one network caused by a nearby transmitter operating in another network. If multiple wireless systems are collocated and controlled by a single host computer, then collaboration is relatively easy to set up. Matters become more complicated if a special wireless link is needed to coordinate operation between two or more wireless systems in different parts of a room or building. Furthermore, this latter means of collaboration is prohibited in many regulatory domains.

General Coexistence Solutions

Several general techniques can be used to improve coexistence between wireless networks. Some of these are as follows:

Clear channel assessment technique   By listening for all signals sharing a particular operating channel, a node will avoid activating its transmitter while the channel is busy, even if the existing transmission originates from another network.

Modulation method   DSSS and FHSS provide some interference immunity through processing gain.

Link-quality indication   This provides an assessment of the level of interference in received packets, allowing other factors to be adjusted as necessary.

Reduced duty cycle   By increasing the data rate and data compression, messages can be conveyed more quickly, reducing the fraction of time the channel is busy.

Lower TX power   This reduces the range of interfering signals generated by a particular node.

Channel alignment   Selecting the least populated channel within the available set of operating channels obviously enhances coexistence. Furthermore, implementing channel agility, where a network moves out of a channel in which interference appears, is a way of adapting to changing channel conditions.

Reserved times for neighbor network operation   Independent networks competing for channel resources can employ a *time-division multiplexing* (TDM) method so that each operates in an environment that is interference-free from the other network.

Antenna isolation   The directivity inherent in all antennas can be exploited to reduce interaction between competing networks.

Packet structure   Packet length and error control processes can be adjusted to maximize reliability.

Retransmission algorithm   When a packet of data is received in error, the number of retransmission attempts, as well as the delay between successive attempts, can be adjusted to reduce interference to other users.

We will address these coexistence solutions as they apply to the various WLAN and WPAN interference situations that occur during typical usage scenarios.

Interference Modeling

Recently, the number of studies examining coexistence among independent WLAN systems, independent WPAN systems, and between different WLAN and WPAN networks has grown considerably. One of two approaches is generally used for modeling and analyzing coexistence. The first is the *analytical* approach. This can consist of a detailed *theoretical analysis*, where the coexistence problem is modeled mathematically and a result is expressed as a function of input variables. For example, the input variables might be the size of the room and the number of Bluetooth piconets, with the result being the average *packet error rate* (PER) or average throughput in such a situation.

Another form of analytical approach is the *computer simulation*. This involves creating a network topology model (such as placing users in a room) and running a PER simulation again and again, each time moving the users around to various locations to determine average performance. The computer simulation is especially useful when a strict theoretical analysis is too complex to be practical. Sometimes the two analytical methods are combined, and a theoretical analysis is completed, followed by a computer simulation to verify the results. The analytical approach lends itself well, for example, to modeling Bluetooth-on-Bluetooth and Wi-Fi–on–Wi-Fi interference, because the networks being analyzed have the same signal structure and timing.

Coexistence can also be examined through the *empirical* approach, where wireless devices are actually deployed and measurements taken. The most straightforward empirical method is to equip some laptop computers with WLAN nodes, others with WPAN nodes, and move them about the room, checking data rates and packet error probabilities in a *real-world* situation. A difficulty with this method is that the measurements can be influenced by additional variables such as unknown third-party interference or a room that doesn't exhibit typical RF path characteristics. To avoid these problems, the empirical analysis can be done using a *laboratory simulation*, where signal generators and protocol analyzers are connected with cable in such a way that coexistence can be tested in a controlled environment. The empirical approach is often used for modeling coexistence between dissimilar networks because it's usually easier to observe their interaction directly than perform a theoretical analysis on the vastly different signaling methods and timing used by each.

IEEE Standards Activity for Coexistence

The IEEE has sanctioned several different committees and task groups to address ongoing issues in both the WLAN and WPAN arenas. These areas include updating the standards for higher data rates, developing new standards for new market applications, and, of course, addressing coexistence. Both the IEEE and the various alliances have recognized the need for modeling coexistence issues and recommending solutions, and they are working closely with each other in these endeavors. For example, the IEEE 802.15 Task Group 2 (802.15 TG2, also called 802.15.2) has published several documents modeling coexistence, as well as material to coordinate solutions among the various manufacturers and other groups. Furthermore, the IEEE is beginning to include coexistence information within wireless standards as well.

The Bluetooth SIG Coexistence Working Group operates within the Bluetooth community and with cross-industry groups such as the IEEE to quantify the detrimental effect of interference on Bluetooth and other systems. Aside from developing solutions, the information also assists manufacturers in providing realistic performance expectations to their customers and provides valuable information to regulatory bodies. The group publishes *best practices* white papers, along with improvements to the specification and profiles as coexistence issues are solved.

802.15.2

In January 2000, the IEEE formed the IEEE 802.15 WLAN/WPAN Coexistence Task Group 2, which released the 802.15.2 recommended practice in August 2003 to address the capability of 802.15.1 Bluetooth and 802.11b Wi-Fi to operate together in the same vicinity. The group also recommends modifications to additional 802.15 standards in order to enhance their coexistence with other wireless devices. A computer model of the mutual interference between Bluetooth and Wi-Fi is part of the published work. Several collaborative and noncollaborative coexistence mechanisms have been defined within the 802.15.2 recommended practice.

Four types of WLAN-WPAN interference have been addressed by the recommended practice, which are

- 802.11 FHSS WLAN in the presence of 802.15.1 WPAN interference
- 802.11b WLAN in the presence of 802.15.1 WPAN interference
- 802.15.1 WPAN in the presence of 802.11 FHSS WLAN interference
- 802.15.1 WPAN in the presence of 802.11b interference

Of these four interference scenarios, we will only address the second and fourth ones in this book. The reason for this is that few 802.11 FHSS networks exist due to the superior data rate provided by 802.11b Wi-Fi and the newer, faster 802.11a and 802.11g renditions. We will also analyze Wi-Fi–on–Wi-Fi and Bluetooth-on-Bluetooth coexistence, neither of which is part of the 802.15.2 draft recommended practice. However, 802.15.2 presents comprehensive solutions to both collaborative and noncollaborative coexistence between 802.11b and 802.15.1, and much of Chapter 8 is devoted to this issue.

802.19

Until recently, the IEEE has focused its short-range wireless coexistence efforts within the 802.15 WPAN group. Bluetooth resides here, but Wi-Fi does not, so the IEEE formed a new group within the 802 umbrella to scrutinize coexistence within all WLAN and WPAN operations. The creation of the 802.19 Coexistence *Technical Advisory Group* (TAG) was approved by the 802 Sponsor Executive Committee in July 2002. This TAG will "develop and maintain policies defining the responsibilities of 802 standards developers to address issues of coexistence with existing standards and other standards under development." [Lan02] As we'll discover in Chapter 9, many other wireless products besides Wi-Fi and Bluetooth will be competing for access to the limited spectrum offered by the unlicensed bands, extending coexistence issues well beyond merely how WLAN and WPAN can affect each other. One of the primary concerns of 802.19 is the impending proliferation of UWB devices and their coexistence challenges.

802.11h

The 802.11 *Task Group h* (TGh) was formed to explore the possibility of harmonizing the European and U.S. 5.x GHz unlicensed bands so that the same network hardware could be used in both domains. Although primarily an interoperability issue, several items addressed by TGh were directly applicable to coexistence. The result was IEEE Standard 802.11h, published in October 2003, which adds spectrum management services to the 802.11 standard. These services consist of TPC and DFS in parts of the 5.x GHz band. Power capability and supported channel fields were added to some of the MAC frames to coordinate operations. More information on the 8021.11h standard can be found in Chapter 8.

Conclusion

Unlike wired communication systems, wireless networks have much more interference potential due to their signals being transmitted in a relatively haphazard fashion in an attempt to be heard by a receiver that may be in an unknown location. Having access to tetherless communications is so convenient, however, that it's important to solve these coexistence issues.

Some of the most popular bands in which wireless networks are placed are the ISM and U-NII bands because these frequencies can be used without obtaining a license. The 2.4 GHz ISM band is essentially for worldwide use, due in part to the existence of microwave ovens that transmit high powers at a center frequency of about 2.45 GHz. All WLAN and WPAN specifications designate some or all of their operations within the 83.5 MHz width of the 2.4 GHz ISM band, and thus have the potential to interfere with each other.

The OSI model has proven useful for designing a wireless communication system due to its layered approach. Ideally, each layer can be developed independently, and changes made to one layer won't affect the operation of other layers. The 802.11 standard upon which Wi-Fi devices are based covers only the PHY and MAC layers of the OSI model, so the Wi-Fi Alliance developed an interoperability test plan to insure that 802.11b products would operate together, regardless of manufacturer. Conversely, the Bluetooth specification addresses all layers of the OSI model, so an attempt was made to build interoperability into the specification itself. Other wireless alliances are working toward ensuring interoperability of their respective systems.

WLAN and WPAN are aimed at separate markets; the former is best at transferring data relatively quickly in a fixed or semifixed environment, whereas the latter is more suited to replacing cables between computers and peripherals in an ad hoc topology. Hence, the potential exists for both systems to routinely be operating together within the same room, or perhaps even on the same computer. Coexistence, or the capability to operate together with minimal degradation, is critically important under these conditions.

The capability of networks to coexist can be assessed through an analytical procedure in which the network is modeled on a computer or through an empirical process with actual network equipment. Each has its advantages, but similar coexisting systems (such as Bluetooth-on-Bluetooth) lend themselves to the analytical approach, and differing coexisting systems (such as Wi-Fi–on–Bluetooth are often better examined through an empirical method. The IEEE and the various alliances are working diligently to address these coexistence issues and recommend solutions.

References

[Bah02] Bahl, V., "ZigBee and Bluetooth—Competitive or Complementary?" ZigBee Alliance white paper 02/054 , September 2002.

[Bar00] Barr, J., "Bluetooth SIG and IEEE 802.15," presented at the Bluetooth Developers Conference, San Jose, CA, December 2000.

[Bla01] Blaney, T., and Nelson, B., eds., "ZigBee Alliance Market Requirements," ZigBee Alliance white paper, August 20, 2001.

[Cok01] Cokenias, T., and Judge, B., "FCC and European Certifications: In by 9, Out by 5?" presented at the Wireless Symposium, San Jose, CA, December 2001.

[IEEE802.11] IEEE 802.11-1999, "Wireless LAN Medium Access Control (MAC) and Physical Layer (PHY) Specifications," August 20, 1999.

[IEEE802.11a] IEEE 802.11b-1999, "High-Speed Physical Layer in the 5 GHz Band," September 16, 1999.

[IEEE802.11b] IEEE 802.11b-1999, "Higher-Speed Physical Layer Extension in the 2.4 GHz Band," September 16, 1999.

[IEEE802.11g] IEEE 802.11g-2003, "Amendment 4: Further Higher Data Rate Extension in the 2.4 GHz Band," June 27, 2003.

[IEEE802.11h] IEEE 802.11h-2003, "Amendment 5: Spectrum and Transmit Power Management Extensions in the 5 GHz Band in Europe," October 14, 2003.

[IEEE802.15.1] IEEE 802.15.1-2002, "Wireless Medium Access Control (MAC) and Physical Layer (PHY) Specification for Wireless Personal Area Networks (WPAN)," June 14, 2002.

[IEEE802.15.2] IEEE 802.15.2-2003, "Coexistence of Wireless Personal Area Networks with Other Wireless Devices Operating in Unlicensed Frequency Bands," August 28, 2003.

[IEEE802.15.3] IEEE 802.15.3/D17, "Wireless Medium Access Control (MAC) and Physical Layer (PHY) Specifications for High Rate Wireless Personal Area Networks (WPAN)" draft standard, February 2003.

[IEEE802.15.4] IEEE 802.15.4-2003, "Wireless Medium Access Control (MAC) and Physical Layer (PHY) Specifications for Low-Rate Wireless Personal Area Networks (LR-PANS)," October 1, 2003.

[Lan02] Lansford, J., excerpt from 802.19 TAG Statement of Purpose, quoted in general message, July 26, 2002.

[Lip03] Lipset, V. "Group Pushes UWB for Low-Power Networks," www.ultra-widebandplanet.com article, July 22, 2003.

[Oha99] O'Hara, B., and Petrick, A., *IEEE 802.11 Handbook: A Designer's Companion*, New York: IEEE Press, 1999.

[Rob02] Roberts, Rick, et al., "IEEE P802.15.SG3a Five Criteria," August 30, 2002, revised.

[Rob03] Roberts, Glynn, "An Introduction to Wi-Media," Wi-Media Alliance white paper, 2003.

[Sta88] Stallings, W., *Data and Computer Communications,* 2nd ed. New York: Macmillan, 1988.

[Xtr01] Gandolfo, P., "The Optimum MAC/PHY Combination for Multimedia Consumer Applications," XtremeSpectrum white paper, www.xtremespectrum.com, September 2001.

[Zig03] "Questions and Answers about the ZigBee Technology," retrieved from www.zigbee.org, August 19, 2003.

Indoor RF Propagation and Diversity Techniques

Modeling the propagation of radio waves in an indoor environment can be one of the most complex endeavors that any engineer can attempt. *Radio frequency* (RF) energy travels from transmitter to receiver over many different paths, each of which acts differently. Therefore, multiple signals arrive at the receiver with different amplitude, phase, and distortion characteristics. Furthermore, these characteristics change as obstacles, including people, change position. It's impossible, then, to determine how the incoming signal appears at a wireless receiver with great precision; instead, various approximations must be made to produce a tractable propagation model. Once that is achieved, we can then proceed to determine the approximate range over which the communication link is reliable. If no interfering signals are present, the maximum range is a function of how weak the received signal can be before the *bit error rate* (BER) rises to an unacceptable level. As the wireless bands become more crowded, the maximum range will instead be determined more by the ratio of desired signal powers to the interfering ones.

In this chapter we will present some models that usually yield good approximations to large-scale losses over the communications link, or path, which can be used to determine the average strength of the signal the receiver is trying to detect. This signal is called the *desired signal*, and the associated receiver is called the *desired receiver*. The propagation models presented here can also produce accurate results for the various interfering signals at the desired receiver, thus giving us the ability to find the *carrier-to-interference ratio* (C/I) at the desired receiver. By combining these results with receiver performance, we can discover what effect interfering signals will have on the capability of the desired receiver to accurately detect the desired signal.

Indoor RF Propagation Mechanisms

A radio wave can take several routes between the transmitter and receiver. These routes generally fall into four different classes, as shown in Figure 2-1. Only one direct path exists, and this is called *line of sight* (LOS) if no obstructions occur between the transmitter and receiver; otherwise, the path is called *obstructed line of sight* (OLOS) or *nonline of sight* (NLOS). In most indoor communication situations, the direct path is OLOS due to the close proximity of people, furniture, and the communication equipment itself to the antenna. The direct path often, but not always, provides the strongest signal at the receiver. By definition, though, the direct path signal arrives at the receiver first.

Signals that propagate via one of the other three path classes will travel a longer distance between the transmitter and receiver, so these arrive at the receiver at a later time than the signal on the direct path. Also, each of these three path classes can contain several signal copies, each taking a different route to the receiver. Diffraction occurs when a radio wave encounters a sharp edge,

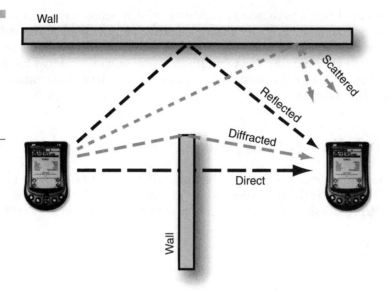

Figure 2-1
Indoor propagation mechanisms can be direct, reflected, diffracted, or scattered.

and some of its energy is bent around the edge. This effect is most common near doorways and other openings. Waves are scattered when they impact a surface that has irregularities that are a significant fraction of the wavelength involved. Under most conditions, though, the most significant of these nondirect paths is the reflected path class. Signals that are in this class produce *multipath*, which can be extremely detrimental to the reliability of signal detection, as we'll discover later in this chapter.

To avoid becoming overwhelmed by attempting to model all these path classes, we will for now consider only the direct path. Next, by retreating from some of the simplifying assumptions, accuracy is improved at the cost of higher calculation complexity. Finally, when multipath is included in the analysis, computation times are increased further but still greater accuracy is achieved.

Large-Scale Propagation Model

The goal of a large-scale RF propagation model is to find the average signal power at the receiver given various characteristics at the transmitter, receiver, and along the path. Signal strengths calculated in this manner are called *large scale* because the power of any signal at a receiver, called the *received signal strength indication* (RSSI), is considered to be averaged over a circular or linear track of about 10 wavelengths (about 3 m at 900 MHz, 1 m at 2.4 GHz, and 0.5 m at 5.x GHz) for each value [Kim96]. Thus, RSSI values, either calculated or measured, don't take into account the smaller-scale variations from multipath effects. It may be argued that multipath is so prevalent in the indoor environment that it's useless to find average RSSI values. Although it's true that RSSI can vary considerably due to multipath, tremendous progress has been made in finding ways to mitigate its effect, resulting in actual RSSI values with a much smaller variance.

Link Budget Equation

The classic free-space link budget equation is used extensively in satellite and LOS terrestrial applications, but it can be suitably modified to model an obstructed indoor or outdoor environment and expressed as [Rap96]

$$P_{r(dBm)} = P_{t(dBm)} + G_{t(dB)} + G_{r(dB)} + 20 \log\left(\frac{\lambda}{4\pi}\right) + 10n \log\left(\frac{1}{d}\right) \quad (2.1)$$

where P_r is the received signal power, P_t is the transmitted power, G_t is the gain of the *transmit* (TX) antenna in the direction of the *receive* (RX) antenna, and G_r is the gain of the RX antenna in the direction of the TX antenna. The carrier wavelength is $\lambda = c/f$, where c is the speed of light at 3×10^8 m/s and f is the frequency in Hz. Wavelengths for carriers in the unlicensed bands are given in Table 2-1. Also, a factor n enables the effect of obstructions to be incorporated into the link budget equation. The value for n will be determined later. Because *decibels* (dB) and *decibels relative to 1 milliwatt* (dBm) are used so often, we will drop the subscripts in the variables and express powers and gains in dBm and dB, respectively, most of the time.

Equation 2.1 has an important restriction: The RX antenna must be far enough away from the TX antenna that the wave is essentially planar so normal

Table 2-1

Carrier wavelengths in the 900 MHz, 2.4 GHz, and 5.x GHz unlicensed bands

Band	Center frequency wavelength
902–928 MHz	0.33 m
2400–2483.5 MHz	0.12 m
5.15–5.25 GHz	0.058 m
5.25–5.35 GHz	0.057 m
5.470–5.725 GHz	0.054 m
5.725–5.825 GHz	0.052 m

propagation physics applies; that is, the RX antenna must be in the *far field*. Although some disagreement exists as to what the minimum far field distance should be [Ban99], the formula is reasonably accurate when [Bes01]

$$d > \frac{2D^2}{\lambda} \qquad (2.2)$$

where D is the longest antenna dimension in meters. For most portable wireless applications, the antenna length is effectively $\lambda/2$, so Equation 2.1 applies for distances greater than half a wavelength from the TX antenna. The formula clearly breaks down for very small values of d, because P_r approaches infinity as d approaches zero for any valid P_t.

Path Loss in the ISM and U-NII Bands

Equation 2.1 can be simplified further by keeping only those terms associated with the path between the transmitter and receiver. As such, the antenna gains are assumed to be 0 *decibels relative to an isotropic source* (dBi) for now. Furthermore, we will change terminology slightly and focus on the *path loss* (PL) between the transmitter and receiver, which is related to the ratio of actual TX and RX signal powers. The PL in dB becomes

$$PL = 20 \log\left(\frac{4\pi}{\lambda}\right) + 10n \log(d) \qquad (2.3)$$

This PL formula forms the basis of what we call the *distance-dependent model*, because the main independent variable in Equation 2.3 is the distance between the transmitter and receiver once the values for n and λ are inserted.

Equation 2.3 produces a loss instead of a gain, so the terms are inverted with respect to those in Equation 2.1. As a result, PL is a positive dB quantity because both terms on the right side are positive for wavelengths shorter than 4π m, and for $d > 1$ m, which is the case for all of our calculations. Also, if both TX and RX antennas have a gain of 0 dB, then Equation 2.3 is just the ratio of P_t to P_r in dB. This insight will become useful when we make our first attempt at calculating *wireless local area network* (WLAN) or *wireless personal area network* (WPAN) maximum range and C/I values.

The Effect of Wavelength on PL

The value of the first term on the right side of Equation 2.3 is influenced by the square of the RF carrier wavelength, and this term becomes larger as λ

decreases. If other factors remain unchanged, doubling the carrier frequency (halving the wavelength) increases PL by 6 dB, which is a factor of 4. This is because the physical size of a particular type of antenna is directly proportional to its operating wavelength. When the wavelength decreases, the antenna becomes smaller for a fixed gain and it therefore captures proportionally less of the available RF wavefront at a given distance d from the TX antenna. Because power is directly proportional to the square of the field strength, an antenna that physically shrinks by half will lose 6 dB of captured signal power. This effect will become obvious as we substitute the appropriate λ for each *industrial, scientific, and medical* (ISM) band into the PL equation.

For the 900 MHz ISM band, substituting $\lambda = 0.12$ m for a frequency of 915 MHz into Equation 2.3, the equation simplifies to

$$PL = 32 + 10n \log(d) \tag{2.4}$$

By substituting $\lambda = 0.12$ m for a frequency of 2.4 GHz into 2.3 we obtain

$$PL = 40 + 10n \log(d) \tag{2.5}$$

Likewise, a substitution of $\lambda = 0.055$ m for the average wavelength at 5.x GHz produces

$$PL = 47 + 10n \log(d) \tag{2.6}$$

Because only one value of the PL exponent is used in the previous equations, we will call this distance-dependent method the *simplified PL model*. For a given distance d, the PL at 2.4 GHz is about 7 dB lower than at 5.x GHz, and about 8 dB higher than at 900 MHz. Also, the first term in each of the three PL equations represents the signal loss at a distance of 1 m from the TX antenna. At a carrier frequency of about 900 MHz, the PL 1 m from the transmitter is 32 dB. To put this in perspective, the popular RG-59 coaxial cable has a loss at 900 MHz of about 30 dB per 100 m [Hal91].

PL Exponent and Typical Values

The dimensionless quantity n in Equations 2.1 and 2.3 to 2.6 is called the *PL exponent*, and it can be adjusted to account for the amount of clutter in the path between the TX and RX antennas. For example, if free-space propagation applies, then $n = 2.0$. The resulting inverse square law states that the signal power drops in proportion to the square of the distance between the antennas. It often helps to visualize these changes in a "per octave" sense (doubling d) or in a "per decade"

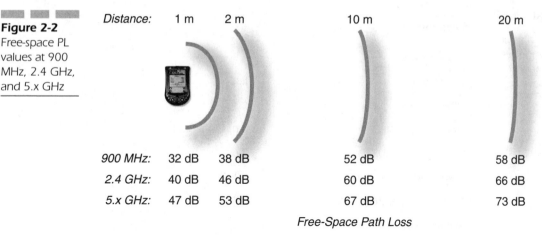

Figure 2-2
Free-space PL
values at 900
MHz, 2.4 GHz,
and 5.x GHz

Distance: 1 m   2 m   10 m   20 m

	1 m	2 m	10 m	20 m
900 MHz:	32 dB	38 dB	52 dB	58 dB
2.4 GHz:	40 dB	46 dB	60 dB	66 dB
5.x GHz:	47 dB	53 dB	67 dB	73 dB

Free-Space Path Loss

sense (multiplying d by 10). Total PL as a function of distance therefore increases by 6 dB per octave, or 20 dB per decade, when free-space propagation applies. This is shown in Figure 2-2 for the three ISM/*Unlicensed National Information Infrastructure* (U-NII) bands.

Equations 2.4 through 2.6 can become even more useful when we substitute different values of n to account for various non-LOS situations. If clutter exists between the transmitter and receiver, which is common in both indoor and outdoor environments, this can be modeled by raising the PL exponent to a value greater than 2.0. Surprisingly, in certain occasions n can be less than 2.0. This commonly occurs when the transmitter and receiver are located in a long hallway, so multiple reflections off the walls, floor, and ceiling cause the RF wave to sustain itself as if it were in a crude waveguide. The PL in these cases may be as low as 10 dB per decade, dramatically increasing its useful range.

Selecting the PL Exponent   Selecting the proper PL exponent value can be done in a number of ways. An empirical method is to place a transmitter of known output power somewhere in the building and measure the RSSI throughout the area of interest. After compensating for the antenna gains, a curve-fitting technique (such as least squares) can be used to find the best estimate for the PL exponent value. Of course, the RSSI measurements apply to the transmitter located only at that specific position, as it would be for a fixed WLAN *access point* (AP) or a fixed Bluetooth-equipped device such as a printer or desktop computer, but most WPAN piconets and some WLANs are more dynamic. Consequently, n must be estimated based on the type of environment the network is located within and by performing many RSSI measurements with the transmitter

Table 2-2

Typical same-floor indoor PL exponent values

Location	n
Grocery store	1.8
Retail store	2.2
Office, moveable walls	2.4
Office, fixed walls	3.0
Textile factory, LOS	2.0
Textile factory, OLOS	2.1
Metalworking factory, LOS	1.6
Metalworking factory, OLOS	3.3

placed in several different locations. Fortunately, much of that work has already been accomplished, and some of the results are summarized in Table 2-2 [And94].

These measurements were made on the same building floor at frequencies ranging from 900 to 1,500 MHz, but n isn't particularly dependent on the carrier frequency, even for frequencies somewhat beyond the 5.x GHz band [Rap96]. For example, the textile factory LOS measurements at 1.3 GHz are accurately modeled for $n = 2.0$, but at 4.0 GHz, n increases only slightly to 2.1. The typical standard deviations of the set of actual PL measurements against the average values found by using Equations 2.4 to 2.6 are between 6 and 10 dB.

As clutter along the direct RF path increases, the PL exponent increases to a value of 3.3 in the difficult non-LOS environment in a metalworking factory. In the same factory, the LOS path can have a PL exponent less than free space due to the waveguide effect. The PL exponent in a typical office environment is about 3.0 when operating through fixed walls, but if the communicating nodes are in the same room, or when cubicle walls are the only obstructions, the PL exponent is usually slightly above 2.0.

When the transmitter and receiver are located on different floors of a building, the PL exponent can be significantly higher, as shown in Table 2-3 for measurements taken at 914 MHz [Sei92]. An RF link often requires transmitter powers in the tens or hundreds of *milliwatts* (mW) to be reliable throughout a large part of a building, especially when the communicating nodes are on different floors. Keep in mind, though, that these high floor attenuations can be advantageous from a coexistence point of view because a low-power network on one floor of a building will probably produce little interference to networks on other floors, nor will it be particularly susceptible to interference from these other networks.

Table 2-3

Average PL exponent values in multi-floor buildings

Number of Floors	n
Same floor	2.8
Through one floor	4.2
Through two floors	5.0
Through three floors	5.2

Estimating Range Using Maximum Allowable Path Loss (MAPL)

To estimate the useful range for wireless devices operating within the ISM and U-NII bands, it's necessary to determine the *maximum allowable PL* (MAPL) that the desired receiver can tolerate and still function. This is given by

$$\text{MAPL} = \text{TX power} - \text{RX sensitivity} - \text{other losses} \qquad (2.7)$$

with all quantities expressed in dB (dBm for TX power and RX sensitivity). We'll assume that the other losses are 0, at least for now. For instance, a perfectly efficient omnidirectional antenna has a gain of 0 dBi and thus won't contribute to other losses.

Example 2-1:
A Wi-Fi AP transmits with a power of +20 dBm, and the RX sensitivity of a *station* (STA) in the network is −80 dBm. What is the MAPL?
Solution:

$$\text{MAPL} = 20 - (-80) = 100 \text{ dB, ignoring other losses}$$

The most common TX power for 802.15 devices is 0 dBm, and 802.11 nodes have typical TX power levels up to +20 dBm. Most commercial receivers for WLAN and WPAN applications have sensitivities between −70 and −90 dBm, so the MAPL ranges from about 70 to 110 dB in the majority of applications.

To determine the maximum range using the simplified PL model, the value for n that best matches the path clutter is placed into the appropriate PL equation, and then d is solved for the required MAPL. These results are summarized in Table 2-4. Because WLAN and WPAN links must be bidirectional, the transmitter and receiver characteristics given in Table 2-4 must be present at both ends

Table 2-4

Range estimates in the ISM and U-NII bands using the simplified PL model

Type of clutter	n	TX power (dBm)	RX sensitivity (dBm)	MAPL (dB)	Range (m) 900 MHz	Range (m) 2.4 GHz	Range (m) 5.x GHz
None	2.0	0	−70	70	79	31	14
(free space)	2.0	0	−80	80	251	100	45
	2.0	+20	−70	90	794	316	141
	2.0	+20	−80	100	2512	1000	447
Light	2.5	0	−70	70	33	16	8
	2.5	0	−80	80	83	40	21
	2.5	+20	−70	90	209	100	52
	2.5	+20	−80	100	525	251	132
Moderate	3.0	0	−70	70	18	10	6
	3.0	0	−80	80	40	22	13
	3.0	+20	−70	90	86	46	27
	3.0	+20	−80	100	185	100	58
Heavy	4.0	0	−70	70	9	6	4
	4.0	0	−80	80	16	10	7
	4.0	+20	−70	90	28	18	12
	4.0	+20	−80	100	50	32	21

of the link for the listed ranges to be valid. For an asymmetrical link, the maximum range is determined by the link direction that can tolerate the lowest PL.

Example 2-2:

A Bluetooth node has a TX power of 0 dBm and an RX sensitivity of −80 dBm. The RF path has moderate clutter. If it is communicating with another node with the same radio specifications and ignoring other losses, what is the estimated range?

Solution:

First, use Equation 2.7 to calculate MAPL: $0 - (-80) = 80$ dB. Next, substitute the MAPL into Equation 2.5 and solve for d:

$$PL = 40 + 10n \log(d) \rightarrow MAPL = 40 + 10(3.0) \log(d) \rightarrow$$

$$80 = 40 + 30 \log(d) \rightarrow d = 22 \text{ m}$$

Outdoor Point-to-Point Range Estimates at 5.x GHz

For fixed point-to-point WLAN operations within the U-NII band at 5.725 to 5.825 GHz, FCC rules allow a TX power up to +30 dBm, feeding a directional antenna with a gain up to +23 dBi. The most common use for such a system is extending 802.11a operation between buildings within a city. Paths are often LOS (or nearly so) under such conditions, so the equivalent PL exponent is 2.0 to 2.5.

Table 2.5 shows range figures for some typical point-to-point links using either an omnidirectional or a gain antenna. As before, we assume that identical conditions exist at both ends of the link. We also assume that the directional antenna enhances both TX power and RX sensitivity by an amount equivalent to the antenna's gain in dBi. This "double-enhancement" property is a good reason to employ a gain antenna, rather than a higher TX power, to increase the maximum range whenever both ends of the link are fixed.

It's immediately apparent from Table 2-5 that the range is enhanced significantly when a gain antenna is placed at each end of the link. These antennas achieve their gain by concentrating power in a particular direction, so they must be aimed more carefully as their gain increases. In return for this effort, we obtain both the increased range for the desired link and the increased interference rejection from directions other than those in which the antenna is aimed.

When a gain antenna is used, a relatively small increase in n from 2.0 to 2.5 results in a drop of the maximum range by a factor of about 10, so it would be

Table 2-5

Range estimates for point-to-point links at 5.7 GHz using the simplified PL model

n	TX power (dBm)	RX sensitivity (dBm)	Antenna gain (dBi)	MAPL (dB)	Range
2.0	+20	−80	0	100	447 m
2.0	+30	−80	0	110	1.4 km
2.0	+20	−80	+23	146	89 km
2.0	+30	−80	+23	156	282 km
2.5	+20	−80	0	100	132 m
2.5	+30	−80	0	110	331 m
2.5	+20	−80	+23	146	9.1 km
2.5	+30	−80	+23	156	23 km

prudent not to underestimate the value of the PL exponent when deploying such systems. Even with an omnidirectional antenna, the maximum range drops by a factor of about 4 for a similar increase in n.

IEEE Breakpoint Path Loss Model

Suppose two members of a WLAN or WPAN are both located in the same room. This will probably be a common occurrence, and the PL exponent in such a situation may be close to that of free space. If one of the network members moves to another room, the adjoining wall may cause the PL exponent to increase to about 3.0 for these longer separation distances. This leads to the possibility of adjusting n as the distance increases in an attempt to improve accuracy without adding undue complexity.

For several types of indoor network path analysis, the *Institute of Electrical and Electronics Engineers* (IEEE) uses a PL model that assumes free-space propagation out to 8 m from the transmitter. For greater distances the PL exponent increases to 3.3 [IEEE802.15.2]. At a carrier frequency of 2.4 GHz, the total PL is found by using a set of formulas given by

$$PL = 40.2 + 20 \log(d), \qquad 0.5 \le d \le 8\,\text{m}; \tag{2.8}$$

$$PL = 58.5 + 33 \log\left(\frac{d}{8}\right), \quad d > 8\,\text{m} \tag{2.9}$$

The first term in Equation 2.9 represents the PL in dB at 8 m with $n = 2.0$, and the second term is the additional loss beyond 8 m using $n = 3.3$. This form of distance-dependent model is called the *breakpoint model*. For 900 MHz carriers, the first terms in Equations 2.8 and 2.9 should be reduced by 8 dB, and for 5.x GHz the first terms should be increased by 7 dB. Figure 2-3 depicts some path loss values in the three unlicensed bands at various distances using the IEEE breakpoint model.

Despite its complex mathematical appearance, it's easy to estimate distance given a PL, or vice versa, using the breakpoint model. Starting with a value of 40 dB at 1 m, the PL drops by 10 dB each time the distance increases by a factor of about 3 to a distance of 9 m, at which the total PL is about 60 dB. Now the higher PL exponent of 3.3 means that the PL drops 10 dB each time the distance doubles. These distance estimates, compared to the values obtained using Equations 2.8 and 2.9, are given in Table 2.6 for a PL up to 100 dB. Although we use a breakpoint at 9 m instead of 8 m, our estimates are extremely close to the values given by the IEEE model.

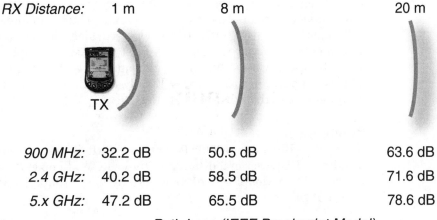

Figure 2-3

PL values at 900 MHz, 2.4 GHz, and 5.x GHz at various distances using the IEEE breakpoint model

RX Distance:	1 m	8 m	20 m
900 MHz:	32.2 dB	50.5 dB	63.6 dB
2.4 GHz:	40.2 dB	58.5 dB	71.6 dB
5.x GHz:	47.2 dB	65.5 dB	78.6 dB

Path Loss (IEEE Breakpoint Model)

Table 2-6

Estimated PL at 2.4 GHz compared to values from the IEEE breakpoint PL model

	Distance (m)	
Path Loss (dB)	**Estimated**	**IEEE model**
40	1	0.98
50	3	3.1
60	9	8.9
70	18	17.9
80	36	35.9
90	72	72.0
100	144	145

Estimating C/I Using the Simplified and Breakpoint PL Models

Up to this point we have estimated the maximum range of a wireless link based strictly upon the desired signal RSSI. As we'll discover in Chapter 5, "Radio Performance," RSSI and the *signal-to-noise ratio* (SNR) at the receiver are related, and the minimum RSSI needed to obtain a certain performance level is specified by the manufacturer. In an environment in which interference plays a major role, receiver performance will be limited by the C/I instead, which is the ratio of the desired signal RSSI to the power in all interfering signals. Receiver C/I performance under various situations is also provided by the manufacturer.

Estimating C/I using either PL model is straightforward, especially if there's just a single interfering transmission. First, the desired signal's RSSI is computed using the appropriate values in the PL equation, and then the interfering signal RSSI is computed in a similar way. Finally, the ratio of the two numbers, given by the difference in their dB values, is the C/I.

Example 2-3:

A desired Bluetooth piconet is operating with the users separated by 5 m, and each has a TX power of 0 dBm. An interfering Wi-Fi node is located 30 m away and has a TX power of +15 dBm. According to the simplified PL model, what is the C/I at the desired receiver?

Solution:

First, substitute d = 5 m into Equation 2.5, which gives a PL of 61 dB. Subtracting the PL from the TX power produces a desired signal RSSI of −61 dBm at the desired Bluetooth receiver. Similarly, substituting d = 30 m into Equation 2.5 yields a PL of 84 dB, so the interfering signal RSSI at the same Bluetooth receiver is −69 dBm. Finally, the C/I is 8 dB, −61 − (−69).

Example 2-4:

Repeat the calculations in Example 2-3 using the IEEE breakpoint model.

Solution:

Substitute d = 5 m into Equation 2.8, which gives a PL of 54 dB. Subtracting the PL from the TX power produces a desired signal RSSI of −54 dBm at the desired Bluetooth receiver. Similarly, substituting d = 30 m into Equation 2.9 yields a PL of 77 dB, so the interfering signal RSSI at the same Bluetooth receiver is −62 dBm. Finally, the C/I is 8 dB, −54 − (−62).

The C/I value calculated in the previous example represents the ratio of desired to interfering signal powers prior to any filtering by the Bluetooth receiver. The estimated C/I values using either PL model are essentially identical.

If multiple interferers are affecting the performance of a desired receiver, the total interfering signal power can be approximated as the sum of the individual interfering signal powers. This quantity is obtained by first converting their respective dBm values to linear values, adding them together, and then converting back to dBm. For example, suppose two interfering signals are present at the desired receiver, one at −50 dBm and the other at −53 dBm. Converting to linear values, we get 10 nW and 5 nW, respectively, so the total interfering power is 15 *nanowatts* (nW), equivalent to −48 dBm.

Evaluation of Distance-Dependent PL Models

With its free-space propagation assumption for TX-RX separation distances up to 8 m, the IEEE breakpoint propagation model yields the average RSSI without including much pessimism in the calculations. The simplified PL model, on the other hand, is biased toward pessimism by assuming a PL exponent of 3.0 for all separation distances. Comparing Equation 2.5 to Equations 2.8 and 2.9, both models begin with a 2.4 GHz PL of about 40 dB at a distance of 1 m, which is the free-space value. As the distance increases, the simplified PL model gradually becomes more pessimistic compared to the IEEE breakpoint model until, at a distance of 8 m, they differ by about 9 dB. At distances greater than 8 m, the PL determined by the IEEE breakpoint model increases slightly faster than that from the simplified PL model. For example, at 100 m the two methods differ by 5.3 dB.

The IEEE breakpoint model can, under most circumstances, claim a higher accuracy than the simplified PL model for finding average RSSI values. The latter model is slightly easier to use, because it's not required that the PL exponent be changed based on distance. The simplified PL model also has a built-in pessimism of 5 to 9 dB at normal short-range wireless separation distances to account for other losses that often creep into a real-world scenario.

Improving Accuracy of the Distance-Dependent Models

Generally, four different problems can lead to errors in the WLAN and WPAN ranges and in the C/I calculations that are based upon the simplified and breakpoint PL models given by Equations 2.4 through 2.6 and 2.8 through 2.9. These four sources of error are as follows:

- Variations in clutter along the direct path
- The presence of reflected signals from multipath
- Antenna directivity and inefficiencies
- Interference variations

The first three of these will be discussed later in this chapter, and the last will be addressed throughout the remainder of this book. It would be helpful, though, to summarize their effects on the accuracy of simple PL models.

Variations in the amount of clutter will have an effect on the actual PL that depends on where the transmitter and receiver are located within a room or building. The actual range, then, could be significantly different from the average predicted value.

Because many rooms are highly reflection-prone to RF signals, several copies of a transmitted wave can arrive at the receive antenna at different times and with different phase relationships. As we will soon discover, the resulting multipath can cause signal dropouts (fades) of up to 30 dB, equivalent to a TX power reduction by a factor of one thousand. Fortunately, elementary techniques are available to reduce multipath fading to 10 dB or less, but its effect on system performance can still be significant.

The antenna in a portable wireless device is often designed to have nearly omnidirectional characteristics, at least in the azimuth plane, in order to free the user from having to aim the device in a particular direction. Various factors such as poor matching to the transmitter or receiver, poor efficiency, and RF absorption often result in a loss of 3 to 5 dB per antenna. Both TX and RX antennas together can reduce RSSI, and hence MAPL, by perhaps 10 dB. That relates to a 50 percent range reduction in a moderately cluttered path.

Finally, with the large number of wireless devices competing for access, especially within the 2.4 GHz band, interference is a major cause of reduced performance. The reliability of outdoor cellular systems, for example, has long been interference limited rather than signal-to-noise limited, and indoor systems will soon follow suit.

Primary Ray Tracing for Improved Accuracy

Up to now we have lumped all clutter between the transmitter and receiver into a quantity called the PL exponent. By selecting n carefully and using Equations 2.4, 2.5, or 2.6, we can quickly find a corresponding PL for a given distance, or a corresponding distance for a given PL. Measured PL standard deviations can exceed 10 dB due to clutter variations as the transmitter and receiver are moved about the operating area. This can cause significant link outage periods near the range limit [Mor02]. As we've already discussed, incorporating pessimism into the PL calculations can produce range or C/I values that are likely to be exceeded in actual network operations.

It is possible to improve accuracy substantially by directly modeling partition losses along the RF link rather than trying to incorporate all these obstructions into the PL exponent. By placing these obstructions into a *computer-aided design* (CAD) drawing of a building floor plan, then each path of interest can be traced on the drawing and a corresponding PL can be calculated based upon the actual set of obstructions between transmitter and receiver. This process is called *primary ray tracing*.

Direct Modeling of Partition Losses

The direct RF path between the transmitter and receiver is characterized by both distance and the type and number of obstructions through which the radiated energy must pass. In a typical building in which wireless devices are used, signals may pass through people, computers, and furniture within a room, as well as through walls, doors, and/or windows if the piconet users are located in different rooms. It's generally reasonable to calculate the PL from partitions that tend to remain fixed over long periods of time, less so from objects that are easily moved such as people, tables, and chairs. Because these movable objects are undoubtedly present in the office environment, a bit of pessimism may be in order here as well.

The concept behind finding total PL values using primary ray tracing employs the following rules [Sei94]:

- Signal strength drops suddenly as it passes through a partition, and the amount of drop depends on the type of partition.
- Free-space PL occurs between partitions.
- Losses between floors exhibit special behavior and are modeled separately.

These rules can be shown mathematically as

$$\text{PL} = 20 \log\left(\frac{4\pi d}{\lambda}\right) + \sum_i [(P_i)(AF_i)] + FAF \qquad (2.10)$$

We'll first customize Equation 2.10 for the ISM and U-NII bands and then explain the meaning of each term. For the 900 MHz band, we can substitute $\lambda = 0.33$ m into 2.10 and obtain

$$\text{PL} = 32 + 20 \log(d) + \sum_i [(P_i)(AF_i)] + FAF \qquad (2.11)$$

For the 2.4 GHz band, the equation becomes

$$\text{PL} = 40 + 20 \log(d) + \sum_i [(P_i)(AF_i)] + FAF \qquad (2.12)$$

and for the 5.x GHz band we get

$$\text{PL} = 47 + 20 \log(d) + \sum_i [(P_i)(AF_i)] + FAF \qquad (2.13)$$

The first term on the right side of Equations 2.11 through 2.13 represents the free-space PL at 1 m for the frequency of interest. The second term includes the additional free-space PL at a distance d beyond 1 m. These values are the same as those given by Equations 2.4 through 2.6 for $n = 2.0$.

The third term models additional losses from the partitions through which the signal must pass. A partition of type i is given an *attenuation factor* AF_i, and P_i partitions of type i are in the path. Floor-ceiling partitions are not included here, but instead are incorporated into the last term, called the *floor attenuation factor* (FAF). This represents an additional attenuation if a signal must pass through one or more floor-ceiling partitions en route to the receiver.

To summarize, primary ray tracing looks at propagation along a vector from transmitter to receiver. RF energy leaving the transmitter experiences free-space path loss until arriving at the first obstruction, which then causes a sudden loss equal to its AF as the wave passes through it. Free-space loss occurs again until the next obstruction is encountered, and so on, until the wave arrives at the receiver.

Table 2-7 shows some typical attenuation factors for various partitions found in a home or office environment [Rap96]. Although most of these measurements were taken at 1.3 GHz, their values are much more dependent upon construction materials than upon carrier frequency.

Example 2-5:

A transmitter and receiver are separated by two fixed walls, one door, and one metal file cabinet. According to Equation 2.12 and Table 2-7, what is the total partition attenuation?

Solution:

Partition attenuation is 2 fixed walls at 3 dB each, one door at 2 dB, and one metal partition at 5 dB, for a total of 13 dB.

Table 2-7

Typical partition attenuation factors

Partition	Loss (dB)
Moveable walls	1.4
Doors	2.0
Windows	2.0
Fixed walls	3.0
Metal partitions	5.0
Exterior walls	10.0
Basement walls	20.0

The partition attenuations in Table 2-7 are average values measured in many different buildings. Losses through a specific partition can be found by using a *continuous wave* (CW) transmitter at the desired carrier frequency, and a receiver with RSSI capability. First, the transmitter is operated at a known distance from the receiver and the RSSI is noted. Next, the partition of interest is placed between the transmitter and receiver, which are kept at the same distance from each other as before, and the new (weaker) RSSI is recorded. Both of these measurements are best accomplished outdoors or in an anechoic chamber to minimize the influence of multipath. The partition attenuation factor is the difference between the two RSSI values in dB.

Signal attenuation through the floors/ceilings in a building exhibits some unusual behavior. The increases in FAF taper off as the signal passes through multiple stories of a building, as shown in Table 2-8 [Sei94]. The results in the table were measured in a specific multistory building. The loss through one floor is quite high, at 13 dB, which is to be expected from a partition constructed from thick, reinforced material. Furthermore, the PL standard deviation measured through the single floor is also somewhat high at 7.0 dB. Adding a second floor increases the FAF by only 6 dB, and the standard deviation is significantly reduced. Finally, the FAF through four or more floors becomes constant at about 27 dB, with a small standard deviation. Researchers believe that this behavior is due to RF finding an alternate path around the high attenuation through floor-ceiling partitions along the direct path. The total PL still increases in proportion to the number of floors separating the communicating nodes because the path is longer and the number of other partitions along the path is probably higher as well, both of which are accounted for in Equations 2.10 through 2.13.

Example 2-6:

Refer to the two 2.4 GHz RF links depicted in Figure 2-4. The PDA TX power is +10 dBm. Phone A is 5 m from the PDA, and Phone B is 20 m from the PDA. Use the partition AF values in Table 2-7, and assume that all communication occurs on the ground floor level. What is the estimated RSSI at phone A and at phone B?

Table 2-8

Typical floor attenuation factors

Number of floors	Loss (dB)	σ(dB)
1	13	7.0
2	19	2.8
3	24	1.7
4 or more	27	1.5

Figure 2-4
Path topology
for Example
2-6

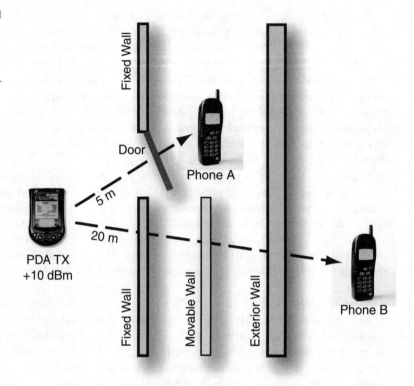

Solution:

Partition attenuation at Phone A is one door at 2 dB, so inserting this value, along with $d = 5$ m into Equation 2.12, we obtain a PL of 56 dB. Subtracting this value from the PDA TX power of +10 dBm produces an estimated RSSI at Phone A of −46 dBm.

Partition attenuation at Phone B is from one fixed wall (3.0 dB), one movable wall (1.4 dB), and one exterior wall (10 dB) for a total AF of 14.4 dB. For a distance of 20 m, Equation 2.12 produces a PL of 80 dB, which equates to an estimated RSSI at Phone B of −70 dBm.

Figure 2-5 shows a plot of RSSI contours (traces of constant RSSI values) on one floor of a building with a centrally located transmitter. Figure 2-5a represents PL contours using the distance-dependent model given by Equation 2.5. Figure 2-5b plots contours given by Equation 2.12 using the primary ray-tracing model. The contours in the latter model are no longer concentric circles; instead, signals are depicted as weaker in the rooms and stronger in the hallways, as we would expect RSSI to be during actual use. Directly modeling the partition losses requires a large number of calculations for each contour, but one of the benefits is an improvement in PL accuracy to a standard deviation of 3 to 5 dB.

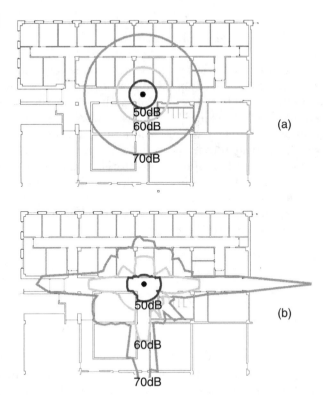

Figure 2-5
PL contours on one floor of a building with a centrally located transmitter. The distance-dependent model in Equation 2.5 produces concentric circle contours (a), and primary ray tracing given by Equation 2.12 shows more realistic irregular contours (b). (Courtesy Wireless Valley Communications, Inc.)

(a)

(b)

50dB
60dB

70dB

50dB

60dB

70dB

Range and C/I Estimation Using Primary Ray Tracing

It should be obvious that primary ray tracing can easily accommodate range calculations by establishing a PL contour corresponding to the weakest useable signal at the receiver. Because the number of partitions changes with distance, doing these calculations by hand is a tedious, iterative process. Computer-aided modeling is a much more practical way of plotting RSSI contours using primary ray tracing.

Finding C/I at a particular point in the building is determined by analyzing two primary rays, one from the desired transmitter and the other from the interfering transmitter, both ending at the point of interest. This is easily done with hand calculations. However, discovering locations within the building where the C/I is at some limit is difficult without a computer. Also, it's clear that this model requires that the number and type of obstructions in the RF path be known, and thus the results apply only to a particular area in which the nodes are located.

For many of the examples in this book, we will use the simplified PL (distance-dependent) model for general results instead.

Primary ray tracing works best for determining the range and C/I when at least one of the nodes in the wireless network is relatively fixed, such as an AP in a Wi-Fi deployment, or a desktop computer or printer in a Bluetooth piconet. PL contours can be plotted from this location to determine the region where RSSI and/or C/I is adequate. WLAN and WPAN are two-way communication systems, though, so what about the reverse link, where the fixed device is the receiver and a mobile device is the transmitter? Fortunately, *reciprocity* usually applies to the direct path, which states that signal characteristics in one direction are nearly the same as signal characteristics in the reverse direction. That is, if both devices use the same transmit power, have the same receive sensitivities, and use the same antennas for transmit and receive, and if the link is good in one direction, then it will also be good in the other direction.

Multipath from a Coexistence Perspective

Multipath can play a significant role from a coexistence point of view. Even if only one wireless network is operating in an area, the multipath components generated by a transmission can produce significant self-interference, even to the point of causing the BER to increase to unacceptable levels. In other words, the wireless network actually has difficulty coexisting with itself. Unfortunately, changing the TX power may not improve the BER since all signal components will be equally affected. When other networks are present, their multipath components can significantly increase the total power in the *co-channel interference* (CCI) components.

When a radio wave strikes the surface of a typical obstruction, part of the wave's energy passes through the obstruction and continues on its way. This forms the direct wave studied earlier. For example, Table 2-7 shows that half the signal's power continues through a typical office wall for a loss of 3 dB. What happens to the other half? Some of the power is absorbed by the wall and some is reflected, either specularly (like a mirror) or through scattering. To simplify the analysis, we will assume that the direct path angle is unaltered as it passes through the obstruction, and that all reflections are specular, so that the angle of incidence equals the angle of reflection (see Figure 2-6).

If a CW wave is transmitted into the multipath channel, the multiple copies add vectorally at the RX antenna. Figure 2-7 shows two of the many possible amplitude-phase orientations of one direct and one reflected path. The magnitude of the vector is the signal's amplitude, and the angle of the vector with respect to the real (Re) axis is the phase. If the phases of the two signals are

Figure 2-6
Radio waves
striking most
indoor
obstructions
are partially
directed,
partially
absorbed, and
partially
reflected.
Diffraction and
scattering are
ignored.

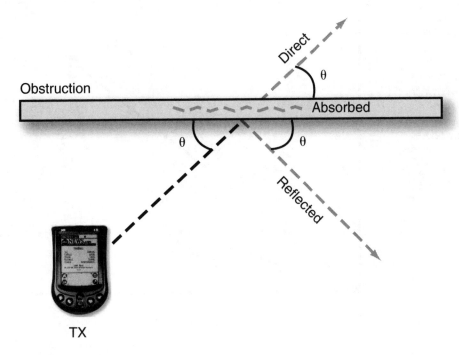

approximately aligned, the resultant RSSI is higher than either of the two signals individually, and if the phases are approximately opposed then the resultant can be lower than either.

These vector additions can change markedly when the receiver or transmitter changes its position by only a fraction of the RF carrier wavelength. Figure 2-8 gives a plot of a standing wave, or interference, pattern provided by a transmitter located near a perfect reflector without accounting for PL [Poo98]. The minimum distance between signal peaks and nulls is $\lambda/4$ along a line segment drawn from the transmitter to a point perpendicular to the reflecting surface. If the surface is partially reflecting and PL is taken into account, the interference pattern itself is similar, but the signal peaks and valleys will be smaller relative to each other. Because the amount of fade depends upon the position of the receiver in the room, fading here is a function of *space*.

For a moving receiver and/or transmitter, the resulting RSSI fluctuations can occur quite rapidly at carrier frequencies in the 2.4 or 5.x GHz band, even when moving at a walking speed, because the carrier wavelength is relatively short. As such, these fades are called *small scale* to differentiate them from the large-scale changes in PL caused by signal attenuation from different obstructions. When motion is involved, then, fading becomes a function of *time*.

Relative to a particular point in the room depicted in Figure 2-8, if the TX frequency changes, then the vector magnitude and phase also change for each path,

Figure 2-7
Resultant signal
magnitude and
phase at a
receiver when
two com-
ponents are
nearly phase
aligned (a) and
nearly phase
opposed (b)

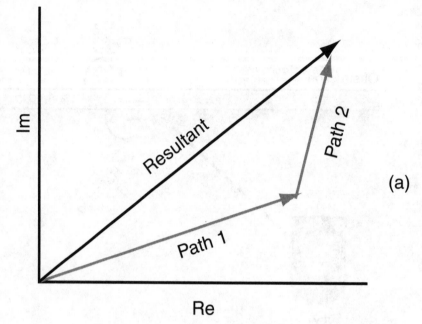

(a)

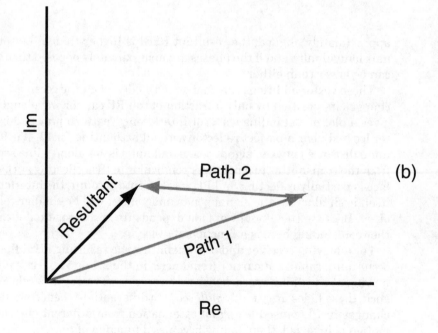

(b)

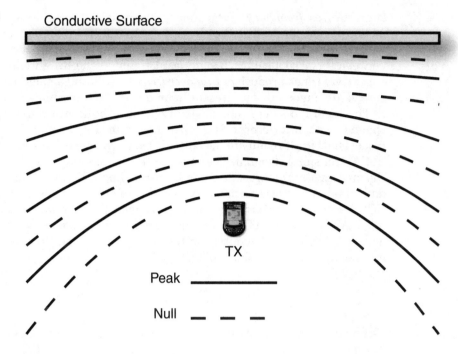

Figure 2-8
Standing wave pattern from a CW transmission near a reflecting surface with no path loss

thus changing the amplitude and phase of the fade. If the signal bandwidth is relatively wide, different parts of the frequency domain representation may be faded differently, so fading is in this case a function of *frequency*.

All three of these fading functions—space, time, and frequency—can cause significant increases in the BER of a communication system. On the other hand, each fading function can be exploited to varying degrees to reduce the BER, sometimes significantly. The exploitation of multipath characteristics using various levels of signal processing is called *diversity*, and this diversity can take place in space, time, and/or frequency.

We continue our study of multipath by quantifying the two major characteristics of the multipath channel, Doppler spread and delay spread. These characteristics are applied to the wireless channel to determine its fading properties. Finally, methods of multipath mitigation are presented and some examples given. Our study of multipath will continue in the next chapter when we show how BER changes when multipath is taken into account.

Fading Characteristics

The reciprocal relationship between fading characteristics in frequency and in time are manifested in both Doppler spread and delay spread. *Doppler spread* characterizes *frequency shifts* among the various signal components due to motion somewhere within the physical communication channel. These frequency shifts appear as *fading over time*. *Delay spread* characterizes the *time shifts* between the arrivals of each significant signal at the RX antenna. These different arrival times produce *fading over frequency*. Both Doppler spread and delay spread are almost always present in an indoor RF environment.

The general discrete-time model for a noisy, multipath-prone communication channel with interference is shown in Figure 2-9 [Poo98]. The discrete-time model lends itself readily to the portrayal of a typical *pulse-amplitude modulated* (PAM) transmitted signal. The input to the channel is $y(n)$, which is filtered by the process $a(n;k)$ representing multipath fading as a linear, time-varying filter, and an additive noise term $w(n)$ consisting of both receiver front-end noise and CCI. Receiver front-end noise is usually modeled as *additive white Gaussian noise* (AWGN). The AWGN approximation is often used for modeling interference as a simplification. The output of the channel, $r(n)$, can therefore be expressed mathematically as the following [Poo98]

$$r(n) = \sum_k a(n;k)y(n - k) + w(n) \qquad (2.14)$$

The term $a(n;k)$ is the response of the channel at time n to a unit-sample input at time $n - k$.

If the noise term $w(n)$ is AWGN, it can be modeled as a zero-mean, complex-valued white Gaussian process with circular symmetry. The variance is

$$E[|w(n)|^2] = N_0 W_0 \qquad (2.15)$$

where W_0 is the system bandwidth and N_0 is the noise PSD. This noise model is often too simplistic for CCI, as we will discover during DSSS analysis in the next chapter. By changing the structure of $w(n)$ to model interference more closely, interference suppression techniques can take advantage of its nonrandom com-

Figure 2-9
Discrete-time baseband model of a channel that has general time-selective and frequency-selective fading

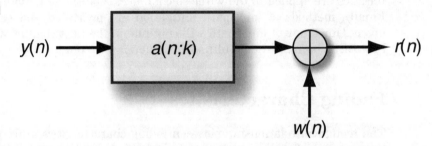

ponents. The accuracy of this model can be enhanced to any desired degree by increasing the resolution of the time intervals k into which the model is partitioned. Furthermore, the discrete model lends itself to the direct application of *digital signal processor* (DSP) diversity solutions.

The continuous time-varying frequency response of the channel $A(\omega;n)$ is related to the first term in Equation 2.15 by

$$A(\omega,n) = \sum_k a(n,k)e^{-j\omega k} \tag{2.16}$$

which represents the effect of the kernel $a(n,k)$ on each frequency component ω of interest. Again, this is a generalized representation of channel fading, and no assumptions are yet made on whether the response varies with time and/or with frequency.

Doppler Spread

Suppose a receiver is placed next to a CW transmitter and then moved away with a constant velocity v. The frequency of the RF carrier along the LOS path will shift lower by

$$f_D = \frac{v}{\lambda} \tag{2.17}$$

where f_D is the Doppler shift in Hz, v is the velocity of the receiver away from the transmitter in meters per second, and λ is the carrier wavelength in meters. If the TX frequency is 2.45 GHz and the receiver is moving away from the transmitter at 1 meter per second (a moderate walking speed), then the TX frequency along the LOS path is shifted down by about 8 Hz. For a 5.x GHz carrier, the frequency shift under these conditions is about 18 Hz. For other signal paths that arrive at different angles, the Doppler shift may be different.

In actual situations, it's not necessary to know the Doppler shift in each of the many paths; instead, a quantity called *Doppler spread*, defined as twice the largest Doppler shift, is used. A signal that fades over time does so because of Doppler spread, and this in turn implies that motion occurs somewhere. The movement can be either the transmitter, the receiver, or any nearby object that can affect RSSI. Figures 2-10a and 2-10b show the fading situation and equivalent Doppler spread when a transmitter and receiver move relative to each other. These motions and the resulting random multipath arrivals cause fades that are often 20 dB, and can even approach 30 dB occasionally. The breadth of the Doppler spread affects the *rate* of signal fading (smaller shifts equate to slower fades), but the *depth* of fading is determined by the relative strengths and phase relationship in the direct and reflected signal paths.

Figure 2-10
Doppler spread
is portrayed in
the time
domain (a and
c) and in the
frequency
domain (b and
d). Fading can
be more severe
if the trans-
mitter and
receiver are
moving relative
to each other
(a and b) than
if their posi-
tions are fixed
(c and d)
[How90].

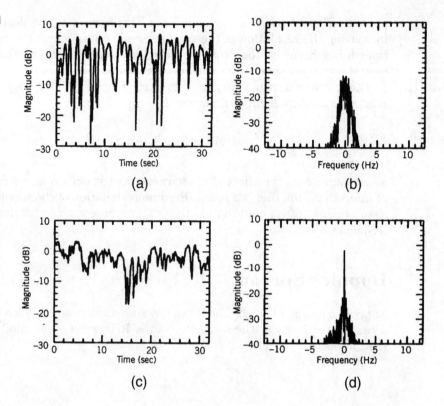

(a)   (b)

(c)   (d)

If the transmitter and receiver are fixed relative to each other, fading can still occur from the movement of nearby objects. This situation is shown in Figures 2-10c and 2-10d. When significant power exists in the (unshifted) LOS path, over-all fades, while still occurring at about the same rate, aren't as deep and reliability is improved.

Doppler Fading and Outage Probability   The probability that a fade will exceed some threshold is easy to calculate provided the following assumptions hold:

■ All frequencies within the bandwidth of the transmitted signal fade together.

■ All incoming paths are subject to random Doppler shifts.

The first assumption will be addressed in more detail in the section on delay spread, and the second assumption is satisfied by the situation depicted in Figures 2-10a and 2-10b, where the transmitter and receiver are randomly moving relative to one another and perhaps other nearby objects are randomly moving as well. If these two assumptions hold, the channel exhibits what is called *Rayleigh fading*. The *probability density function* (pdf) of a Rayleigh-distributed random variable a is given by

$$p(a) = \begin{cases} \dfrac{a}{\sigma^2} \exp\left(\dfrac{-a}{2\sigma^2}\right) & a \geq 0 \\ 0 & a < 0 \end{cases} \tag{2.18}$$

where σ^2 is the average power in the received signal. The time domain plot of Rayleigh fading shown in Figure 2-10a can be accurately portrayed using Jakes' model, which simulates the Rayleigh channel by summing a set of complex sinusoids. More information can be found in Jakes [Jak71] and Pahlavan [Pah95].

For a Rayleigh faded channel, the probability Pr(outage) that the signal power has faded below a certain threshold level P_{th} can be expressed in terms of the receiver's average signal power level P_{ave} by the formula [Rap96]

$$\text{Pr(outage)} = 1 - \exp\left(-\frac{P_{th}}{P_{ave}}\right) \tag{2.19}$$

If P_{th} and P_{ave} are expressed in dB, then they must be converted back to linear power units before their ratio is taken.

Example 2-7:
The average RSSI on a certain Wi-Fi link is -70 dBm and the receiver will lose the signal if it falls below -80 dBm. The fading on the channel is Rayleigh distributed. What is the outage probability?

Solution:
The actual ratio of the two powers is 0.1. According to Equation 2.19 the probability of outage in a Rayleigh faded channel is about 0.095, so the channel will be unusable about 9.5 percent of the time.

To reduce the probability of outage, the RF link is usually designed with a *fade margin*, which is additional signal power above the MAPL such that some fading can occur without the resulting path loss exceeding the MAPL. In terms of P_{th} and P_{ave} expressed in dB, the fade margin is the difference between the two. The exponential nature of Equation 2.19 makes it relatively easy to estimate outage probability given the fade margin in dB. For a fade margin of 10 dB (power ratio of 10), the outage probability is about 10 percent, and for a fade margin of 20 dB (power ratio of 100) the outage probability is about 1 percent.

The situation depicted in Figures 2-10c and 2-10d is somewhat different. Because the transmitter and receiver are fixed relative to each other, the direct (LOS or OLOS) path between them has no random Doppler shift and the direct signal is strong enough to dominate the Doppler-spread multipath reflections from nearby moving objects. The channel now exhibits *Rician fading*, and as a result the fades, although they still occur about as often, aren't as deep as they were in the Rayleigh case. The Rician distribution is given by [Rap96]

$$p(a) = \begin{cases} \dfrac{a}{\sigma^2} \exp\left(\dfrac{-(r^2 + A^2)}{2\sigma^2}\right) I_0\left(\dfrac{A_r}{\sigma^2}\right) & a \geq 0, r \geq 0 \\ 0 & \text{otherwise} \end{cases} \tag{2.20}$$

The parameter A is the peak amplitude of the dominant signal and $I_0(x)$ is the modified, zero-order Bessel function of the first kind. A factor K, called the *Rician factor*, is defined as the ratio between the deterministic signal power and the multipath variance. As such, K completely specifies the Rician distribution and is given by

$$K(\text{dB}) = 10 \log\left(\dfrac{A^2}{2\sigma^2}\right) \text{dB} \tag{2.21}$$

As $A \to 0$, it follows that $K \to -\infty$, which in turn means that the dominant signal path approaches zero and the Rician distribution degenerates to a Rayleigh distribution. Conversely, as A becomes large, the channel behaves as if only one dominant path exists without any significant multipath components.

Fading characteristics in a Rician channel clearly depend on the ratio of the power in the dominant path compared to the power in the Doppler-spread paths. If the ratio is high, then fades are relatively shallow, and if the ratio is low then fades are relatively deep. In general, Pr(outage) in a Rician channel will be less than that in a Rayleigh channel given the same average signal power at the receiver, so Equation 2.19 can be used to find an upper bound on this probability.

Coherence Time During the time a channel symbol is undergoing detection at the receiver, the BER is reduced if the symbol's amplitude remains relatively constant over its duration. This is especially true if information is encoded into the signal's amplitude. Doppler spread can be used to determine a channel's *coherence time*, which is the maximum time during which fades exhibit some dependency. If two points in time are separated by less than the coherence time, then the associated signal amplitudes fade mostly together; if the two points are separated by more than the coherence time, then the signals fade mostly independently. If the symbol duration is much shorter than the coherence time, and the channel is frequency nonselective (which we discuss in the next section), then symbol amplitude is essentially constant during its reception.

The coherence time is related to the reciprocal of the maximum Doppler shift, but the channel is continuously changing when fading is present, so a small amount of fading will occur even during a time period less than the coherence time. The amount of tolerable fading will dictate the formula that is used for calculating the coherence time. One that has gained acceptance is given by [Rap96]

$$T_C = \frac{0.4}{f_D} \tag{2.22}$$

where T_C is the coherence time and f_D is the maximum Doppler shift. In other words, the coherence time is about 40 percent of the reciprocal of the maximum Doppler shift.

Now we can relate coherence time to the data symbol rate to determine whether the channel fades significantly during the transmission of a data symbol. Two outcomes are possible:

■ **Slow fading**   The coherence time is greater than the symbol period, so the baseband signal (the symbols representing the data bits) varies faster than the channel does. This means that each symbol has nearly constant amplitude while it is being received, and distortion from fading is negligible.

■ **Fast fading**   The coherence time is less than the symbol period, so the channel varies faster than the baseband signal does. This means that each symbol has a high probability of experiencing a fade while it is being received, and distortion from these fades may be significant.

To reduce distortion from Doppler fading, a slow-faded channel is preferred. This can be achieved by transmitting a relatively fast symbol rate for a given set of fading conditions. Keep in mind, though, that even with a slow-faded channel the possibility exists that deep fades will occur from time to time, resulting in outage. This outage could affect several consecutive channel symbols or data packets when fading is slow.

With the increasing speeds at which data can be sent over the wireless channel, packet transmission times are becoming shorter. If the coherence time is longer than the time required to send an entire packet of data, then the channel is called *quasi-stationary*. For a quasi-stationary channel, the outage probability and the packet error rate are identical if fading is the only source of errors.

Delay Spread

The second major way that multipath can affect communication performance is by the channel's *delay spread*. This is a measure of the spread in the time it takes for reflected signals of significant amplitude to arrive at the receiver. Unlike Doppler spread, delay spread does not require any motion and is strictly a function of arrival times. Of course, if motion occurs, the delay spread characteristics will probably change with time.

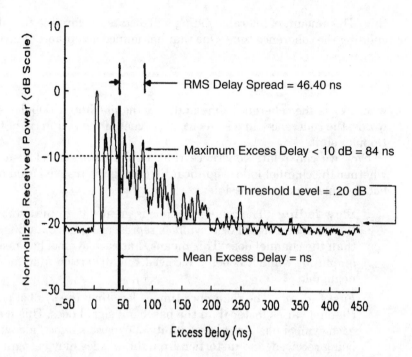

Delay spread can be measured, at least theoretically, by transmitting a brief pulse and plotting the arrival times of all incoming copies of the pulse at the receiver. Figure 2-11 shows an example of this in an indoor situation [Rap96]. The first arrival is set to a time of 0 *nanoseconds* (ns) and an amplitude of 0 dB. Additional pulses that reflect off the various partitions then arrive later with their own amplitudes, eventually tapering off into the noise floor or threshold level of the receiver.

Delay spread can be quantified in one of several different ways. Two of these are *maximum excess delay* and *root mean square* (rms*) delay spread*. Maximum excess delay is the time between the first arrival and the last arrival that occurs above a particular threshold, usually 10 dB below the strongest. The quantity is easy to determine by simply looking at the excess delay plot. In Figure 2-11, for example, a total of six arrivals are present within 10 dB of the power in the strongest arrival, and the last occurs 84 ns after the first. Therefore, the maximum excess delay is 84 ns. This quantity has limited use, though, because it provides no indication of the number or relative amplitudes of the various arrivals within the 10 dB window. Consequently, two channels with the same excess delays can behave very differently.

The rms delay spread σ_τ can alleviate some of these problems at the expense of a more complex calculation method. This is done by finding the second central moment about the mean excess delay. This quantity adjusts for the number, relative amplitudes, and times of arrival of the various multipath components, and is given by

$$\sigma_\tau = \sqrt{E[\tau^2] - (E[\tau])^2} \tag{2.23}$$

where

$$(E[\tau])^2 = \left(\frac{\sum_k P_r(\tau_k)\tau_k}{\sum_k P_r(\tau_k)} \right)^2 \tag{2.24}$$

and

$$E[\tau^2] = \frac{\sum_k P_r(\tau_k)\tau_k^2}{\sum_k P_r(\tau_k)} \tag{2.25}$$

The quantity $P_r(\tau_k)$ is the received power at time τ_k.

Example 2-8:

Calculate the rms delay spread for the indoor delay profile given in Figure 2-11. Use a threshold value of -10 dB.

Solution:

First, create a list of the approximate arrival times and powers of the components, with the dB values in Figure 2-11 converted to absolute powers. These values are given in Table 2-9. Next, find the sum of the powers at all arrival times

$$\sum_k P_r(\tau_k) = 1.0 + 0.30 + 0.55 + 0.30 + 0.17 + 0.20 = 2.52$$

followed by the other two quantities in Equations 2.24 and 2.25:

$$\sum_k P_r(\tau_k)\tau_k = 1.0(0) + 0.30(20) + 0.55(35) + 0.30(55) + 0.17(75) + 0.20(85) = 71.5$$

$$\sum_k P_r(\tau_k)\tau_k^2 = 1.0(0) + 0.30(400) + 0.55(1225) + 0.30(3025) + 0.17(5625)$$
$$+ 0.20(7225) = 4103$$

Arrival time (ns)	Absolute power
0	1.0
20	0.30
35	0.55
55	0.30
75	0.17
85	0.20

Now find the mean square value using Equation 2.24, which is

$$(E[\tau])^2 = \left(\frac{71.5}{2.52}\right)^2 = 805$$

followed by the second moment given by Equation 2.25:

$$E[\tau^2] = \frac{4103}{2.52} = 1628$$

Finally, use Equation 2.23 to find the result as

$$\sigma_\tau = \sqrt{1628 - 805} = 28.7 \text{ ns}$$

Note that the rms delay spread value given in Figure 2-11 is higher because a threshold of −20 dB was used.

Table 2-10 gives some measured rms delay spreads for some typical office and store environments [Hal00]. As a rule, larger rooms with more obstructed paths have larger rms delay spread values.

Coherence Bandwidth   If data is sent at a fast rate over the multipath channel, delay spread may cause *intersymbol interference* (ISI) at the receiver. Figure 2-12 depicts a transmitter and receiver in which the signal takes two paths: One is LOS, and the other bounces off a highly reflective surface. Suppose the transmitter sends a 0 followed soon afterward by a 1. Furthermore, suppose that the symbol duration is the same as the time difference between signal arrivals on the LOS and reflected paths; that is, if the reflected signal arrives at the receiver

Table 2-10

Measured RMS delay spread values

Location	Mean (ns)	Median (ns)	Maximum (ns)
Single room office	n/a	25	30
Cafeteria	n/a	30	75
Engineering building LOS	42	n/a	65
Engineering building, lightly obstructed	56	n/a	75
Engineering building, heavily obstructed	70	n/a	85
Retail store LOS	23	n/a	74
Retail store, lightly cluttered	74	n/a	97
Retail store, heavily cluttered	84	n/a	100
Office building	n/a	40	150
Shopping center	n/a	105	170
Laboratory	n/a	106	270

Figure 2-12
Delay spread can cause ISI at the receiver, increasing the probability of bit error.

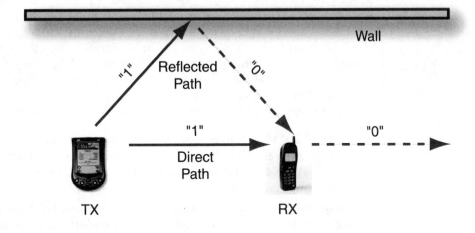

50 ns after the LOS signal, the symbol duration is also 50 ns, corresponding to a data rate of 1/(50 ns), or 20 megabits per second (Mb/s).

First, the 0 appears at the receiver along the LOS path. Next, the 1 along the LOS path and the 0 from the reflected path appear *together* at the receiver, as shown in Figure 2-12. The receiver is attempting to detect a new 1 from the LOS path and the previous 0 from the reflected path at the same time. If the 0 is detected over the 1, then a bit error occurs from the ISI, which leads to *irreducible bit error rates*. Increasing the transmitter power won't improve the situation, because the ratio between the LOS and reflected signal powers remains unchanged.

The degree to which the received channel symbol is vulnerable to ISI is determined by the *coherence bandwidth* of the channel. This is a measure of the range of frequencies over which the channel response is *flat*, meaning that the distortion over the channel that results in ISI is negligible. The formula can be expressed as

$$B_C \approx \frac{0.1}{T_{\text{rms}}} \tag{2.26}$$

where B_C is the coherence bandwidth and T_{rms} is the rms delay spread. The coherence bandwidth is essentially the widest bandwidth that a transmitted signal can have with negligible ISI. The multiplier 0.1 can actually range between 0.02 and 0.2, depending on the amount of correlation desired among the various parts of the transmitted signal bandwidth [Rap96]. Obviously, larger rooms (with larger rms delay spreads) have smaller coherence bandwidths.

The concept of coherence bandwidth can sometimes be more easily grasped by returning to the associated rms delay spread. In these terms, Equation 2.26 states that if the transmitted symbol duration is greater than 10 times the rms delay spread, the ISI will be small. In other words, only the first 10 percent (or less) of the duration of a data symbol will be significantly affected by the previous symbol. The remaining 90 percent or more will be received without ISI because all the significant reflections now contain the same data symbol.

Like its coherence time counterpart, a channel's delay spread and associated coherence bandwidth can also be placed into one of two major categories:

- **Flat fading** The bandwidth of the transmitted signal is less than the channel coherence bandwidth. Equivalently, the symbol period is greater than 10 times the rms delay spread. The transmitted waveform is not significantly altered by ISI.

- **Frequency selective fading** The bandwidth of the transmitted signal is greater than the channel coherence bandwidth. Equivalently, the symbol period is less than 10 times the rms delay spread. The transmitted waveform may be significantly altered by ISI.

To reduce distortion from ISI, a flat-faded channel is preferred. This is accomplished for a given set of parameters by reducing the channel symbol rate or by using some of the diversity techniques we discuss later in this chapter. Incidentally, reducing the channel symbol rate doesn't necessarily equate to reducing the data rate, as we will discover in the next chapter.

Doppler Spread and Delay Spread Channel Models

Let us now return to our generalized channel model given by Equation 2.14. A channel that produces fast, frequency-selective fading requires use of all terms in the equation and is depicted in Figure 2-9. If the channel is slow-faded to the point that no channel-induced amplitude changes exist over an entire data packet, but the fading is still frequency-selective, then $A(\omega;n)$ does not vary with n and $a(n;k)$ represents a linear, time-invariant channel over the symbol duration. As a result, the channel output is represented by a simple convolution on the input signal, shown as [Poo98]

$$r(n) = \sum_k a(k)y(n-k) + w(n) \qquad (2.27)$$

and depicted in block diagram form in Figure 2-13.

For a channel in which all transmitted frequency components are faded with the same amplitude and linear phase (flat fading), but the fading is fast, the model in Figure 2-9 can be modified to make the fading frequency nonselective, given by [Poo98]

$$r(n) = a(n)y(n) + w(n) \qquad (2.28)$$

and shown in Figure 2-14. In this situation the channel acts like an additional modulator.

When the channel is both time nonselective and frequency nonselective, Equation 2.28 simplifies further to [Poo98]

$$r(n) = Ay(n) + w(n) \qquad (2.29)$$

Figure 2-13
Discrete-time baseband model of a channel that has frequency-selective and time-nonselective fading

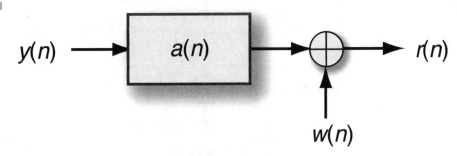

Figure 2-14
Discrete-time
baseband
model of a
channel that
has time-
selective and
frequency-
nonselective
fading

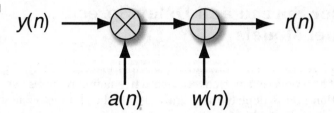

Figure 2-14
Discrete-time
baseband
model of a
channel that
has time-
selective and
frequency-
nonselective
fading

where A is a complex random variable that fixes the amplitude and phase change applied by the channel to the input signal as a whole.

The Two-Ray Rayleigh Fading Model The Rayleigh fading model applies only to flat-faded (frequency-dispersive) communication channels, but accounting for delay spread is becoming more important as data rates and occupied bandwidths increase. Modeling all time- and frequency-dispersive channel components within the summation in Equation 2.9 would be extremely complex and difficult. The two-ray Rayleigh model has been developed to account for both time and frequency dispersion while maintaining tractability. This model is given by [Rap96]

$$h_b(t) = \alpha_1 \exp(j\phi_1)\delta(t) + \alpha_2 \exp(j\phi_2)\delta(t - \tau) \tag{2.30}$$

where α_1 and α_2 are independent, Rayleigh-distributed random variables, ϕ_1 and ϕ_2 are independent phase components uniformly distributed on $[0, 2\pi]$, and τ is a time delay component between the two rays. This channel model is depicted in Figure 2-15. By varying τ it's possible to model channels with different time-dispersive characteristics.

Figure 2-15
Continuous-
time two-ray
Rayleigh fading
model that can
incorporate
time dispersion

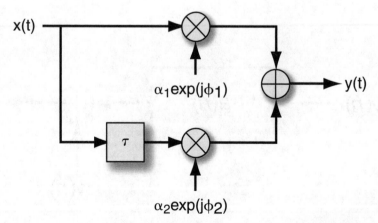

ITU Recommendation for Propagation Modeling

The *International Telecommunications Union* (ITU) has issued a recommendation for modeling propagation data and prediction methods for the indoor channel in the frequency range of 900 MHz to 100 GHz [ITU01]. Extensive research into the behavior of the indoor channel was conducted and several conclusions and recommendations were made as a result.

Path Loss (PL) Model

The PL model can be either site general or site specific. The site-general model's loss equation has this form:

$$L_{total} = 20 \log_{10} f + N \log_{10} d + L_f(n) - 28 \ \ \text{dB} \tag{2.31}$$

where L_{total} is the total PL, f is the carrier frequency in MHz, N is the power loss coefficient, and $L_f(n)$ is the floor penetration loss factor as a function of the number of floors n. (Don't confuse the number of floors n in Equation 2.31 with the PL exponent n we used earlier.)

This equation is simply an expansion of the terms in Equation 2.3, along with the addition of a floor loss term. The power loss coefficient N is our earlier PL exponent multiplied by 10. The values recommended by the ITU for the power loss coefficient are given in Table 2-11 for the frequencies in which we're interested. These closely match the values that were used in earlier examples. Notice once again that the power loss coefficient in the office environment isn't particularly frequency dependent and corresponds to a PL exponent of about 3.0.

The floor penetration loss factor is given in Table 2-12 as a function of the number of floors in different environments. The ITU noted that a limit may exist to the isolation expected from penetrating multiple floors, just as we discovered in the section on primary ray tracing. Standard deviations for the PL values predicted by Equation 2.31 ranged between 8 and 10 dB at 2.4 GHz and were typically about 12 dB in the office environment at 5.2 GHz.

Table 2-11

ITU recommended power loss coefficient

Frequency	Residential	Office	Commercial
1.8–2 GHz	28	30	22
5.2 GHz	—	31	—

Table 2-12

ITU-recommended floor penetration loss factor

Frequency	Residential	Office	Commercial
1.8–2 GHz	$4n$	$15 + 4(n - 1)$	$6 + 3(n - 1)$
5.2 GHz	—	16 (1 floor)	—

Table 2-13

ITU-measured delay spread values

Frequency	Environment	Low (ns)	Average (ns)	High (ns)
1.9 GHz	Indoor residential	20	70	150
1.9 GHz	Indoor office	35	10	460
1.9 GHz	Indoor commercial	55	150	500
5.2 GHz	Indoor office	45	75	150

Delay Spread

The ITU also conducted studies of the rms delay spread in several indoor environments. Table 2-13 lists the typical minimum and average spreads, along with a maximum that was occasionally observed. The maximum value can be used in a worst-case calculation to determine if ISI will be a problem for a given symbol rate in a particular situation. In general, the delay spread tended to increase as the communication path increased in length, and little frequency dependence was found when omnidirectional antennas were used.

Antenna Effects

The ITU discovered that polarization and antenna radiation patterns can both affect indoor propagation characteristics. In a LOS situation, rms delay is reduced by using a directional antenna due to the inherent capability of the antenna to amplify the LOS signal and attenuate signals arriving from directions other than that where it is aimed. Delay is also reduced when a *circularly polarized* (CP) antenna is used instead of one that is *linearly polarized* (LP). This is due to the fact that the handedness (left or right) of the CP wave reverses with each reflection that occurs at an incidence angle less than the Brewster angle. As a result, stronger multipath components from a single reflection are orthogonally

	RX antenna beamwidth (degrees)	Static RMS delay spread (ns)
	Omnidirectional	17
	60	16
	10	5
	5	1

Table 2-14

Effect of antenna directivity on RMS delay spread

polarized to the LOS component and are therefore rejected by the CP antenna. Because all existing building materials have Brewster angles greater than 45 degrees, single reflection multipath components are effectively suppressed by a CP antenna in most room environments. The exception is in long hallways where waves often strike a surface at angles greater than the Brewster angle.

Table 2-14 shows the effect of RX antenna directivity on rms delay spread in an empty 14 × 9 m office room at 60 GHz [ITU01]. The TX antenna is omnidirectional. As the RX beamwidth decreases, which implies higher gain and directivity, rms delay spread drops significantly. The resulting increase in coherence bandwidth keeps the channel flat faded for relatively large TX signal bandwidths.

For OLOS paths, the effect of antenna polarization and directivity is much less due to the random direction from which the strongest signal component arrives. Therefore, if a point-to-point wireless link is to be established between two fixed nodes, then placing these so that an LOS path is achieved and, if necessary, using a directional antenna at each end will help ensure adequate signal strength and low rms delay spread. A significant difficulty for portable wireless equipment, even in LOS situations, is that antenna orientation will change as the device is moved around during use. This alters antenna polarization and makes the use of simple directional antennas impractical.

Object Movements

The movement of persons and objects within a room can cause changes in the signal's propagation characteristics. The variation is slow relative to the data rate so the effect is similar to a slow-faded Doppler channel. The movement of persons or objects outside the room in which communication is taking place has a negligible effect on signal propagation.

Measurements at 1.6 GHz indicate that a person moving into the LOS signal path causes a 6 to 8 dB drop in RSSI, and in some cases a Rician channel will become a Rayleigh channel from attenuation of the LOS path. In OLOS situations people moving near the antenna had little effect on the channel, provided they weren't close enough to a node to disrupt the antenna's impedance match to the transmitter or receiver. Experiments at 900 MHz showed an RSSI drop of 4 to 7 dB when the terminal was held next to a person's waist, and 1 to 2 dB when the terminal was held next to someone's head.

In a typical office lobby environment, RSSI measurements taken in an OLOS path at 37 GHz exhibited fades of 10 to 15 dB, and these fades demonstrated a lognormal distribution. That is, RSSI measurements expressed in dB exhibit a normal (Gaussian) distribution. At a fade depth of 10 dB, the mean duration was 0.11 s with a standard deviation of 0.47 s. At 15 dB, the mean duration was 0.05 s with a standard deviation of 0.15 s.

Modeling the UWB Channel

Modeling the indoor *ultra-wideband* (UWB) channel is in many ways simpler than previous models due to the inherent nature of the wideband signal and its capability to separate individual multipath components. The impulse response of such a channel can be approximated by [Pen02]

$$h(t) = \sum_{k=0}^{N} a_k \delta(t - \tau_k) \tag{2.32}$$

That is, the channel simply replicates the transmitted impulse with different amplitudes a_k and different time delays τ_k. The actual impulse response can be measured in the time domain using a UWB pulse and attempting to resolve the individual components. It can also be measured by using a sweep generator transmitter and a network analyzer receiver and then converting the measurements into the time domain via the inverse Fourier transform.

The IEEE 802.15 TG3a examined several UWB channel models and their associated data, and it was discovered that multipath components tended to arrive in clusters rather than in a continuum characteristic of narrowband systems [Mol03]. This is a result of the fine resolution that UWB signals provide, effectively separating the otherwise blurred components from a single object such as a desk. As such, multipath components can be grouped into cluster arrivals and ray arrivals within each cluster. Four parameters are required to complete this model: the cluster arrival rate, its decay factor, and the ray arrival rate, and its decay factor within each cluster. The amplitude statistics of the path components were found to follow the lognormal distribution. Equation 2.32 can be modified to incorporate both clustering and lognormal shadowing [Mol03].

Several empirical studies of the indoor UWB channel have been accomplished [Mol03], [Gha02], [Pen02], [Siw02], [Win98] and from these we can conclude the following:

- The strongest pulse over the indoor UWB channel follows a PL exponent of $n = 3.0$, which closely agrees with the PL coefficients measured in the ISM and U-NII bands. The strongest pulse is not necessarily the first arriving pulse.

- The total UWB power density falls off with a PL exponent of 2.0 with the usual λ^{-2} wavelength factor given in Equation 2.3.

- The energy in each received path varies by at most 5 dB as the receiver position in the room varies.

- The channel exhibits almost no Doppler fading.

- The delay spread varies depending on the indoor environment being examined, with rms values typically between 5 ns (LOS) and 30 ns (non-LOS) at a distance of 4 to 10 m from the transmitter.

  - One study [Siw02] modeled the UWB rms delay spread as $\tau_{rms} = 3d$ ns, where d is the distance between transmitter and receiver in meters.

  - Another study [Gha02] modeled the LOS rms delay spread as $\tau_{rms} = 3.1 \times d^{0.27}$ and the non-LOS rms delay spread as $\tau_{rms} = 3.8 \times d^{0.4}$.

- The average number of signal components for an 85 percent energy capture over a 0 to 4 m LOS distance in a room with metal studs is 24 [Pen02]. Over a 4 to 10 m NLOS distance the average number is 62 [Mol03].

Over the short range of less than 10 m, which UWB is intended to be used for, a large number of signal components are distributed over a relatively short time at the receiver, with no significant fading over time. If the high-speed signal-processing requirements can be met, then UWB has the potential to deliver extremely fast data rates over short distances.

Diversity for Multipath and Interference Mitigation

The presence of multipath and other forms of CCI can often severely cripple the reliability of wireless devices, especially when the distance between two nodes approaches the range limit. Fading depth from the Doppler spread can be as high as 30 dB, and it's unrealistic from both a cost and a coexistence perspective to simply increase the transmit power by a factor of 1,000 to compensate. ISI can be so severe that effective maximum data rates may be limited to only a few megabits per second. If wireless communications are to meet customer demands for ever-increasing data rates, long battery life, and good coexistence properties, the problems caused by multipath need to be addressed and solved.

A few "brute force" methods for multipath mitigation exist, but many are unsatisfactory for indoor wireless. As an example, the easiest way to compensate for degradation from a channel that has time-selective fading is to just live with it by building a fade margin into the PL equations when estimating the maximum communication range. With Doppler fading as high as 30 dB, subtracting that amount from the MAPL insures that the receiver will still have (barely) adequate signal strength during these deep fades to maintain the link. For low-power indoor wireless, though, not enough signal strength overhead exists to accommodate fade margins that high. Suppose our typical Bluetooth transmitter power is 0 dBm and the associated receiver sensitivity is −70 dBm. Maximum PL with a 30 dB fade margin is only 40 dB, and we know from previous analysis that a 2.4 GHz wave in free space has a PL of 40 dB over a distance of just 1 m. Even Wi-Fi links with TX powers of +20 dBm can't include much fade margin without limiting the range to impractically short distances. Although we can't include fade margins high enough for the deep fades, it certainly makes sense to use perhaps 10 dB or so of "wiggle room" in the PL calculations to account for shallow fades and other inefficiencies, as we've discussed earlier.

Types of Diversity

Multipath, along with other forms of CCI, is mitigated most elegantly through *diversity*, which is a means for exploiting the characteristics of such interference to reduce or eliminate its degradation. Fortunately, self-interference in the form of multipath has a structure that can be accurately estimated at the receiver, so given sufficient DSP capabilities, diversity can even enhance system performance beyond that achieved if multipath were simply canceled. It follows, then, that diversity methods can be categorized in the same way as the particular form of multipath that is being addressed: spatial diversity, temporal diversity, and spectral diversity. Furthermore, for a particular diversity to enhance performance, the equivalent multipath category must also exist. For example, spectral diversity provides no improvement on a channel that is frequency nonselective, and temporal diversity won't improve the BER on a channel that is time nonselective. Implementing diversity requires varying levels of signal processing, depending on the complexity of the problem and the desired performance improvement.

Spatial Diversity *Spatial diversity* exploits the fact that deep fades usually occur over a relatively small physical space, often over a distance of less than half a wavelength in any direction [Jak71]. Therefore, if a receiver employs multiple antennas placed far enough apart, the probability is high that at least some of these antennas will capture strong signals. The receiver could then select the antenna with the strongest signal (*selection diversity*) or simply check the antennas in sequence until one is found where the signal is strong enough to yield the desired performance (*scanning diversity*). Using signals at all the antennas can reduce the BERs still further. The signals are first phase aligned and then their contributions are combined after each has its gain set according to its signal strength (*maximal ratio combining*), or all the antenna elements can be given equal gain (*equal gain combining*) to reduce signal processing requirements. For noncoherent demodulation, where no incoming signal phase information is available, the demodulator outputs can be squared, summed, and sent to the detector (*square-law combining*) instead.

Selection combining is simple to implement and will improve performance in many WLAN and WPAN installations. If L receive antenna elements are used, then outage occurs only if all elements are faded below the minimum RSSI needed for proper operation. For independent Rayleigh fading on each antenna element, outage occurs with probability

$$\Pr(\text{outage}) = \left[1 - \exp\left(-\frac{P_{th}}{P_{ave}} \right) \right]^L \tag{2.33}$$

Angle diversity is a form of spatial diversity that uses a directional antenna to enhance the capability of a receiver to "hear" better in the direction of the strongest signal, and to reject signals arriving from other directions. In this way, the detrimental phase cancellation effect of multipath is reduced. An added benefit is that interfering transmissions from other sources can also be reduced at a receiver using angle diversity. Incorporating angle diversity into a portable wireless device is a challenge, though, because antenna directivity must compensate for the device's change in orientation during use. Multiple antennas can also be used at the transmitter, but sophisticated signal processing is required at the receiver to realize a significant performance improvement from this arrangement.

Another form of spatial diversity is *polarization diversity*, in which each antenna is oriented to enhance its capability to receive an incoming wave with a certain polarization. A wave's polarization is determined by the orientation of its electric field (E-field), which is also the orientation of the long dimension of a simple linear antenna. To implement polarization diversity, one antenna could be oriented horizontally and the other vertically. This provides reasonable RSSI from arbitrarily polarized incoming signals. In practice, though, polarization diversity yields only mild improvements in indoor communications due to the different polarizations already present in the various multipath components.

Temporal Diversity   *Temporal diversity* is based on the assumption that fades last a relatively short time. If a set of data is repeated over time, the probability is high that at least one of the data sets will be successfully received. Rather than blindly transmitting data over and over, a process called *automatic repeat request*

(ARQ) can be used, where only the transmissions received in error are repeated. ARQ works well over a quasi-stationary channel, because entire packets will either be faded or received with adequate RSSI.

To combat both fading and ISI from delay spread, temporal diversity can be implemented through DSSS to enable a receiver to actually discern among different copies of the same transmitted signal separated by only a few nanoseconds in time. As such, the receiver is no longer held hostage to the possibility of multipath phase cancellation and the resulting deep fades. Instead, the receiver can pick out the strongest signal and decode it, or even combine several incoming copies into a composite from which the transmitted data bits can be extracted with high reliability. The latter process is used in a device called a *RAKE receiver*, which is implemented in CDMA cellular phones. Sometimes the RAKE receiver is categorized as a form of spectral diversity, because it relies on the existence of an incoming signal with a sufficiently wide bandwidth to enable the receiver to resolve separate multipath components. CCI transmissions with the same DSSS structure appear as a slight increase in background noise to the desired receiver, and this interference can be further mitigated by eliminating such CCI through a process called *multiuser detection* (MUD). To be effective, though, the bandwidth of all signals must be wide enough to allow the receiver to discern each multipath and CCI arrival separately.

An intense area of research is in *space-time diversity*, where multiple TX and RX antennas are used, each fed with delayed versions of the modulated carrier. The result is a significant drop in BER for a given TX power and receiver sensitivity [Ald02], [Boy00]. Space-time diversity will be discussed in more detail later in this chapter.

Spectral Diversity   *Spectral diversity* works by transmitting the same information using several different carrier frequencies, each with independent fading. The probability is low that all carrier frequencies will experience simultaneously deep fades. A benefit of the Bluetooth *frequency hopping spread spectrum* (FHSS) ARQ process, for example, is that transmissions that aren't received correctly by a piconet member are retransmitted on a different hop frequency and at a later time, thus employing both temporal and spectral diversity. Spectral diversity is also exploited through *orthogonal frequency division multiplexing* (OFDM), where several subcarriers are modulated with relatively slow channel symbols, for a high aggregate data rate. Although frequency selective fading may affect a few subcarriers, error correction coding can often recover from the resulting symbol errors. OFDM is the modulation method used in 802.11a and 802.11g WLAN.

As DSP techniques improve, along with associated cost and power reductions, such processing power will certainly find its way into inexpensive portable wireless devices. We can eventually expect to see more sophisticated diversity techniques being employed to combat interference and multipath. The amount of improvement from several forms of diversity will be explored later in this chapter and elsewhere in this book, but we will preface that with a look at spatial diversity as implemented in a Wi-Fi AP and an STA, and by examining the temporal diversity that stems from Bluetooth FHSS.

Spatial Diversity in Wi-Fi

One of the most powerful methods for combating multipath (and, to a lesser degree, CCI) is with a simple spatial diversity technique that uses two antennas at the receiver. Inside, a circuit monitors the relative signal strengths on each antenna and switches the one with the higher signal level into the receiver front end (see Figure 2-16). This rendition of selection combining is common in such devices as wireless microphones and Wi-Fi WLAN implementations.

A considerable performance improvement is possible with two RX antennas separated far enough in space that they experience independent fades. If path clutter is evenly distributed between the transmitter and receiver, this separation can be as small as half a wavelength [Jak71]. Because a deep fade lasts a short time compared to the time between fades (see Figure 2-10), only a small probability exists that both antennas will be deeply faded at the same time. Figure 2-17 shows typical independent Doppler fades at each of the two antennas, along with a trace of the composite fading that occurs at a receiver using ideal selection combining. The average RSSI is normalized to 0 dB. Notice that, although each antenna occasionally experiences fades in excess of 20 dB, the selection combining process limits fades to about 5 dB.

Example 2-9:
In Example 2-7, the average RSSI on a certain Wi-Fi link was given as −70 dBm and the receiver was assumed to lose the signal if it fell below −80 dBm. For

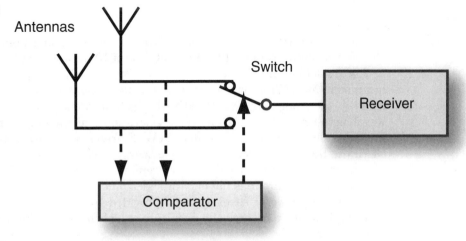

Figure 2-16
A receiver using selection combining with two antennas [Mor02]

■■ ■■ ■■

Figure 2-17
Selection
combining,
even with only
two RX
antennas,
reduces fading
significantly
[Mor02].

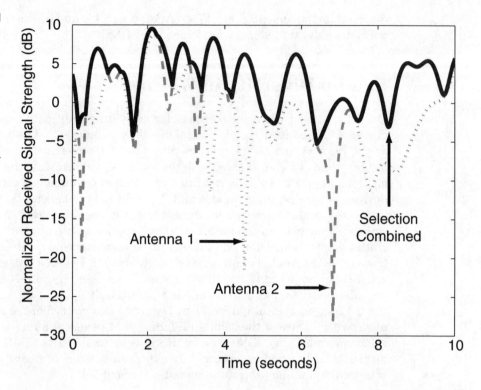

Rayleigh fading on one antenna, the outage probability was 0.095. Suppose now that two antennas are selection combined, and each has independently faded Raleigh channel statistics. What is the new outage probability?

Solution:

The actual ratio of the two powers is 0.1, as before. Using Equation 2.33, the outage probability is now 0.0090, for a factor of 10 improvement.

The implementation of selection diversity is fairly straightforward and is included in most Wi-Fi nodes. Figure 2-18a shows how the two antennas are oriented in the Apple AirPort 802.11b AP. Each antenna is about a quarter of a wavelength long at 2.4 GHz, and their separation is about $\lambda/2$. The relatively long preamble sent at the beginning of each Wi-Fi frame allows the receiving node to check the RSSI from each of the two antennas and select the antenna with the higher signal power for use during the remainder of the frame. The slight angle between the two antennas not only provides mild polarization diversity, but also increases the angle of coverage area provided by the AP. The transmitter in such implementations typically uses only one of these antennas, so diversity is receive-only.

■■ ■■ ■■

Figure 2-18
Selection
combining as
implemented
in the Apple
AirPort
802.11b AP (a)
and the Proxim
Orinoco
802.11a/b
combination
PC card (b)

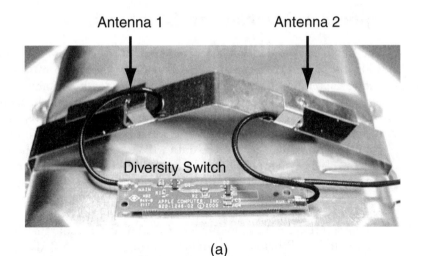

Antenna 1   Antenna 2

Diversity Switch

(a)

Antenna 1

Antenna 2

(b)

    Spatial diversity can improve performance even when the antennas are closer together than $\lambda/2$, which is a necessity within the limited space available on a PCMCIA card. Figure 2-18b shows an X-ray of the Proxim Orinoco 802.11a/b combination card that provides two dual-band antennas for receive selection diversity. Although the antennas are too close to each other to experience independent fades, performance is still enhanced because only rarely will a fade severe enough for an outage occur at both antennas simultaneously.

Spectral Diversity in Bluetooth

The Bluetooth frequency hopping channel set ranges from 2,402 to 2,480 MHz in 1 MHz steps, so a total bandwidth of about 79 MHz is eventually covered during a long hopping sequence. The transmitted bandwidth on any one hop frequency is about 1 MHz. Because Bluetooth uses FHSS, it would be interesting to discover if frequency hopping can be employed successfully as an antimultipath technique similar to spectral diversity. If a Bluetooth hop channel is faded to the point that an outage occurs, the data will be resent on a new hop frequency. This process, then, is actually a combination of spectral and temporal diversity. We need to discover if the new hop channel is sufficiently far from the faded channel that they don't necessarily fade together; in other words, they both fade independently.

The solution can be found by first calculating the coherence bandwidth for the typical indoor Bluetooth channel. According to Table 2-10, the median delay spread is 25 ns in a typical single-room office, which results in a coherence bandwidth of 4.0 MHz according to Equation 2.26. That tells us that portions of the 79 MHz band that are more than 4 MHz apart fade somewhat independently in this environment. Longer delay spreads, such as the maximum of 270 ns found in a laboratory (see Table 2-10), translate to a coherence bandwidth of only 370 kHz, so even adjacent Bluetooth hop channels would be independently faded in these large rooms.

Example 2-10:

A Bluetooth WPAN is operating in a small office that has a coherence bandwidth of 4 MHz. Assume that hopping across the 79 channels, each 1 MHz wide, is completely random. If a particular hop channel is faded, what is the probability that the next hop channel will have independent fading characteristics?

Solution:

If a particular hop channel is faded, the probability is approximately 0.95, derived from $(79 - 4)/79$, that the next hop will not experience the same fading characteristics in the small office. We can conclude that FHSS, as implemented by Bluetooth, successfully exploits spectral diversity as an antimultipath technique.

Incidentally, implementing spatial diversity in the form of selection combining becomes more complex when combined with Bluetooth FHSS. We know that the probability is high that each hop will be independently faded, so antenna selection must be reaccomplished at the beginning of each hop for best performance. However, Bluetooth provides only 4 preamble symbols, sent in just 4 μs, during which time the antenna selection process must be completed. Furthermore, the nominal hop rate is 1,600 per second, which is the same rate at which the selection algorithm must be run. Rather than sacrifice battery life to support signal processing for such an endeavor, chip manufacturers have opted to use a single antenna and rely instead on combined spectral and temporal diversity provided by Bluetooth FHSS and ARQ as an antimultipath technique.

Temporal Diversity Using Error Control

Error control compensates for channel imperfections by providing several ways for the communication link to improve communication reliability by exploiting temporal diversity. In general, error control attaches redundancy to the transmitted bits that can be used to improve the BER at the receiver. Somewhat surprisingly, Shannon's theorems state that this improvement can be made arbitrarily high, provided that arbitrarily long delays can be accepted. Error control can be implemented in three general ways, all of which will be examined in greater detail in the next chapter:

- **Error detection codes** are additional bits that are sent as part of a data string that the receiver uses to check the accuracy of the incoming message. The receiver performs an efficient mathematical test out of which a pass-fail verdict is reached.

- **Error correction codes** are additional bits that are sent as part of a data string that the receiver uses to actually correct errors that may appear in the incoming message. These codes are often substantially more complex than those used for error detection, but their ability to correct bad data makes them more powerful as well. These codes work best when errors are scattered throughout a packet rather than concentrated within a small number of symbols. Catastrophic error events that occur during a collision with another transmitted packet can overwhelm the error correction code, so the utility of such a code is questionable when most packet errors are caused by high levels of interference rather than by random noise.

- **Automatic repeat request**, also called ARQ, is used in conjunction with error detection codes to enable a receiver to request a bad set of data to be retransmitted. If a set of incoming data is good, the receiving node returns an *acknowledgment* (ACK) to the transmitting node; otherwise, the receiving node returns a *negative acknowledgment* (NAK) or makes no response. If an ACK is returned, the transmitting node continues with the next part of the message; otherwise, the last part of the message is sent again.

Bluetooth usually uses all three of these error control methods when piconet members exchange data. First, the error correction code is applied to an incoming data set to correct as many errors as possible; then the data set is checked by the error detection code for any remaining errors. Finally, an ACK is returned if all is well. In Bluetooth, each message, and each ACK or NAK, is transmitted on a different hop frequency, thus combining spectral and temporal diversity.

The Wi-Fi specification is a bit less structured in its treatment of error control. Some error control is placed into the modulation scheme, which we will discuss in the next two chapters. Within their frames, all 802.11-based wireless systems have built-in error detection and ARQ, but only 802.11a and 802.11g include error correction as well. This is to combat the possibility that some OFDM carriers may experience fading or interference levels beyond the receiver's capability to demodulate their data.

Both WiMedia and ZigBee incorporate some error control into their modulation schemes. Within the packets themselves, though, only error detection is included for use with ARQ. These newer specifications treat error detection differently, sometimes by using a single detection sequence for checking the integrity of the entire packet, and sometimes by using separate sequences for checking different parts of a single packet.

Smart Antennas and Space-Time Diversity

In Chapter 1, "Introduction," we noted that significant increases in the maximum range are possible using directional antennas, but because the traditional gain antenna must be carefully aimed, they are more suitable in point-to-point wireless links when both nodes are in fixed locations. There has been tremendous interest in designing directional antennas that have electronically steerable main lobes that can automatically enhance the desired signal while placing nulls in the direction of undesired signals in the form of both multipath and CCI. The operation of such a *smart antenna*, also called an *adaptive antenna*, is depicted in Figure 2-19.

An adaptive antenna consists of a phased array in which the timing of the transmitted signal into each element is altered slightly using DSP techniques to place the main lobe and nulls in the desired position. As with their aimed counterparts, these antennas enhance both TX and RX performance. Smart antennas, with their high DSP demands, are much easier to implement into fixed nodes, such as Wi-Fi APs, for a number of reasons:

- DSP demands require relatively high power consumption and would drain batteries quickly.

- Antenna orientation at the fixed location remains constant, so the antenna needs to follow only the relatively slow-moving mobile unit(s).

- Beam forming is mostly confined to the azimuth plane, with directivity in the elevation plane being limited to a relatively small angle.

For smart antennas to be successful in handheld wireless devices, the cost and power requirements of the associated DSP must be brought to a manageable

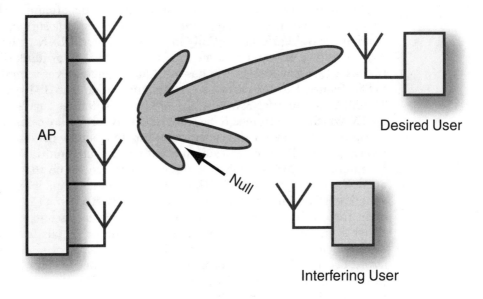

Figure 2-19
A smart antenna can automatically steer its main lobe toward the desired signal and place nulls in the direction of undesired signals.

level. The processing speed and antenna design also must be such that the directional characteristics can change quickly as the device is moved around by the user while communicating. Although difficult, the reward in enhanced reliability and performance is substantial, and smart antennas in portable devices may be the best way to mitigate coexistence problems.

Multiple-Input Multiple-Output Architecture

High-rate wireless access requires the transfer of large amounts of data over a short time and, UWB exempted, within a relatively narrow bandwidth. IEEE 802.15.3 WiMedia is an example of high-rate access at the WPAN level, with data rates of up to 55 Mb/s placed within a bandwidth of only 15 MHz. Although special modulation and coding make this possible in 802.15.3, studies have shown that additional data rate increases of one or two orders of magnitude can be achieved through exploiting spatial diversity using a process called *space-division multiplexing* (SDM) that is accomplished with *multiple-input, multiple-output* (MIMO) antenna systems. Their use reduces the effect of the three major impediments over the wireless channel: fading, delay spread, and CCI. As we've already discovered, simple antenna diversity methods such as selection combining help mitigate fading, and adaptive (smart) antennas mitigate CCI. At the expense of requiring complex signal processing algorithms, SDM can mitigate all three wireless impediments together.

Most of the research into MIMO antennas is built around a communication framework consisting of a *base station* (BS) and a *mobile station* (MS) [Mur02], making the configuration initially more suitable for WLAN rather than WPAN use. Up to four antenna systems can be used. On the downlink (BS to MS), the BS has a TX array and the MS has a RX array, and vice versa on the uplink (MS to BS). Figure 2-20 depicts one such implementation. The BS has M antennas for TX and L antennas for RX, while the MS has N antennas for RX and K antennas for TX, which are assumed to be a shared array for a more compact design.

Figure 2-21 shows MIMO operation from an intuitive point of view. The BS is receiving the MS transmission, and both nodes use multiple antennas. Each antenna at the MS is sending a different data stream on the same carrier frequency, so channel capacity is tripled over a single-antenna implementation. The adaptive RX antenna array at the BS effectively steers a main lobe toward each of the three TX antennas, and the associated nulls are used to reject interference, either from the other antennas at the same MS or signals from other users as necessary for the best performance. This process is sometimes called *spatial filtering*, because the rejection of signals is accomplished in space by using a sophisticated form of angle diversity. A time element can also be included, in which case the processing is called *space time*.

A receiver is assumed to have access to *channel state information* (CSI), which it obtains by examining an incoming, known TX sequence for distortions over the channel. The receiver uses CSI to adjust various parameters in its multiple-antenna processing algorithms for best performance. An example of a space-time antenna array and its associated signal processing is shown in Figure 2-22. The

Figure 2-20
A wireless link between BS and MS, both of which use a MIMO antenna array

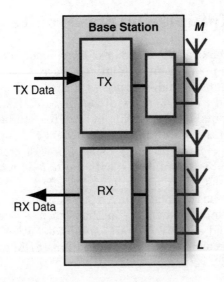
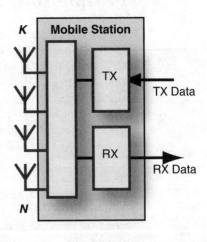

Figure 2-21
MIMO operation over the MS-to-BS uplink

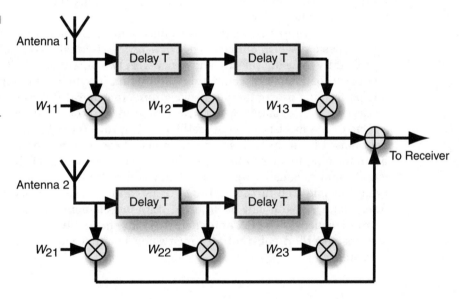

Figure 2-22
An example of space-time processing for two RX antennas

sequence of weighted delays is adjusted to match channel conditions for both fading and ISI characteristics. Both maximal ratio and equal gain combining can be realized with this implementation. Typical *binary phase shift keying* (BPSK) performance improvement for two antennas using maximum ratio combining in a Rayleigh faded channel is about 10 dB.

Providing CSI to the transmitter for use with multiple antennas is more difficult because accurate information must be passed to the sender from the receiver at the other end of the link. Alternatively, the sender can make various CSI parameter assumptions based on its corresponding receiver's CSI or make some other approximation. Consequently, SDM at the TX side will probably be suboptimal.

Summary

WLAN and WPAN implementations are used under vastly varying conditions with both fixed and portable network members. Simple approximations can be used for quick and reasonably accurate estimates of the maximum distance over which two nodes can communicate. Accuracy can be improved by using primary ray tracing, where the path from the transmitter to receiver is analyzed with its obstructions and distance information. This method is most applicable for fixed devices such as a printer, desktop computer, or WLAN AP. The analysis will help determine where a device should be placed within a room or building for best access by its intended users, and what specifications the transmitter, receiver, and antenna should have.

Multipath components can cause fading of up to about 30 dB and ISI that limits the speed at which data can be sent over the channel. Fading can be reduced significantly by using selection diversity with two antennas. Additional improvements can be gained by combining all antenna inputs by aligning their phases and performing proper amplitude weighting. In general, diversity combining can be categorized as spatial, spectral, or temporal, or a combination of these.

One of the most promising methods to improve coexistence among independent networks is through smart antennas. These antenna arrays use electronic beam-forming to steer a high-gain lobe toward the source of the desired signal while simultaneously creating nulls in the direction of interfering signals. Carrying this one step further, MIMO arrays can increase capacity by sending different data streams out of different antenna elements, each using the same carrier frequency. The RX antenna array can effectively demodulate these data streams together, substantially increasing channel capacity while improving coexistence.

References

[Ald02] Al-Dhahir, N., et al., "Space-Time Processing for Broadband Wireless Access," *IEEE Communications Magazine*, September 2002.

[And94] Anderson, J., et al., "Propagation Measurements and Models for Wireless Communication Channels," *IEEE Communications Magazine*, November 1994.

[Ban99] Bansal, R., "The Far-Field: How Far is Far?" *Applied Microwave & Wireless,* November 1999.

[Bes01] Best, S., "Antennas and Propagation for Wireless Communication," Cushcraft Corporation white paper, 2001.

[Boy00] Boyle, B., "Space-Time Codes: A Synergistic Integration of Multiple Communication Disciplines," *MPRG Propagator*, Virginia Polytechnic University, 2000.

[Bre98] Breed, G., "A Primer on Antenna/Human Body Interaction," *Applied Microwave and Wireless,* November/December 1998.

[Car96] Carey, T., "Fading and Multipath Testing in Communications Systems," *Microwave Journal*, November 1996.

[Che98] Cheung, K., et al., "A New Empirical Model for Indoor Propagation Prediction," *IEEE Transactions on Vehicular Technology*, August 1998.

[Che01] Chen, J., "Wireless Comes Home with 5-GHz Technologies," *Electronic Products Supplement*, Winter 2001.

[Cui01] Cuinas, I., and Sanchez, G., "Measuring, Modeling, and Characterizing of Indoor Radio Channel at 5.8 GHz," *IEEE Transactions on Vehicular Technology*, March 2001.

[Dob02] Dobkin, D., "Indoor Propagation Issues for Wireless LANs," *RF Design*, September 2002.

[Dur98] Durgin, G., et al., "Measurements and Models for Radio PL and Penetration Loss in and Around Homes and Trees at 5.85 GHz," *IEEE Transactions on Communications*, November 1998.

[Gha02] Ghassemzadeh, S., "The Indoor Ultra-Wideband Multipath Channel Model," IEEE document 802.1502/283 r1-SG3a, July 2002.

[Hal91] Hall, G., ed., *The ARRL Antenna Book*. Connecticut: American Radio Relay League, 1991.

[Hal00] Halford, K., and Webster, M., "Multipath Measurements in Wireless LANs," Intersil Application Note AN9895, May 2000.

[Han02] Hansen, J., "802.11b/a—A Physical Medium Comparison," *RF Design*, February 2002.

[How90] Howard, S., and Pahlavan, K., "Doppler Spread Measurements of Indoor Radio Channel," *IEE Electronics Letters*, January 18, 1990.

[IEEE802.15.2] IEEE 802.15.2-2003, "Coexistence of Wireless Personal Area Networks with Other Wireless Devices Operating in Unlicensed Frequency Bands," August 28, 2003.

[ITU01] International Telecommunication Union—Radio Communication Sector, "Propagation data and prediction methods for the planning of indoor radiocommunication systems and radio local area networks in the frequency range 900 MHz to 100 GHz," Recommendation ITU-R P.1238-2, 1997-1999-2001.

[Jak71] Jakes, W., "A Comparison of Specific Space Diversity Techniques for Reduction of Fast Fading in UHF Mobile Radio Systems," *IEEE Transactions on Vehicular Technology*, November 1971.

[Kim96] Kim, S., et al., "Pulse Propagation Characteristics at 2.4 GHz Inside Buildings," *IEEE Transactions on Vehicular Technology*, August 1996.

[Lee00] Lee, D. J. Y., and Lee, W. C. Y., "Propagation Prediction in and Through Buildings," *IEEE Transactions on Vehicular Technology*, September 2000.

[Mol03] Molisch, A., et al., "Channel Models for Ultrawideband Personal Area Networks," *IEEE Wireless Communications Magazine*, December 2003.

[Mor99] Morrow, R., "Site-Specific Engineering for Indoor Wireless Communications," *Applied Microwave & Wireless*, Vol. 11, No. 3, March 1999.

[Mor02] Morrow, R., *Bluetooth Operation and Use*, New York: McGraw-Hill, 2002.

[Mur02] Murch, R., and Letaief, K., "Antenna Systems for Broadband Wireless Access," *IEEE Communications Magazine*, April 2002.

[Pah95] Pahlavan, K., and Levesque, A., *Wireless Information Networks*, John Wiley & Sons, 1995.

[Pen02] Pendergrass, M., "Empirically Based Statistical Ultra-Wideband (UWB) Channel Model," IEEE document 802.1502/295 SG3a, July 2002.

[Poo98] Poor, V., and Wornell, G., eds., *Wireless Communications: Signal Processing Perspectives*. New Jersey: Prentice Hall, 1998.

[Rap96] Rappaport, T., *Wireless Communications: Principles & Practice,* 1st ed. New Jersey: Prentice Hall, 1996.

[Sch98] Schweber, B., "RF-Channel Simulators," *EDN*, September 11, 1998.

[Sei92] Seidel, S., and Rappaport, T., "914 MHz PL Prediction Models for Indoor Wireless Communications in Multifloored Buildings," *IEEE Transactions on Antennas and Propagation*, February 1992.

[Sei94] Seidel, S., and Rappaport, T., "Site-Specific Propagation Prediction for Wireless In-Building Personal Communication System Design," *IEEE Transactions on Vehicular Technology*, November 1994.

[Siw02] Siwiak, K. "UWB Propagation Phenomena," IEEE document 802.1502/301 r2, July 2002.

[Ste97] Stein, J., "Indoor Radio WLAN Performance. Part II: Range Performance in a Dense Office Environment," Harris Semiconductor White Paper, 1997.

[Tan95] Tang, Y., and Sobol, H., "Measurements of PCS Microwave Propagation in Buildings," *Applied Microwave & Wireless*, Winter 1995.

[Tol98] de Toledo, A., and Turkmani, M., "Estimating Coverage of Radio Transmission into and Within Buildings at 900, 1800, and 2300 MHz," *IEEE Personal Communications Magazine*, April 1998.

[Wel02] Welborn, M., and Siwiak, K., eds, "Ultra-Wideband Tutorial," IEEE document 802.1502/133 r1, March 2002.

[Win98] Win, M., and Scholtz, R., "On the Robustness of Ultra-Wide Bandwidth Signals in Dense Multipath Environments," *IEEE Communications Letters*, February 1998.

[Val97] Valenzuela, R., et al., "Estimating Local Mean Signal Strength of Indoor Multipath Propagation," *IEEE Transactions on Vehicular Technology*, February 1997.

Basic Modulation and Coding

Since the purpose of the radio wave is to convey information from one point to another, this chapter will examine some of the ways this information can be coded and modulated onto the carrier, and we'll also look at the relative performance of each method. Coding attempts to take raw data and create a bit stream that has certain desirable characteristics beyond just the data itself. Examples would include compressing data into a shorter bit stream for faster transmission or adding error detection, and perhaps correction, to prevent corrupted data from being accepted at the receiver. Modulation is the process of placing data, coded or uncoded, onto the carrier.

In this chapter we'll take a quick look at basic modulation principles, followed by the modulation techniques used in *wireless local area network* (WLAN) and *wireless personal area network* (WPAN) specifications. That is, we won't discuss modulation methods other than those used in the various 802.11 WLAN and 802.15 WPAN renditions.

Elementary Modulation Methods

In digital communication systems, the data (consisting of binary digits, or *bits*) is represented in some way as electronic signals sent from transmitter to receiver. Perhaps the simplest representation is to let the binary 1 be a "high" voltage, and a 0 be a "low" voltage. These voltages can be sent one by one down a wire from the transmitter to the receiver. A sequence of raw data represented in this way is called the *baseband signal*. If communication is via wireless, the baseband signal is used to alter the *radio frequency* (RF) carrier accordingly.

An RF carrier is represented mathematically as a sinusoid having amplitude, frequency, and phase. Any of these three quantities, or a combination of them, can be controlled by the baseband data signal. Because a data bit can be either a binary 1 or 0, a set of two different signals composes the simplest possible modulation scheme. One of these signals is called $s_1(t)$, corresponding to a 1, and the other is called $s_0(t)$, corresponding to a 0. In this scheme, the *symbol rate*, which is the rate at which new transmitted signals are sent, is equal to the *bit rate*, which is the rate at which new data bits are created. Some examples of data with their corresponding baseband signal, along with waveforms produced by each of these elementary modulation techniques, are given in Figure 3-1.

The receiver attempts to determine which of these was actually sent by processing a copy of the transmitted signal that often becomes extremely weak and

Figure 3-1
Amplitude shift
keying (ASK),
frequency shift
keying (FSK),
and phase shift
keying (PSK)
modulation
waveforms

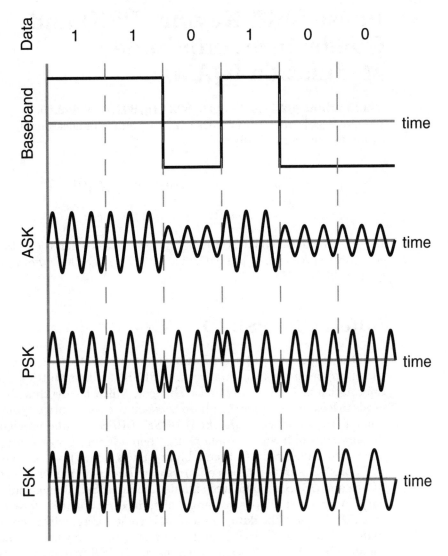

noisy during its journey to the receiver. Occasionally, the noise will be such that the receiver will mistake $s_0(t)$ for $s_1(t)$ or vice versa, and a bit error occurs. The bit error probability, also called the *bit error rate* (BER), is an extremely important measure of the performance of a digital communication system.

Phase Shift Keying (PSK) and Quadrature Amplitude Modulation (QAM)

Application: 802.11a, 802.11b, 802.11g, 802.15.3, 802.15.4

The simplest form of PSK occurs when the data changes the phase of the carrier, expressed mathematically as

$$s_1(t) = A \cos\left[2\pi f_c t + \phi_1(t)\right] \tag{3.1}$$

$$s_0(t) = A \cos\left[2\pi f_c t + \phi_0(t)\right] \tag{3.2}$$

For example, the phase could be $\phi_0 = 0$ degrees when the data bit is a 0, and $\phi_1 = 180$ degrees when the data bit is a 1. This is called *binary phase shift keying* (BPSK or 2PSK).

Differential PSK (DPSK)

Both the original 802.11 specification and the 802.11b enhancement require the use of *differential phase shift keying* (DPSK) for transmitting data at 1 Mb/s. Information is found in the phase changes, which means that the receiver only needs to look for whether the phase between two consecutive symbols changes or remains the same. For *differential BPSK* (DBPSK), if the previous symbol and the new symbol have the same phase, then a binary 0 was sent; if they change phase by π radians (180°), then a binary 1 was sent. This is shown in Table 3-1.

Figure 3-2 depicts a representative data stream along with its corresponding BPSK- and DBPSK-modulated signals. Unlike BPSK, if an incoming DBPSK signal is interrupted, there's no danger of an incorrect phase reference during reacquisition. Instead, the data stream will become correct within one or two symbol changes after the desired signal appears again. A *signal-to-noise ratio* (SNR) penalty about 1 to 3 dB occurs when using DBPSK instead of BPSK, depending on the receiver detection method. That is, the DBPSK signal needs to be about 1

Table 3-1

DBPSK phase shift encoding in 1 Mb/s 802.11

New bit	Phase shift from previous signal
0	0
1	π

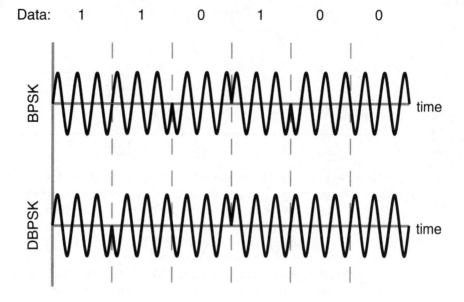

Figure 3-2
Ordinary BPSK
compared to
DBPSK
modulation
waveforms

to 3 dB stronger at the receiver to achieve a BER performance equivalent to that of ordinary BPSK. A slightly different form of differential encoding is used in 802.15.4 for data rates of 20 and 40 kb/s. In this method, a new differentially encoded bit is determined by taking the previously encoded bit and performing an *exclusive OR* (XOR) operation on it using the new raw data bit.

One potential problem with DBPSK is that the receiver can lose track of bit boundaries if a stream has too many consecutive binary zeros. With no phase transitions available (and no phase lock on the incoming signal), clock drift between transmitter and receiver could result in the receiver adding or dropping data bits during demodulation. To avoid this, bit streams are scrambled, or *whitened*, before transmission, and then dewhitened at the receiver after detection. The whitening process uses an algorithm to randomize the data stream so that long runs of either 1's or 0's are prevented. The receiver uses the same algorithm to derandomize the detected stream and restore the original data.

Quadrature Phase Shift Keying (QPSK)

For a given symbol rate, the data rate can be doubled within the same transmitted bandwidth by using *quadrature phase shift keying* (QPSK) instead of BPSK. The transmitted signal set now has phases 0, $\pi/2$, π, and $3\pi/2$ radians, so that two bits can be encoded per transmitted symbol. QPSK can also be coded differentially as *differential QPSK* (DQPSK), which is used for 802.11b *Wireless*

Table 3-2

DQPSK phase shift encoding

New bit pair	Phase shift from previous signal
00	0
01	$\pi/2$
11	π
10	$3\pi/2$

Fidelity (Wi-Fi) at 2 Mb/s and 802.15.3 WiMedia at 22 Mb/s. The transmitted signal is determined by taking the data stream to be sent and arranging it in pairs of bits. The four different possibilities determine the phase shift of the new symbol relative to the previous symbol, as shown in Table 3-2.

At first glance, it seems strange that the bit pairs in Table 3-2 progress from 00 to 01 to 11 to 10, instead of the usual binary counter sequence of 00, 01, 10, 11. The former type of binary counting (00, 01, 11, 10) is called the *Gray code*, and this is used to attempt to prevent multiple bit errors from occurring in response to a single symbol error. If noise corrupts one of the symbols, the probability is high that an adjacent symbol will be mistakenly selected by the receiver, and an adjacent symbol error will result in only a single bit error.

Offset QPSK   Because all transmitters have nonlinearities inherent in their circuitry, out-of-band emissions can be reduced by preventing a phase transition of π, which momentarily requires the signal envelope to pass through zero. Instead, if the two data components are offset by one bit period (half a symbol period), then phase transitions occur more often but are limited to at most $\pi/2$. An example of *offset QPSK* (OQPSK), a version of which is used in 802.15.4 ZigBee, is shown in Figure 3-3.

Signal Constellations

Instead of using only mathematical formulas to represent the various PSK signals, we will turn once again to the phasor diagram. Phasor diagrams make it easy to visualize the different phase shifts and amplitude variations of complex signal sets that are sometimes used for sending several data bits per transmitted symbol. When plotted on a phasor diagram, this collection of transmitted symbols is called the *signal constellation* due to its similarity to the arrangement of stars in an astronomical constellation.

Figure 3-4 shows the signal constellations for both DBPSK and DQPSK used in low-data-rate 802.11b Wi-Fi transmissions. The two axes are labeled *I* for the *in-phase* and *Q* for the *quadrature* components of the signals. In Wi-Fi, DBPSK

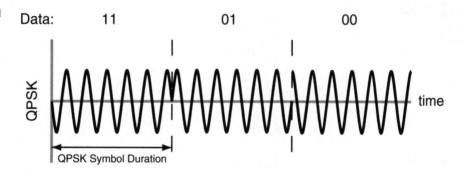

Figure 3-3
OQPSK
prevents the
transmit (TX)
signal
envelope from
passing
through zero
by limiting
phase changes
at most to $\pi/2$.

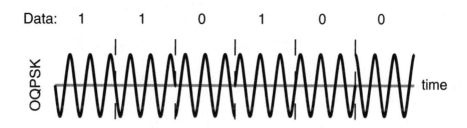

results in a raw transmitted data rate of 1 Mb/s, and DQPSK has a raw data rate
of 2 Mb/s. These rates were part of the original 802.11 *direct sequence spread
spectrum* (DSSS) specification, and they are also included in 802.11b to provide
backward compatibility with 802.11 DSSS. The diagrams here are a bit oversim-
plified, though, because a special DSSS spreading code is superimposed on each
channel symbol, as we discuss in the next chapter. However, the BER performance in *additive white Gaussian noise* (AWGN) is approximately the same with
or without the spreading code. As we will soon discover, though, some types of
DSSS can enhance performance over a channel susceptible to interference.

There is really no theoretical limit to the number of signals that can be part of
a constellation, but it should be obvious that the more crowded the constellation
becomes, the higher the symbol error rate, because it becomes more difficult for
the receiver to discern the differences among symbols in a noisy or interference-
prone environment. The number of bits per symbol k is related to the number of
symbols M in the signal constellation by the formula

$$k = \log_2 M \qquad (3.3)$$

or, equivalently, $M = 2^k$. For example, if 16 signals are in the constellation, then
each contains 4 bits of information.

Figure 3-4
Signal
constellations
for DBPSK and
DQPSK
modulation

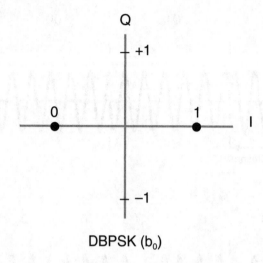

DBPSK (b_0)

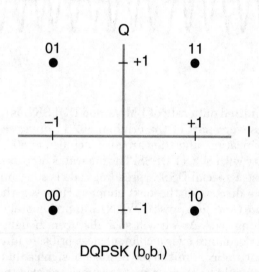

DQPSK (b_0b_1)

Quadrature Amplitude Modulation (QAM)

QAM combines both ASK and PSK in such a way that several bits can be transmitted per symbol while keeping adjacent symbols separated enough that the BER is lower than if ASK or PSK were used alone. Each symbol in the signal constellation is now represented by a unique amplitude and phase. A QAM constellation containing M symbols is called M-ary QAM, where, as before, M is usually

Figure 3-5
Signal
constellation
for 16-QAM

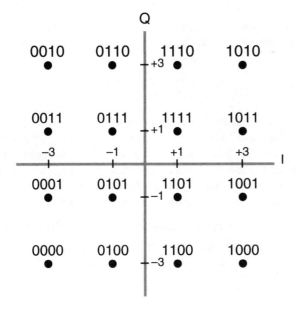

16-QAM ($b_0 b_1 b_2 b_3$)

an integer power of 2 and is related to k by Equation 3.3. BPSK can be thought of as 2-QAM and QPSK as 4-QAM. As a trivial aside, using two different amplitude values for 4-QAM has no performance advantage over specifying the same amplitude for all four signals in its QPSK rendition [Pro94].

802.11a, 802.11g, and 802.15.3 allow BPSK, QPSK, 16-QAM, or 64-QAM, depending upon the desired data rate. 802.15.3 also includes 32-QAM. The 16-QAM constellation is depicted in Figure 3-5, which also shows how an adjacent symbol error still results in a single data bit error within that symbol.

PSK and QAM Performance in AWGN

Finding the BER of various digital modulation methods in AWGN without accounting for fading is, for the most part, a mature area of research and is extensively covered in many communication textbooks. It will be interesting here to compare the BER plots for the various modulation methods used by most of the wireless networks examined in this book. The BER will be plotted against the bit energy per unit noise *power spectral density* (PSD), defined as $\dfrac{E_b}{N_0}$, or SNR/bit, which is related to received SNR by

$$\frac{E_b}{N_0} = \frac{S}{N}\left(\frac{W}{R}\right) = \frac{S}{N}(WT) \tag{3.4}$$

where S/N is the received SNR, R is the rate at which data is sent in bits/second, T is the bit duration, and W is the channel bandwidth in Hz [Skl88]. In this way we can easily discern the SNR penalty associated with attempting to transmit multiple bits per channel symbol in an effort to increase the overall data rate.

Several idealistic assumptions are made during the derivation of most of the BER formulas in this section. Among these are the following:

- All impediments to perfect data reception can be modeled as AWGN.

- Optimum detection is used for a given modulation method.

- Incoming data is random with approximately the same number of 1s and 0s.

Of course, no actual link operates under these conditions. Indeed, real-world performance can often be significantly worse due to cost constraints on transmitter and receiver design. However, by making these idealistic assumptions, we can compare the relative performance of the various WLAN/WPAN signal structures. Furthermore, by assuming that interfering signals act as AWGN at the desired receiver, useful BER results can be obtained without resorting to additional complex interference modeling.

BPSK and QPSK Bit Error Rates   Straight BPSK and QPSK produce good BER values for a given γ_b at the receiver. Consequently, these plots can serve as a standard by which others can be compared. In one of those mysterious quirks of probability theory and how it relates to communication system performance, the BER of BPSK and QPSK under identical operating conditions is the same. Although the symbol error rate for BPSK is twice that of QPSK, QPSK transmits symbols at only half the rate of BPSK when both are sending data at the same bit rate, so their BER performances are equivalent. Achieving identical BER values for BPSK and QPSK in practice is difficult, however, due to the more stringent timing requirements placed upon QPSK modulation and demodulation to achieve zero crosstalk between the in-phase and quadrature signal components.

For BPSK and QPSK that are detected coherently, meaning that the receiver must track the phase of the incoming signal, we have

$$P_b = Q\left(\sqrt{\frac{2E_b}{N_0}}\right) \tag{3.5}$$

where P_b is the bit error probability, which is the same as the BER. The so-called *Q-function* is the area under the tail of a Gaussian curve and is used to measure the part of the probability density function on the wrong side of the receiver detector's threshold. This function is given by

$$Q(x) = \int_x^\infty \frac{1}{\sqrt{2\pi}} \exp\left(\frac{-u}{2}\right) du \qquad (3.6)$$

Unfortunately, this integral cannot be solved in closed form, but tables of values for this function are readily available, and it can be solved numerically. Most mathematics computation programs contain this function, or a closely related cousin called the *error function*, in their libraries. The error function is given by

$$\text{erf}(x) = \frac{2}{\sqrt{\pi}} \int_0^x \exp(u^2) \, du \qquad (3.7)$$

and is related to the Q-function by

$$Q(x) = \frac{1}{2}\left[1 - \text{erf}\left(\frac{x}{\sqrt{2}}\right)\right] = \frac{1}{2}\text{erfc}\left(\frac{x}{\sqrt{2}}\right) \qquad (3.8)$$

where $\text{erfc}(x) = 1 - \text{erf}(x)$ is the *complementary error function*.

For DBPSK and DQPSK, the receiver can use noncoherent techniques because it only needs to determine whether or not a phase change occurs at the symbol boundaries, and there's no need to track the phase itself. A noncoherent receiver detecting DBPSK has its BER given by

$$P_b = \frac{1}{2}\exp\left(-\frac{E_b}{N_0}\right) \qquad (3.9)$$

Unlike coherently detected BPSK and QPSK, DQPSK and DBPSK don't perform identically. Instead, the BER of DQPSK is approximately equal to that of QPSK with a 3 dB SNR penalty. In other words, for a given noise PSD, DQPSK requires twice the signal power at the receiver to produce the same BER as coherently detected QPSK [Pro94].

QAM Bit Error Rates The performance of QAM depends greatly on how the signal constellation is structured and how the bits are mapped to each constellation point. The rectangular constellation pattern is convenient because a transmitter and receiver can be designed fairly easily to produce and detect these signals. For convenience, the number of signals in the QAM constellation is almost always represented by an integer power of 2.

As with other M-ary modulation schemes, channel errors affect the entire QAM symbol rather than individual bits. If almost all symbol errors result in, at most, a single bit error from the erroneous detection of an adjacent symbol, the BER is approximately the *symbol error rate* (SER) divided by k. Due to the close proximity of QAM symbols in the constellation, however, this assumption may be too optimistic at times, especially when symbol errors are caused by occasional high levels of interference. If all symbol errors are equally likely, then the BER for a given SER, designated as P_s, can be found by noting that $2^k - 1$ possible error patterns are in a k-bit sequence. After accounting for the number of bad bits in each pattern, we discover that the BER and SER are related by

$$P_b = \left(\frac{2^{k-1}}{2^k - 1} \right) P_s \tag{3.10}$$

For large k, $P_b \rightarrow P_s/2$, because on average, for equally distributed symbol error probabilities, half the bits in a bad symbol will be in error.

Finding the SER for QAM in AWGN is quite difficult, but fortunately a tight upper bound has been discovered [Pro94]. This is given by

$$P_s \leq 4Q\left(\sqrt{\frac{3kE_b}{(M-1)N_0}} \right) \tag{3.11}$$

where $M = 16$ and $k = 4$ for 16-QAM, and $M = 64$ and $k = 6$ for 64-QAM.

A BER plot is shown in Figure 3-6 for the modulation methods we've discussed so far. For QAM, we assume each symbol error contains a single bit error, but additional pessimism can be accommodated through Equation 3.10 if desired. As more and more bits are encoded per channel symbol, the SNR for a given BER must be higher so the receiver can discern the differences among a larger number of points on the signal constellation.

One of the great strengths of Wi-Fi operation is its ability to automatically adjust its modulation to match channel conditions. When an 802.11b link is first established, for example, it operates with DBPSK or DQPSK for transmitted data rates of 1 or 2 Mb/s, respectively. If channel conditions and available hardware can support higher data rates, the system will automatically switch to speeds up to 11 Mb/s using *complimentary code keying* (CCK). For 802.11a and 802.11g, data rates can be as low as 6 Mb/s using BPSK, or as high as 54 Mb/s using 64-QAM. All these higher data rates are achieved by encoding more data bits per transmitted symbol, and this in turn dictates a higher SNR at the receiver for reliable symbol detection. If channel conditions later deteriorate due to increased noise or interference levels, reliability is maintained by encoding fewer data bits per channel symbol.

Because the symbol duration remains fixed regardless of the data rate for 802.11a/g, the bit energy is reduced for these higher rates. Equation 3.11 provides the basis for deriving the increase in average *received signal strength indication* (RSSI), in dB, needed to maintain a given level of performance for QAM as the number of signals M in the constellation changes. The required RSSI increase is given by $10\log[2(M-1)/3]$ [Pro94].

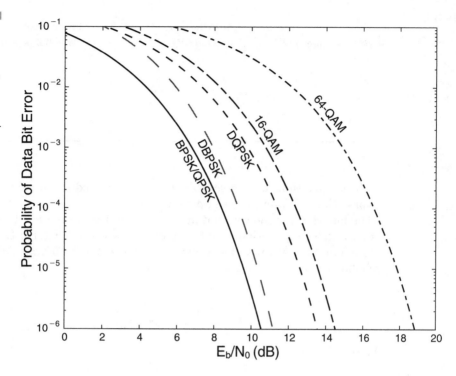

Figure 3-6
BER plots for BPSK/QPSK, DBPSK, DQPSK, 16-QAM, and 64-QAM in AWGN

As an example, suppose a reliable 50 m link is established using BPSK. For the same BER using other modulation methods, the required SNR (that is, RSSI) must increase, which equates to a reduced range as shown in Table 3-3 using the simplified *path loss* (PL) model and constant TX power. The range decreases significantly as more bits are encoded per symbol while maintaining a constant symbol rate. The trend shown in Table 3-3 is typical of actual 802.11a/g performance, where the slowest data rate has a range about twice that of a moderate data rate and triple that of the highest-speed data rate.

PSK Performance in Rayleigh Fading and Interference

As discussed in Chapter 2, "Indoor RF Propagation and Diversity Techniques," fading is almost always present in an indoor communication system. Rayleigh fading occurs when no *line of sight* (LOS) or *obstructed line of sight* (OLOS) path exists, which can occur in many WLAN and WPAN usage scenarios. The

Table 3-3

Range versus modulation method, path loss exponent 3.0

Modulation method	Additional RSSI needed (dB)	Estimated range (m) for fixed TX power
BPSK	0	50
16-QAM	10.0	25
64-QAM	16.2	15

resulting variations in *receive* (RX) signal power produces an average BER that is worse than if fading weren't present.

The last chapter showed that antenna diversity can reduce the probability of outage and by implication can also reduce the average BER. For BPSK, if one TX antenna and L RX antennas employ *maximal ratio combining* (MRC), then the probability of bit error is given by [Pro94]

$$P_b = \frac{K_L}{(4E_b/N_0)^L} \qquad (3.12)$$

where

$$K_L = \frac{(2L - 1)!}{L!(L - 1)!} \qquad (3.13)$$

and the desired signal on each RX antenna is assumed to have independent Rayleigh fading. Because E_b/N_0 at any RX antenna varies with time, the BER is given as a function of average E_b/N_0.

Figure 3-7 shows the performance of BPSK over a Rayleigh fading channel with no diversity and with MRC using two RX antennas and one TX antenna. The BER plot when no fading exists is also shown for comparison. MRC using two RX antennas provides about 10 dB of improvement in BER compared to using only one antenna with no diversity combining in a Rayleigh channel. DBPSK performance is about 3 dB worse than BPSK performance in most situations.

When interference is present on the channel, it can have a profound effect on the BER of signals modulated with various forms of PSK. Quantifying this effect accurately requires knowing the appropriate *carrier-to-interference ratio* (C/I) values, the structure of both the desired and interfering signals, and receiver performance criteria. As mentioned earlier, the process is facilitated by treating interference as AWGN and using standard formulas that determine the BER as

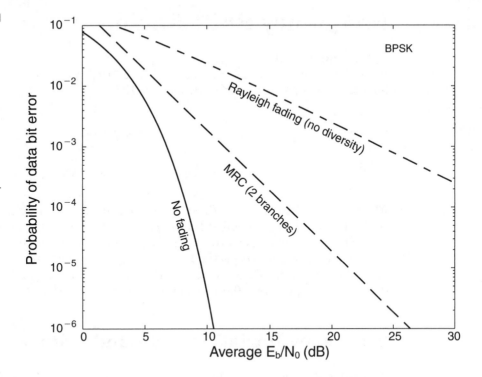

a function of SNR/bit. C/I values can be found using the techniques in Chapter 2, and receiver performance figures are published by the manufacturer.

Interference analysis is facilitated further by assuming that if an interfering signal brings the C/I of the desired receiver below some specified limit, the incoming desired symbols are randomized. The result is a useless channel while the interference is present, and the desired packet or frame is destroyed. To combat this, many WLAN and WPAN implementations use a listen-before-talk channel access method that allows transmissions to occur only when the channel is clear. As discussed in Chapter 6, "Medium Access Control," network throughput can slow to a crawl when large numbers of nodes compete for access to the channel. Because the unlicensed bands, especially at 2.4 GHz, are quickly becoming crowded, additional interference mitigation is necessary for reliable performance. This takes the form of more sophisticated modulation such as DSSS, *frequency hopping spread spectrum* (FHSS), or *orthogonal frequency division multiplexing* (OFDM), along with error control to prevent corrupted data from being accepted by the receiving node.

Frequency Shift Keying

Application: 802.15.1

If the data bits change the frequency of the carrier instead of its amplitude, we have for *binary frequency shift keying* (BFSK)

$$s_1(t) = A \cos \left[2\pi (f_c + f_d)t + \phi \right] \tag{3.14}$$

$$s_0(t) = A \cos \left[2\pi (f_c - f_d)t + \phi \right] \tag{3.15}$$

where the amplitude is now fixed and the frequency varies with the bit being sent. The quantity f_d is called the *frequency deviation*, which determines the amount of shift up or down from the nominal carrier frequency f_c that occurs when a 1 or 0 is sent, respectively. For example, a carrier frequency of 1 MHz might be shifted up slightly to 1.01 MHz to represent a 1 and down slightly to 0.99 MHz to represent a 0. In this case f_d equals 0.01 MHz.

Frequency Deviation and Bandwidth

Frequency deviation is different from the data rate, so how should f_d be selected? A larger f_d gives the receiver a better chance of deciding whether $s_0(t)$ or $s_1(t)$ was transmitted (which is good), but it is also obvious that a larger f_d equates to a greater transmitted bandwidth for a given data rate (which is bad).

One way to help quantify the relationship between the frequency deviation f_d and the data rate R (or the bit duration $T = 1/R$) is through the *modulation index* β, defined as

$$\beta = \frac{2f_d}{R} = 2f_d T \tag{3.16}$$

assuming that one data bit is sent per symbol. Many digital communication systems that use BFSK (including Bluetooth) define a value, or range of values, for the modulation index. The data rate R is usually fixed, so Equation 3.16 can be used to find the deviation f_d given the other two quantities. In general, inexpensive, noncoherent BFSK transmitters use a modulation index of around 1, equivalent to the frequency deviation equal to half the data rate, for reasonable receiver performance. If the more sophisticated coherent (phase-tracked) BFSK detection is used, the modulation index can be as small as 0.5, which is the minimum possible for the signals $s_0(t)$ and $s_1(t)$ to remain *orthogonal*, or completely separated, at an ideal detector. For this reason, the aforementioned process is called *minimum shift keying* (MSK). Orthogonality exists between signals $s_0(t)$ and $s_1(t)$ when the following is true:

$$\int_0^T s_0(t)s_1(t)\,dt = 0 \qquad (3.17)$$

The bandwidth of an FSK signal is theoretically infinite, but the energy density in the transmitted waveform becomes insignificant for frequencies far removed from the carrier frequency. For an unfiltered baseband signal used in a BFSK modulator, the bandwidth B_T of the transmission is approximated by *Carson's rule*, which says that

$$B_T \approx 2(f_d + R) \qquad (3.18)$$

so the transmitted bandwidth in Hz is approximately twice the sum of the deviation and the data rate.

When Carson's rule is applied to the Bluetooth data rate of 1 Mb/s, coupled with a deviation of 500 kHz for a modulation index of 1, the bandwidth B_T is 3 MHz. However, Bluetooth was developed when the FCC restricted transmitted bandwidth to 1 MHz per hop channel, so more work must be done to maintain a 1 Mb/s FSK data rate within this bandwidth.

Reducing FSK Bandwidth with Gaussian Baseband Filtering The simple FSK modulation scheme transitions abruptly between $s_0(t)$ and $s_1(t)$ when the baseband data signal changes from 0 to 1 and likewise from $s_1(t)$ to $s_0(t)$ when a transition from 1 to 0 occurs. Fourier transform theory states that fast transitions in time correspond to wide spans in frequency, so Bluetooth uses a technique called *Gaussian-filtered frequency shift keying* (GFSK) to remove the abruptness of these transitions.

GFSK first passes the baseband signal though a low-pass filter that has a Gaussian response curve in the frequency domain (Figure 3-7c). The Gaussian filter's response $H(f)$ as a function of frequency f can be expressed mathematically as

$$H(f) = \exp\left(\frac{-1.4f^2}{B^2}\right) \qquad (3.19)$$

where B is the -3 dB bandwidth of the filter. The bandwidth B equals 500 kHz for the Bluetooth Gaussian filter. This filter smoothes the sudden transitions in

Figure 3-8
An unfiltered baseband signal (a) passed through a Gaussian baseband filter (c) produces a filtered base-band signal (b). Although the Gaussian filter response is shown here in the frequency domain, its time domain impulse response is also Gaussian [Mor02].

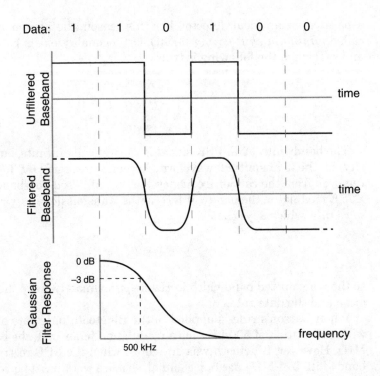

the baseband signal so that the FSK-modulated RF carrier also smoothly transitions between a frequency of $f_c + f_d$ and $f_c - f_d$, thus reducing the bandwidth. Figure 3-8 shows representative examples of the unfiltered and filtered baseband signal.

Even though the transmitted bandwidth is reduced by this filtering process, the price for this reduction is the introduction of intersymbol interference by the modulation process itself. Refer again to Figure 3-8. After filtering, the shape at the baseband of a particular bit is influenced by its neighboring bits. For instance, a binary 1 with 1's on either side of it has maximum amplitude (corresponding to maximum deviation of the modulated carrier) for the duration of the symbol period; however, a 1 surrounded by 0's has time periods on either side of its midpoint where the amplitude is less than the maximum. ISI increases as B in the Gaussian filter decreases for a given symbol rate and will increase with an increasing symbol rate for a constant B [Mur81]. Because the waveform in Figure 3-8 depends on both filter bandwidth B and symbol duration T, it is customary to express the Gaussian filter's characteristics as a BT product. As a general rule, ISI is fairly small if the BT product is 0.5 or greater [Rap96]. The BT product for the Bluetooth Gaussian filter is 0.5, so the modulation is called 0.5 GFSK.

Summary of Bluetooth Modulation Requirements

To receive Bluetooth qualification, a radio must have the following modulation characteristics:

- GFSK with *BT* at 0.5
- Symbol rate of 1 *megasymbols per second* (Ms/s), corresponding to a data rate of 1 Mb/s
- Modulation index β between 0.28 and 0.35
- Binary 1 with a positive f_d and 0 with a negative f_d
- Symbol timing better than ± 20 *parts per million* (ppm)
- Zero crossing error not greater than 1/8 of a symbol period
- The f_d corresponding to a 1010 sequence to be at least 80 percent of f_d corresponding to a 00001111 sequence
- Minimum f_d equal to 115 kHz

Most of the previous modulation characteristics are self-explanatory, except perhaps the last two. If the modulation index is between 0.28 and 0.35, then Equation 3.18 tells us that the corresponding f_d must be between 140 and 175 kHz. This can be considered a "steady-state" f_d for a long sequence of binary 1's or 0's. Due to the ISI inherent in GFSK, it's possible that f_d may not reach its maximum value for a binary sequence that alternates between 1 and 0. The specification requires that f_d under these conditions reach at least 115 kHz, which is about 80 percent of the minimum steady-state f_d of 140 kHz.

A typical Bluetooth signal power spectrum within a single hop channel is shown in Figure 3-9 [Sho01]. The -20 dB bandwidth is 1 MHz, as stated in the specification, and this value is almost universally used for coexistence analysis.

FSK Performance in AWGN

Ordinary noncoherent BFSK has a BER in AWGN given by [Skl88]

$$P_b = \frac{1}{2} \exp\left(\frac{-E_b}{2N_0}\right) \tag{3.20}$$

and thus its performance is 3 dB worse than that of DBPSK in Equation 3.9. For Equation 3.20 to be valid, the two transmitted BFSK signals must be orthogonal. This implies a modulation index of at least 1.0 for noncoherent FSK detection, or 0.5 for MSK. The Bluetooth ISI penalty for 0.5 GFSK and its lower modulation index is only about 0.2 dB, which means that a BFSK transmission with no ISI

Figure 3-9
Spectrum of
the Bluetooth
signal [Sho01]

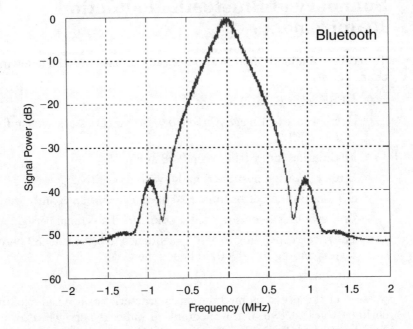

requires an SNR/bit about 0.2 dB less for the same BER performance. The BER performance of the Bluetooth GFSK signal in AWGN is shown in Figure 3-10, with the performance of BPSK/QPSK added for comparison. GFSK performs about 4 dB worse than BPSK for nearly all E_b/N_0 values. Also, it's important to remember that the energy E_b per bit varies depending on the value of adjacent bits, so care must be taken when using Figure 3-10 to determine Bluetooth system performance. For example, the BER for the sequence 10101010 is higher than for the sequence 11110000 because E_b is lower due to increased ISI when the bit pattern alternates between 1's and 0's.

For a more accurate rendition of the BER of a GFSK signal, we must account for the modulation index, especially because this index has a wide range of 0.28 to 0.35 in the Bluetooth specification. A smaller modulation index results in a higher BER due to the smaller deviation. The formula for incorporating the modulation index into the BER calculation is quite complex and is omitted here [Pro94].

FSK Performance in Rayleigh Fading and Interference

Like PSK, FSK also shows significant performance degradation when operating in a channel that exhibits Rayleigh fading. For BFSK, if one TX antenna and L

Figure 3-10
Approximate
BER versus SNR
plots for
Bluetooth
GFSK in
AWGN, with
BPSK/QPSK
shown for
comparison

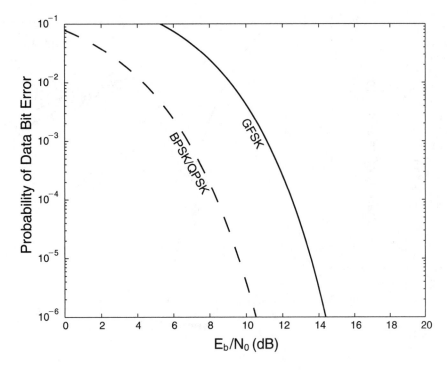

RX antennas employ *square-law combining* (SLC), then the probability of bit error is given by [Pro94]

$$P_e = \frac{K_L}{(E_b/N_0)^L} \tag{3.21}$$

where K_L is given by Equation 3.13. As before, each RX antenna has a desired signal with independent Rayleigh fading, and the BER results are shown as a function of average E_b/N_0.

Figure 3-11 shows the performance of BFSK over a Rayleigh fading channel with no diversity, and with square-law combining using two RX antennas and one TX antenna. The BER plot when no fading exists is also shown for comparison.

FSK can be more robust than PSK and its variants when CCI is present due to a phenomenon called *capture*. In analog *frequency modulation* (FM), the *capture ratio* is the minimum power ratio that must exist between two competing signals for the stronger one to completely dominate the weaker. That is, only the stronger FM broadcast would be heard. The equivalent term used in digital communications is *co-channel C/I*, which is the ratio between the desired and

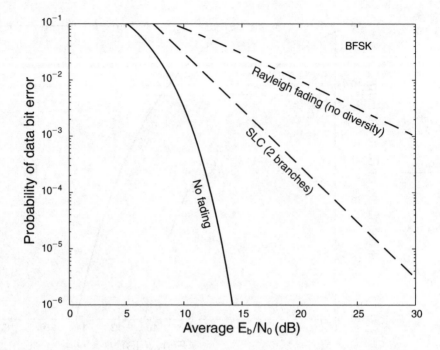

interfering signal powers at which the BER of the desired signal is at or below
some designated value. For example, the Bluetooth specification requires that
the BER cannot exceed 0.001 when the co-channel C/I is 11 dB. Because the FSK
modulation process used in Bluetooth can tolerate reasonably high interference
levels without significant BER degradation, nodes in the piconet don't use a listen-before-talk channel access. Instead, transmissions are made during regular
intervals regardless of whether or not the channel is otherwise occupied. It
appears at first glance that Bluetooth violates the good neighbor policy outlined
in Chapter 1, "Introduction." However, Bluetooth also uses FHSS for interference
avoidance, with the added benefit that interference by a Bluetooth node affects a
particular frequency only occasionally. So it's a (reasonably) good neighbor after
all. (For more information on channel access techniques, see Chapter 6.)

Error Control Coding

An understanding of error control is essential before any wireless network's
packet throughput figures can be calculated in a noisy or interference-prone
environment. An 802.11b receiver, for example, is required to have a *frame error
rate* (FER) of 8×10^{-2} or less for a 1,024-byte packet sent at 2 Mb/s when the RSSI

is -80 dBm. For Bluetooth certification, the receiver must have a raw BER of 10^{-3} or less at an RSSI level of -70 dB m. These error rates are much too high for reliable and efficient file transfers, so a means must be employed to reduce the effective FER or BER by several orders of magnitude. This is accomplished through *error control coding*. We will look first at error control in general and then examine the details of the various WLAN and WPAN implementations both in this and subsequent chapters.

An error control code can be either an *error detection code* for finding whether or not errors occur within a block of data, or an *error correction code* that can both find and correct a limited number of errors in a block or stream of data. Both codes work by adding redundant patterns of bits to the data. The sending node runs an algorithm to generate the redundant bits associated with the error control process, and these bits are transmitted along with the data. The receiving node may first process the incoming packet to try to correct any bit errors, and then it checks the packet for any remaining uncorrected errors. If additional errors remain that cannot be corrected, the receiving node requests a retransmission of the same information from the sending node. 802.15.1 Bluetooth, 802.11a and 802.11g versions of Wi-Fi, and 802.15.3 WiMedia have the capability to use all these methods to reduce errors. The 802.11b version of Wi-Fi and 802.15.4 ZigBee use error detection and repeat transmissions for reliable communication, but no error correction.

Several sources are available that provide detailed construction methods for error control codes [Lin70], [Bla83]. We will provide a glimpse of the mathematics behind error detection and then move on to a discussion of how the various codes perform. Generally, when noise causes the occasional bit error within a data packet, the error correction code, if it is present, will correct these errors. Next, the error detection code claims the packet is good, and it will be accepted at the higher protocol layers without any further action. On the other hand, if interference is present to such a degree that part or all of a desired packet is randomized, error correction cannot recover the data, so error detection will identify the packet as bad and a retransmission occurs. Under these conditions, error correction code overhead is a hindrance, and efficiency is improved without it. This will become more apparent as we quantify network throughput under both noisy and interference-prone conditions.

Error Detection

The *cyclic redundancy check* (CRC) consists of a group of j bits (typically 8, 16, or 32 bits) appended to the end of a *frame*, or block, of k bits of data. This group of bits is sometimes called the *frame check sequence* (FCS). The FCS bits are calculated such that the resulting $k + j$ bits are exactly divisible by some predetermined number. The receiver then divides the block of received data by the same number and checks the remainder. If it is zero, the block is considered to be error-free. If not, the receiver asks for a retransmission of the data block.

One of the first questions that probably comes to mind is this: Do circumstances exist in which the CRC will find nothing, but the data block actually contains errors? This situation can be extremely serious because it could result in erroneous data being accepted by the WLAN or WPAN recipient. To find the answer, we must first describe mathematically how the CRC is calculated [Sta88].

Mathematical Description of the CRC Process   A string of bits can be represented as a *polynomial*, also called a *vector*, in terms of a dummy variable X, such that each exponent corresponds to the position of each 1 in the string. For example, the binary string 11001 can be represented as $1X^4 + 1X^3 + 0X^2 + 0X^1 + 1X^0 = X^4 + X^3 + 1$. By doing this, we can perform special mathematical operations by using familiar polynomial algebra. Multiplication follows the usual rule of adding the exponents, such as $X^4 \times X^2 = X^6$. Addition is done modulo-2, which is the same as the XOR function: $0 + 0 = 0$, $1 + 0 = 1$, $0 + 1 = 1$, and $1 + 1 = 0$. The first three of these are the same as ordinary addition, but in the last equation the 1's add together to produce 0 with no carry bit. For more information on finite field arithmetic, see the literature [Lin70], [Bla83], [Sta88].

If the message in polynomial form is represented as $M(X)$ and the bit pattern $G(X)$ is the polynomial divisor (the *generator polynomial*), the associated quotient $Q(X)$ and remainder $R(X)$ have the following relationship:

$$\frac{X^j M(X)}{G(X)} = Q(X) + \frac{R(X)}{G(X)} \tag{3.22}$$

The quantity X^j is a left-shift operator by j bits to make room at the end of the message for the FCS $R(X)$. If $R(X)$ has j bits, $G(X)$ must have $j + 1$ bits. Now the sending unit transmits the sequence

$$T(X) = X^j M(X) + R(X) \tag{3.23}$$

As an example, suppose $M(X)$ equals $X^9 + X^7 + X^3 + X^2 + 1$ (1010001101) and $G(X)$ equals $X^5 + X^4 + X^2 + 1$ (110101). After performing the operation in Equation 3.22, we obtain $R(X) = X^3 + X^2 + X$ (1110), and the resulting $T(X)$ equals $X^{14} + X^{12} + X^8 + X^7 + X^5 + X^3 + X^2 + X$, equivalent to 101000110101110. Note that the FCS is actually 01110 to make it the required 5 bits long.

If the sequence $T(X)$ is received error-free, then

$$\frac{T(X)}{G(X)} = 0 \tag{3.24}$$

and if errors occur then the result should not be 0. Therefore, the receiver performs the division in Equation 3.24 and asks for a retransmission if the result isn't 0.

CRC Error Detection Capability   Returning to our original question, we can now conclude that if $T(X)$ can be corrupted by a certain pattern of errors, such that division by $G(X)$ produces 0 anyway, the CRC process fails and the WLAN or WPAN host could accept bad data. It turns out that all of the following bit patterns are *not* divisible by $G(X)$ and thus will be detected as errors in the data [Pet61]:

- All single-bit errors
- All double-bit errors as long as $G(X)$ has a factor with at least three terms
- Any odd number of errors as long as $G(X)$ contains the factor $X + 1$
- Any burst error that has a length less than the length of the FCS
- Most burst errors of longer length

A burst is defined as a binary sequence beginning and ending with 1 and having, at worst, randomized bits in between. If the received data block contains a burst error longer than the FCS length, it's possible that the block will be divisible by $G(X)$ with no remainder, and thus the burst error will go undetected. For a burst error of length r, where r is greater than the length of the FCS, the probability that the CRC will fail is approximately equal to 2^{-r}. As an example, suppose the FCS is 16 bits long. All burst errors of 15 bits or less will be detected by the CRC with certainty, but if the burst is, say, 20 bits long then the CRC will erroneously indicate that the data is good with a probability of 2^{-20}, or about 1 in a million. This probability may seem high, but remember that the chance of only 20 bits in a row being randomized from a noise burst is usually low to begin with. If the randomization occurs from interference, the run is usually far longer than 20 bits, with a corresponding reduction in the probability that the CRC will fail to discover a bad packet. Also, many data files contain their own error-checking mechanism independent of that included in the wireless *medium access control* (MAC) layer, so the end result is an exceedingly tiny chance that bad data will be accepted.

WLAN and WPAN Error Detection   CRC codes used for error detection in WLAN and WPAN systems can have a length field of 8, 16, or 32 bits, depending on the application. Short packet fields such as headers are usually protected by a short FCS, and long fields such as data payloads by a long FCS. Precisely where these FCS fields appear within the WLAN and WPAN packets will be discussed in later chapters.

The 8-bit FCS is used to protect WPAN header fields less than a few dozen bits long. The generator polynomial is

$$G(X) = X^8 + X^7 + X^5 + X^2 + X + 1 \qquad (3.25)$$

This FCS detects all errors that are odd in number, all double-bit errors, burst errors of 7 bits or less, and most other error patterns.

WLAN and WPAN packet fields containing a few thousand bits are protected by the 16-bit FCS called CRC-CCITT, also known as CCITT-16, given by the following generator:

$$G(X) = X^{16} + X^{12} + X^5 + 1 \tag{3.26}$$

As with the 8-bit FCS, this CRC detects all odd errors and all double-bit errors. Burst errors of 15 bits or less and most other error patterns will also be discovered.

Both 802.11 Wi-Fi and 802.15.3 WiMedia packets can have payloads exceeding 20,000 bits of data, and these long fields are protected by a 4-byte FCS called CRC-32 that uses the following generator:

$$G(X) = X^{32} + X^{26} + X^{23} + X^{22} + X^{16} + X^{12} + X^{11} + X^{10} + X^8 + X^7 + X^5$$
$$+ X^4 + X^2 + X + 1 \tag{3.27}$$

This FCS will detect all odd-numbered errors, all double-bit errors, all burst errors of 31 bits or less, and most other error patterns.

Automatic Repeat Request (ARQ)

In most WLAN and WPAN networks, communication occurs between two endpoints in almost all operations, so the systems are ideally suited for the implementation of a simple ARQ process for repeating the transmission of bad packets. An ARQ works as follows:

1. The sending node transmits a data packet.
2. The receiving node performs any applicable error correction and then checks packet integrity via the FCS.
3. The receiving node returns an *acknowledgment* (ACK) packet if the incoming packet is good; otherwise, it returns a *negative acknowledgment* (NAK) packet or remains silent.
4. The sending node transmits the next data packet if an ACK is returned; otherwise, it repeats the same packet.

The ACK/NAK information is usually sent as a separate packet, but it can sometimes be combined with other data. Successive 802.15.1 Bluetooth packets are transmitted on different hop frequencies, so if one of the frequencies is unusable because of interference, then the resulting NAK, the packet retransmission, and the following ACK probably won't use this particular frequency again. Wi-Fi, WiMedia, and ZigBee have somewhat less protection from such interference because they typically use a single, fixed carrier frequency for operations, but the intermittent nature of most nondeliberate interference generally results in only slight performance degradation. Furthermore, most of these specifications allow some type of frequency agility, where members of the network can change to a new carrier frequency either periodically or when the present channel contains unacceptable interference levels.

The ARQ process guarantees that a file will eventually be transferred completely error-free provided the following criteria are met:

- The FCS will never fail to discover a packet containing errors.
- The ACK/NAK process always operates correctly.
- The channel BER is lower than 0.5.

We've already made the reasonable assumption that the FCS will almost always work and that additional error control at higher protocol levels is usually present as a backup. For the ACK/NAK process to be reliable, we must assume that a NAK is never mistaken for an ACK, which the FCS will prevent. Furthermore, we must also assume that the sending node never fails to receive an incoming ACK. If an incoming ACK is missed, then the sending node will retransmit the previous packet, but the receiving node will think that it's getting a new packet. A corrupted file will then result.

So what happens when the sender misses an ACK? In Bluetooth, the baseband packet header also contains a *sequence* (SEQN) bit that is toggled (changed from binary 1 to 0, or from 0 to 1) for each new packet transmitted by the sending node. The receiving node can now check for duplicate packets by examining their SEQN bits and discarding any that arrive due to a lost ACK. The other WPANs and Wi-Fi use a more complex packet numbering system, but the result is that the receiving node can recognize and discard duplicate packets at the MAC level. We can now conclude that the probability is extremely low that a corrupted packet will be accepted as good.

Error Correction

Error *detection* is essential for a wireless communication link to operate with an effective BER low enough that file transfers can be accomplished essentially error-free. Unfortunately, the only way to recover from a bad packet so identified is though its retransmission using ARQ, and this can increase delay and decrease throughput markedly when the raw BER across the communication channel is only moderately high.

An alternative to simply placing a "good" or "bad" verdict on a packet of data is to include an *error correction code* along with the packet's data bits. This code is generated by an algorithm at the sending node that creates additional bits that the receiving node can use in another algorithm to correct up to a certain number of errors per block of data. Because the recipient performs error correction

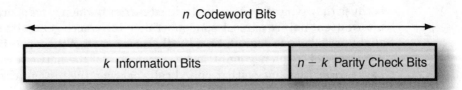

Figure 3-12
The structure of a systematic block FEC code

without any further assistance from the transmitting node, the process is called *forward error correction* (FEC).

FEC can be extremely useful when errors are relatively few and are independent from bit to bit within the communication channel. This situation occurs, for example, when AWGN predominates as the error-causing mechanism. When significant interference is present, though, bit errors may no longer be independent but instead often manifest themselves by relatively long strings of bit randomizations. These can easily overwhelm any FEC code that may be in use. For this reason, not all WLAN and WPAN implementations include error correction coding in their specifications.

Block FEC Codes   Bluetooth uses a type of FEC called a *systematic block code*, where a block of data with a length of k bits (usually smaller than the length of the packet itself) has an additional $n - k$ *parity check bits* appended to it such that the total codeword block length is n (see Figure 3-12). This code is called an (n, k) code. (The term "parity check" is a bit misleading because these bits are used for correcting errors, not just checking for errors.) The *rate* of the code is a measure of the fraction of the block devoted to actual user data and is given by the following equation:

$$r = \frac{k}{n} \tag{3.28}$$

The number of bit errors that an FEC code can correct is determined by the *minimum Hamming distance* associated with its set of codewords of length n. The Hamming distance is simply the number of bits between two equal-length strings of binary data that differ. For example, the strings 11010 and 10100 have a Hamming distance of three because the middle three bits are different between the two strings. The Hamming distance between arbitrary codewords in an error correction code may vary depending on which codewords are selected. The minimum Hamming distance is the smallest distance between any two codewords in the set. If this distance is called d, the maximum number of errors t that the code can correct is given by

$$t = \left\lfloor \frac{d - 1}{2} \right\rfloor \tag{3.29}$$

where the special brackets mean "integer part of." As an example, if the minimum Hamming distance of a particular code is three, it can correct one error per encoded block of data.

Block FEC Operation The mathematics behind systematic block error correction codes is similar to error detection codes. A generator polynomial $G(X)$ is used on the message vector $M(X)$ to create a remainder $R(X)$, as shown in Equation 3.22. The parity check bits that comprise $R(X)$ are appended to the end of the message, just like the FCS bits are for CRC, and the resulting codeword vector $T(X)$ is transmitted. Because FEC is more complex than CRC, the FEC generator $G(X)$ is usually longer than that used for the FCS.

Decoding the FEC codeword is also more involved because we must somehow identify where the errors are within the received binary string. Suppose the received binary string is written as a polynomial vector $V(X)$. If no errors exist, then $V(X)$ equals $T(X)$. However, if $V(X)$ contains errors, we would like to create another polynomial called the *error vector* $E(X)$ that contains terms that identify the error locations in $V(X)$. Remember, a nonzero term in a polynomial represents a binary 1, so if we simply perform a term-by-term XOR operation between $V(X)$ and $E(X)$, the result will be $T(X)$. Finding $E(X)$ requires finite-field matrix algebra, which is beyond the scope of this discussion. However, you can find a practical discussion of this subject in several of the references: [Lin70], [Skl88], [Bla83].

Bluetooth FEC: Binary Repetition Code The binary repetition code is one of the simplest FEC codes. The transmitting node simply sends each data bit n times, where n is odd. At the receiver, a majority vote is taken to determine the actual bit that was sent. The (3,1) binary repetition code used by Bluetooth has two codewords: a 1 is encoded as 111 and a 0 is encoded as 000. The minimum Hamming distance (in fact, the only Hamming distance) for this code is three, so according to Equation 3.29 the code can correct a single-bit error within each block.

If a double- or triple-bit error occurs in a codeword block, the decoder will fail and output a data bit error. This situation is rare if the bit errors are independent and the overall BER is low, but a burst of errors caused by high noise or interference could cause consecutive errors to occur within a single codeword. Therefore, if data integrity is critical, an error detection scheme such as a CRC is still required to check for uncorrected errors after the FEC process is completed.

The binary repetition code is not particularly efficient because the highest rate possible with this code is only $1/3$. Since the blocks are quite short, it runs the risk of failing even when a short burst error occurs. For example, two consecutive bit errors have a $2/3$ chance of occurring within one three-bit block, and three consecutive bit errors will always cause one block to be decoded in error. Therefore, this code has limited utility when interference causes a string of bit errors. The advantage of this code is that the encoding and decoding algorithms are extremely simple and fast, so little delay exists between the time the signal arrives at the receiver and when the original information can be extracted. The code is also fairly powerful with its ability to correct one bit error for every three bits received. Both of these features are important during baseband packet reception because the payload immediately follows the header, so the header must be decoded with minimal delay.

Bluetooth FEC: Shortened Hamming Code   An (n, k) Hamming code has a structure given by [Bla83]

$$(n, k) = (2^m - 1, 2^m - 1 - m) \tag{3.30}$$

For example, if m equals 5, then the codeword length n is 31 bits and the corresponding data block length k is 26 bits with an additional 5 bits for error correction. This code can correct one bit error, or detect two bit errors, anywhere within the 31-bit codeword.

Now suppose the first 16 bits of the 26-bit data block are always 0. Because the systematic codeword consists of the data itself plus five extra bits for error correction, it will be 31 bits long, but the first 16 bits will always be 0. By dropping these extraneous bits, we've created a (15,10) "shortened" Hamming code, which is one of the FEC codes available for use on Bluetooth baseband data packet payloads. The generator polynomial for this code is

$$G(X) = X^5 + X^4 + X^2 + 1 \tag{3.31}$$

which is used to generate the correct five parity check bits for any 10-bit message vector. This shortened code can actually be more powerful than the original (31,26) code because the one correctable error or two detectable errors can occur over a shorter 15-bit field. In other words, a higher BER can exist before this new code becomes overwhelmed.

Figure 3-13 gives examples of how this code operates. The transmitting node uses its error correction encoder on 10 data bits to create a 15-bit codeword that is sent over the air. The receiver feeds the 15 bits into its error correction decoder. Provided that the incoming codeword has at most one error, the decoder will output the correct data sequence. If two errors exist, the receiver's decoder will set an "error alarm" to inform the Bluetooth controller that the data block is bad, but the location of the errors is unknown and hence can't be corrected. A NAK (or no response at all) is returned and the sender retransmits the packet. Finally, if three or more errors occur in the 15-bit block, the decoder may output the incorrect data block without necessarily triggering the error alarm. All is not lost, however, because the CRC is still available to check data integrity after the FEC decoder submits its output for all blocks within a single baseband packet. Once again, a NAK (or silence) is returned and the sender retransmits the packet.

Let's summarize the error control process as it applies to a Bluetooth baseband packet payload. The transmitting node first appends a CRC to a set of data,

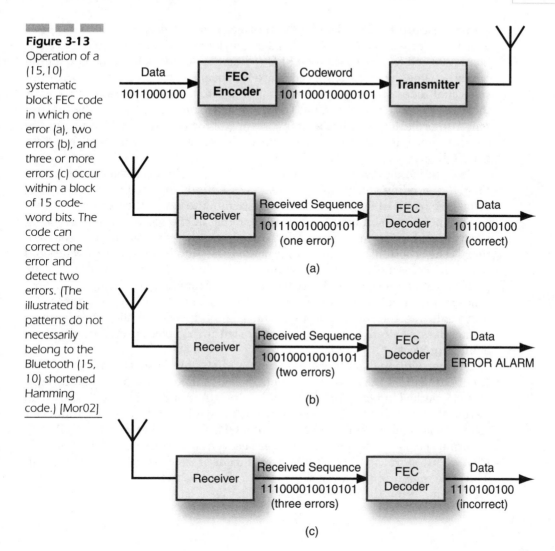

Figure 3-13
Operation of a (15,10) systematic block FEC code in which one error (a), two errors (b), and three or more errors (c) occur within a block of 15 codeword bits. The code can correct one error and detect two errors. (The illustrated bit patterns do not necessarily belong to the Bluetooth (15, 10) shortened Hamming code.) [Mor02]

divides it into blocks of k bits, and runs the FEC encoding algorithm for each block. (If the last block is shorter than k bits, the encoder appends binary 0's to the string until its length is k bits.) Next, the sequence of codewords, each of length n, is concatenated and transmitted as a packet payload. The receiving node separates the payload from the rest of the packet and divides the sequence into blocks of n bits each. The node uses its FEC decoder on each block to obtain k bits of data. These are then concatenated to form the (hopefully) original data set. Finally, the CRC is checked to assess data integrity and a retransmission requested if necessary.

Convolutional Codes   Unlike the Bluetooth block codes, Wi-Fi uses a continuous FEC process called convolutional coding for its 802.11a and 802.11g renditions. (802.11b frames contain no FEC coding.) These codes have memory in the coding scheme, whereas block codes are independent from block to block. Convolutional codes still map k information bits into n bits to be transmitted over the channel, for a rate of k/n, but the n bits are determined not only by the present k information bits, but also by previous information bits. As a result, the encoder is a finite state machine and can be implemented with shift registers and XOR gates.

Convolutional codes are sometimes selected over block codes for two reasons. First, the encoding and decoding process is essentially continuous, so that no long delays are required for a block of data to accumulate before the encoding process can begin, nor must a long block finish arriving at the receiver before decoding commences. Furthermore, with a few exceptions, the performance of a convolutional code is better than a block code of the same rate. That is, the convolutional code can successfully correct incoming bits having a slightly higher BER. One problem with a convolutional decoder is that an incoming burst error event sometimes results in a longer burst of errors at its output before it once again settles into producing the correct data stream. The block code, on the other hand, is memoryless; as a consequence, burst errors within one block cannot propagate to a different block.

The convolutional encoder for 802.11a and 802.11g is shown in Figure 3-14. Input data enters the six shift registers bit by bit, and each time the shift registers are clocked, a new pair of output data bits, A and B, are created. These can be concatenated for transmission as two BPSK symbols or be used separately for the in-phase and quadrature portions of a QPSK or QAM modulator. Each pair of output bits is a function of the values of seven consecutive input data bits, six of which are stored in the shift registers and one of which is the current input bit. Thus, the *constraint length* of this code is 7. Because two output bits are created for each input bit, the rate of this code is ½.

Figure 3-14
Convolutional encoder for 802.11a and 802.11g

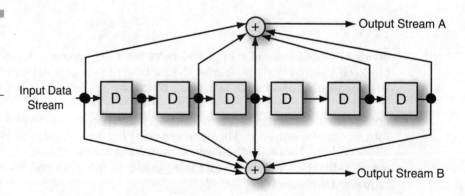

As a brief aside, Figure 3-14 depicts a *nonsystematic, nonrecursive* convolutional encoder. That is, the original data stream doesn't appear in either of the two output streams, and the code is generated without the use of any digital feedback mechanism. A second method of convolutional code generation is *systematic*, whereby one of the output streams is the data itself, and *recursive*, which is accomplished by feeding the output of at least one of the registers into a previous register's input. A subset of these encoders produces *turbo codes*, which are superior to convolutional codes in several ways. Significant research activity is taking place in modeling and designing efficient encoders and decoders of turbo codes [Ber03].

When the communication channel is less error-prone, the 802.11a and 802.11g specifications optionally allow the use of convolutional codes with rates ⅔ and ¾ for a more efficient transmission of information with less FEC overhead. Rather than requiring new encoding and decoding hardware for these lower rates, the rate ½ code is still used; only some of the output bits are removed to form a *punctured code*. These removed bits are replaced at the receiver's convolutional decoder with binary placeholders called *erasures*, and the decoding process proceeds normally. By carefully selecting the bits to puncture, the original bit stream can still be retrieved by the receiver in its entirety as long as the BER is low enough that the FEC isn't overwhelmed. In general, the rate ½ convolutional code can be punctured to produce a new code with the rate $(k - 1)/k$ for any integer k [Vit89].

Decoding a convolutional code usually takes place through the *Viterbi algorithm*. This is a somewhat involved process computationally (and conceptually!) so we won't cover the details here. Generally, the decoder considers several possible data streams given the incoming codeword stream and selects the data stream that best fits the codeword stream. Of course, if no errors occur, then only one perfect fit exists, but if some of the codeword bits are received in error the data stream with the best fit is selected.

Trellis-Coded Modulation (TCM)   Traditional FEC codes operate at the bit level, and the resulting codeword is separately modulated onto the carrier for transmission. Greater coding gains can sometimes be realized by combining both processes into a single operation. One of the most successful of these is *trellis-coded modulation* (TCM), where a convolutional encoder directly controls a nonbinary ASK, PSK, or QAM modulator [Ung87a], [Ung87b].

Figure 3-15 shows how TCM signals are created. A set of $k - 1$ parallel data bits enters the rate $(k - 1)/k$ convolutional encoder, each at rate R b/s. The output of the encoder is k parallel bits, and these are sent to a QAM modulator for selecting one of $M = 2^k$ symbols for transmission [Vit89]. Because $k - 1$ data bits are sent per symbol, the symbol rate is R per second.

TCM has built-in error control due to the fact that not every QAM symbol is available for transmission at any particular time. The presence of the rate $(k - 1)/k$ convolutional encoder in Figure 3-15, for example, means that, once a

Figure 3-15
A trellis-coded
modulator

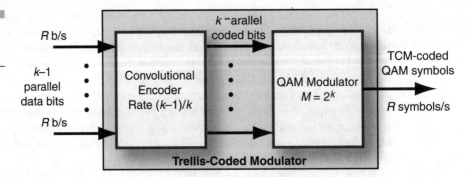

Figure 3-16
Example of
allowable
transitions
within a QAM
signal set
when TCM is
used

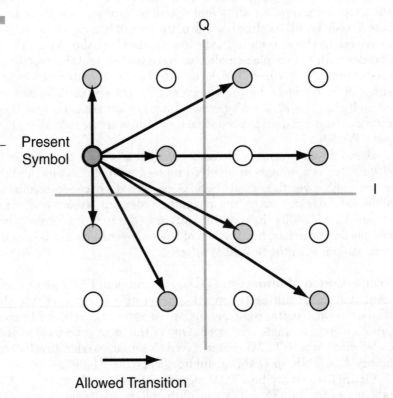

particular symbol is sent, the follow-on symbol is selected from a subset containing $M/2$ of the M symbols in the QAM signal set. Therefore, the effective distance between QAM symbols is increased, resulting in a typical coding gain of 3 to 4 dB. This concept is shown in Figure 3-16.

In Rayleigh fading, TCM can lower the BER significantly compared to ordinary QAM [Cav92]. This is due to the exploitation of time diversity provided by the memory elements in the convolutional encoder. As a function of SNR, the BER on a TCM channel decreases faster than the simple inverse dependence commonly found on Rayleigh faded channels.

TCM is used in 802.15.3 WiMedia for data rates ranging from 11 Mb/s using a QPSK constellation to 55 Mb/s with a 64-QAM constellation. The symbol rate is always 11 Mbaud, so the number of bits per second is always one fewer than the QAM constellation can support without TCM. For example, 64-QAM can support 6 uncoded bits per symbol, but TCM uses only half the available signal set for each symbol transition. Therefore, only 5 bits per symbol are actually sent for a data rate of 55 Mb/s.

Summary

Information can be placed on an RF carrier by changing its amplitude, frequency, or phase in response to the data stream. The maximum rate at which this data can be sent over an imperfect channel is given by the channel capacity, which is based on both the SNR and the bandwidth of the signal at the receiver.

Wi-Fi transmissions are modulated with DBPSK, DQPSK, BPSK, QPSK, 16-QAM, or 64-QAM, depending on the application and channel conditions. The BER in AWGN is higher for the more complex modulation methods, reducing the range for a given TX power. In general, moderate data rates have about half the range of low data rates, and the highest data rates operate over about one-third the distance of the lowest rate.

The Bluetooth radio modulates data onto a carrier using FSK as a compromise between cheaper but less capable on-off keying and better performing but more expensive PSK. The Bluetooth baseband signal is Gaussian filtered for a higher data rate within a 1 MHz bandwidth, but at the expense of a slight increase in intersymbol interference.

The 802.15.3 WiMedia PHY, with its high-speed streaming applications over very short range, uses QAM constellations with up to 64 signal points and incorporates TCM for moderate error control. The symbol rate is always 11 Mb/s, with 1, 2, 3, 4, or 5 bits per symbol.

The inevitable noise and interference on the RF channel can be partially mitigated using error control. Error detection places an FCS at the end of each packet that can be used by the receiver to determine if any errors exist. If so, the receiving node requests a retransmission. Error-correction coding has the ability to correct the occasional error, with additional redundancy needed for more powerful error-correcting abilities. Bluetooth uses block FEC, whereas 802.11a and 802.11g nodes use convolutional codes and WiMedia employs TCM.

References

[Ber03] Berrou, C., "The Ten-Year-Old Turbo Codes Are Entering into Service," *IEEE Communications Magazine*, August 2003.

[Bla83] Blahut, R., *Theory and Practice of Error Control Codes*, Massachusetts: Addison-Wesley, 1983.

[Cav92] Cavers, J., and Ho, P., "Analysis of the Error Performance of Trellis-Coded Modulations in Rayleigh-Fading Channels," *IEEE Transactions on Communications*, January 1992.

[Lin70] Lin, S., *An Introduction to Error-Correcting Codes*, New Jersey: Prentice-Hall, 1970.

[Mor02] Morrow, R., *Bluetooth Operation and Use*, New York: McGraw-Hill, 2002.

[Mur81] Murota, K., and Hirade, K., "GMSK Modulation for Digital Mobile Radio Telephony," *IEEE Transactions on Communications*, July 1981.

[Pet61] Peterson, W., and Brown, T., "Cyclic Codes for Error Detection," *Proceedings of the IRE*, January 1961.

[Pro94] Proakis, J., and Salehi, M., *Communication Systems Engineering*, New Jersey: Prentice Hall, 1995.

[Sho01] Shoemake, M., "Wi-Fi (IEEE 802.11b) and Bluetooth: Coexistence Issues and Solutions for the 2.4 GHz ISM Band," Texas Instruments white paper, Version 1.1, February 2001.

[Skl88] Sklar, B., *Digital Communications: Fundamentals and Applications*, New Jersey: Prentice Hall, 1988.

[Sta88] Stallings, W., *Data and Computer Communications*, 2nd ed., New York: Macmillan, 1988.

[Ung87a] Ungerboeck, G., "Trellis-Coded Modulation with Redundant Signal Sets, Part 1: Introduction," *IEEE Communications Magazine*, February 1987.

[Ung87b] Ungerboeck, G., "Trellis-Coded Modulation with Redundant Signal Sets, Part 2: State of the Art," *IEEE Communications Magazine*, February 1987.

[Vit89] Viterbi, A., et al., "A Pragmatic Approach to Trellis-Coded Modulation," *IEEE Communications Magazine*, July 1989.

Advanced Modulation and Coding

Each *wireless local area network* (WLAN) and *wireless personal area network* (WPAN) implementation uses its own form of advanced modulation and coding to enhance its operation in a noisy and interference-prone environment, as well as to reduce its level of interference to other users operating in the same band. IEEE 802.11b Wi-Fi and 802.15.4 ZigBee use variations of the *direct sequence spread spectrum* (DSSS), and 802.15.1 Bluetooth uses the *frequency hopping spread spectrum* (FHSS). Modulation and coding are combined in 802.15.3 WiMedia through the use of *trellis coded modulation* (TCM). Both 802.11a and 802.11g employ *orthogonal frequency division multiplexing* (OFDM), which uses multicarrier modulation coupled with error control coding as a way to speed the data rate and increase reliability, especially in a channel that experiences frequency selective fading and/or narrowband interference. Finally, *ultra-wideband* (UWB) modulation is being considered for very fast (greater than 100 Mb/s) data rates as an alternate *physical* (PHY) layer called 802.15.3a and 802.15.4a.

Direct Sequence Spread Spectrum (DSSS)

Application: 802.11b, 802.15.4

The original 802.11 specification required spread spectrum transmission to be used for *radio frequency* (RF) transmissions at 1 and 2 Mb/s data rates. Both FHSS and DSSS were included in the specification, and their use allowed higher *transmit* (TX) powers under FCC rules in effect at the time. By deliberately spreading the bandwidth of the transmitted signal, additional immunity from interference (but not from *additive white Gaussian noise* [AWGN]) was gained. This method also reduced the average TX power per Hz of bandwidth, which in turn lowered the *co-channel interference* (CCI) provided by 802.11 to many other wireless devices. The 802.11b enhancement allowed data rates of 5.5 and 11 Mb/s through *complimentary code keying* (CCK), which has a lot in common with DSSS, and backward compatibility with the lower-data-rate DSSS was retained. The vast majority of Wi-Fi networks are compliant with the 802.11b specification.

DSSS Signal Construction

In ordinary digital communications such as those we've studied up to now, the data modulates a carrier such that the transmitted signal's bandwidth is proportional to the channel symbol rate. We also learned that if a sufficient *signal-to-noise ratio* (SNR) exists at the receiver, then several data bits can be encoded into each symbol for an increased data rate without a proportional increase in bandwidth.

DSSS takes the opposite approach to enhancing performance by deliberately increasing the TX signal's bandwidth by superimposing a much faster signal, called the *spreading code,* or *pseudonoise* (PN) sequence, onto the data. To demonstrate the principle, we'll start with the *binary phase shift keying* (BPSK) signal in Equations 3.1 and 3.2 and combine these into a single equation like this:

$$s(t) = Ab(t)\cos(2\pi f_c t) \qquad (4.1)$$

where $s(t)$ is the transmitted signal and $b(t)$ is the data stream represented as a time series of $+1$ and -1 values, each with duration T. After a little thought, it should become apparent that when $b(t) = +1$ the transmitted signal has a phase of 0 and when $b(t) = -1$ the transmitted signal has a phase of π, just like before. Therefore, this is just a BPSK transmission in which a binary 0 is represented by $+1$ and a binary 1 is represented by -1 in the data stream $b(t)$. Equation 4.1 can be easily modified to depict *differential BPSK* (DBPSK) by changing the sequence $b(t)$ accordingly.

Now suppose that we create another stream of $+1$ and -1 values called $a(t)$, but this stream switches values much more rapidly than $b(t)$. By placing this sequence into Equation 4.1, we obtain

$$s(t) = Aa(t)b(t)\cos(2\pi f_c t) \qquad (4.2)$$

The bandwidth of the new $s(t)$ is clearly greater than it was in Equation 4.1 due to the faster changing stream $a(t)$ now modulating the carrier along with $b(t)$. The $+1$ and -1 values that comprise the spreading sequence $a(t)$ are called *chips*, and there are N chips in $a(t)$ per data bit in the sequence $b(t)$. Each chip has duration $T_c = T/N$ where N is an integer. The value N also represents the bandwidth increase of the DSSS signal over ordinary BPSK.

It's also possible to employ *quadrature phase shift keying* (QPSK) modulation to the carrier in Equation 4.2, with the quadrature coding being applied to the data stream, spreading stream, or both. It's mathematically convenient to express the QPSK symbols in $a(t)$ and $b(t)$ in the complex plane as $+1$, $+j$, -1, and $-j$. For 802.11b running at 2 Mb/s, DQPSK is used for the data stream, doubling the data rate while retaining the 1 Ms/s symbol rate. The *bit error rate* (BER) performance in AWGN is essentially the same for DSSS as it is for an equivalent nonspread signal.

In a manner similar to the Gaussian baseband filtering used in 802.15.1 Bluetooth, baseband pulse shaping can be employed to reduce the transmitted bandwidth by preventing sudden phase changes in either $a(t)$ or $b(t)$. For 802.15.4 ZigBee, pulse shaping takes place in the spreading sequence $a(t)$ and is done by using a raised cosine instead of a rectangular envelope for each chip. In this way, phase and amplitude transitions are smoothed in the time domain, reducing chip-occupied bandwidth.

Figure 4-1
A DSSS
communi-
cation system

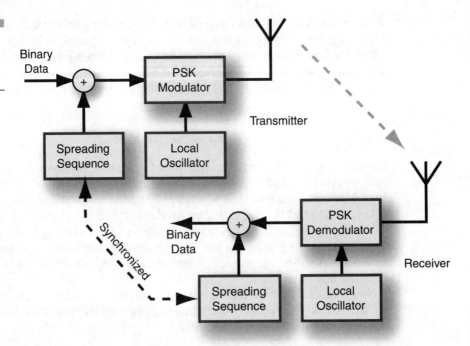

The DSSS transmitter and receiver are shown in Figure 4-1. The transmitter places both the data stream and spreading code onto the carrier, and the receiver multiplies the incoming signal by the spreading code and then extracts the data. (Actual implementations sometimes perform the spreading and despreading operations at RF, but the result is the same.)

Figure 4-2 provides an example of DSSS signal generation and detection. For clarity, we show only the data and spreading sequence associated with the base-band part of Equation 4.2; that is, we depict only the values represented by the streams $a(t)$ and $b(t)$ without including modulation and demodulation. Both the transmitter and receiver use identical spreading codes, so the spreading is removed at the receiver and the data sequence from $b(t)$ appears at the detector if no bit errors are present. The 11-chip spreading code in Figure 4-2 is actually used in 1 and 2 Mb/s Wi-Fi signals and is called a *Barker sequence*. This sequence was selected for its desirable correlation properties to facilitate synchronization at the receiver. Although the 802.11 specification dictates that the sequence is to be transmitted from left to right, we'll write it in the order it appears in the specification, both here and in the following correlator example, rather than reversing the sequence.

The approximate spectrum of the Wi-Fi DSSS signal is shown in Figure 4-3 [Sho01]. The −20 dB bandwidth of this signal is about 16 MHz, but researchers often use the rule that the bandwidth of a DSSS signal without pulse shaping

Figure 4-2
DSSS signal construction and detection for a data bit of +1 and −1. For clarity, the PSK modulation and demodulation processes are omitted.

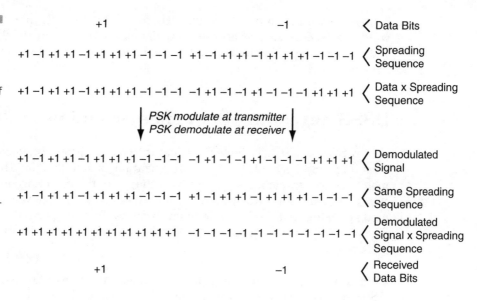

Figure 4-3
Spectrum of the 802.11b signal [Sho01]

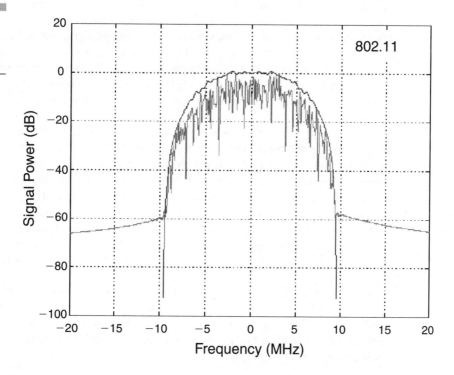

is about twice the chip rate, or 22 MHz for Wi-Fi DSSS. Either of these band-widths is used in the literature for coexistence analysis, but the 22 MHz band-width seems to predominate. That is the bandwidth we will use for coexistence modeling.

DSSS Spreading Sequence Selection

For proper operation, it's important that the receiver properly synchronize on the incoming DSSS signal. Such synchronization is required for the receiver to match its own spreading sequence with that from the incoming signal such that they cancel and the data stream can be extracted. If the receiver's locally gener-ated spreading code isn't in alignment with the incoming signal's code within a fraction of a chip, the signal isn't despread properly and the detector will output gibberish.

Correlator Operation   One of the simplest and most effective ways for the DSSS receiver to synchronize on the incoming signal is by using a correlator, as shown in Figure 4-4. This device consists of a set of N 1-bit memory elements, or flip-flops, into which the desired sequence is loaded. The incoming DSSS signal is entered into an N-stage shift register, chip by chip, and a multiply-and-add operation is performed after each shift. It's easy to see that when the two sequences are aligned the correlator produces a high output level and triggers the receiver's tracking circuitry to maintain chip alignment with the incoming signal. It doesn't matter whether the incoming data bit is a $+1$ or -1 because the correlator output will be a large positive value for the former and a large neg-ative value for the latter. However, if the spreading sequence is misaligned or if the wrong spreading sequence is present, the correlator outputs a low value and no trigger occurs.

For the correlator to operate properly and synchronize on the correct shift of the spreading sequence, this sequence must have good *autocorrelation proper-ties*. This means that shifts other than exact alignment between two identical copies of the spreading sequence should result in correlation values of nearly 0. A generalized Barker sequence has maximum autocorrelation values between $+1$ and -1 at any shift other than perfect alignment, in which case its correla-tion value is $+N$. Unfortunately, only a few Barker sequences exist. Further-more, autocorrelation properties are altered when the data stream inverts some of these sequences and the results are concatenated into a long string of $+1$ and -1 values.

Another method of generating spreading sequences is by using a *linear feed-back shift register* (LFSR) string consisting of n stages with logic that sends some combination of states to its input. The longest possible sequence produced by such an LFSR is $2^n - 1$, and these maximum length m-sequences have the fol-lowing properties [Dix94]:

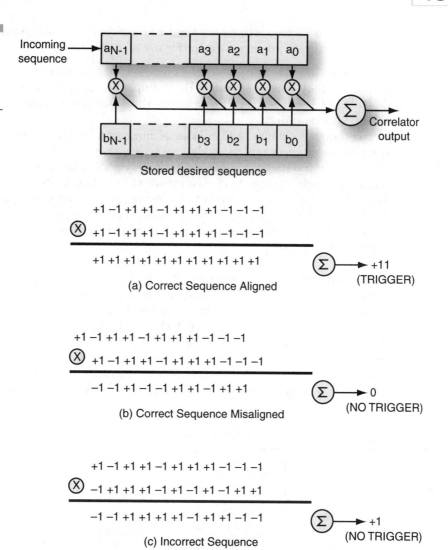

Figure 4-4
Correlator operation at the DSSS receiver

Incoming sequence

a_{N-1} a_3 a_2 a_1 a_0

b_{N-1} b_3 b_2 b_1 b_0

Stored desired sequence

Correlator output

+1 −1 +1 +1 −1 +1 +1 +1 −1 −1 −1
$\times$ +1 −1 +1 +1 −1 +1 +1 +1 −1 −1 −1
_____
+1 +1 +1 +1 +1 +1 +1 +1 +1 +1 +1 $\longrightarrow$ +11 (TRIGGER)

(a) Correct Sequence Aligned

+1 −1 +1 +1 −1 +1 +1 +1 −1 −1 −1
$\times$ +1 −1 +1 +1 −1 +1 +1 +1 −1 −1 −1
_____
−1 −1 +1 −1 −1 +1 +1 −1 +1 +1 $\longrightarrow$ 0 (NO TRIGGER)

(b) Correct Sequence Misaligned

+1 −1 +1 +1 −1 +1 +1 +1 −1 −1 −1
$\times$ −1 +1 +1 +1 −1 +1 −1 +1 −1 +1 +1
_____
−1 −1 +1 +1 +1 +1 −1 +1 +1 −1 −1 $\longrightarrow$ +1 (NO TRIGGER)

(c) Incorrect Sequence

- The number of +1 values and −1 values is equal within one chip. Their statistical distribution is the same and meets various tests for randomness.

- Autocorrelation values are −1 for all shifts other than perfect alignment, where the value is +N.

- A modulo-2 addition of the sequence with a shifted version of itself produces another replica with a still different shift.

■ Every possible state within the LFSR exists during the generation of the associated *m*-sequence except for the all-zeros state.

Because these *m*-sequences are easy to create and have good autocorrelation properties, they are used in many DSSS implementations. For example, 802.15.4 ZigBee uses a four-stage LFSR to create the 15-chip *m*-sequence used for its 20 and 40 kb/s renditions.

The coding efficiency of DSSS can be improved by designating a collection of spreading sequences and using data to select a particular sequence to transmit. When sending data at 250 kb/s, a ZigBee symbol transmits 1 of 16 spreading sequences, each 32 chips long, as a means of encoding 4 bits of data per symbol. These PN sequences are also selected for their good autocorrelation properties for the ease of synchronization, along with good cross-correlation properties (near orthogonality) for reliable symbol detection. The ZigBee modulation details are shown in Figure 4-5.

Complementary Code Keying The higher Wi-Fi data rates of 5.5 and 11 Mb/s were also achieved by placing information into the spreading sequences. Because backwards compatibility with low-rate 802.11 was required, the chip rate of 11 *million chips per second* (Mchips/s) was retained. This meant that one bit of information had to be sent per chip at the highest data rate, making spreading sequence selection and detection more difficult than with ZigBee. To meet this challenge, the designers had to find multiple short spreading sequences with good autocorrelation and cross-correlation properties.

Figure 4-5
802.15.4
ZigBee
modulation
details

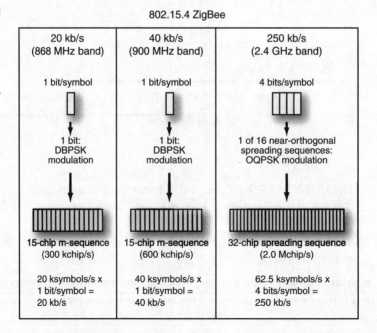

It turns out that such sequences, called *complementary code sequences*, exist, and when they are used as spreading codes, the modulation method is called CCK. Instead of the 11-chip Barker sequence, the Wi-Fi node operating at 11 Mb/s selects one of 64 eight-chip complex CCK sequences and transmits these using QPSK. There are 65,536 (4^8) possible complex 8-bit spreading sequences, out of which only a few have complementary properties. Of these, 64 were chosen for their superior correlation characteristics [Pea00]. The minimum distance between any two sequences is four, which implies a coding gain of 3 dB. At practical error rates, the coding gain is actually about 2 dB [Hee00]. This means that the signal produces BER values at the receiver equivalent to uncoded transmissions that are 2 dB stronger. Six data bits are used to select the particular CCK sequence and two additional data bits are used for DQPSK modulation of the entire symbol; thus, 8 bits are transmitted for every 8-chip CCK sequence. Because the chip rate is 11 Mchips/s, the data rate is 11 Mb/s. The price paid for the faster data rate is a shorter range and higher node complexity.

If channel conditions are such that the receiver has difficulty deciding which of 64 CCK sequences belong to an incoming signal, the transmitting node can instead limit the sequence set to one of only four sequences, giving the receiver a better chance of picking the correct one. The selected sequence is QPSK modulated for a raw data rate of 5.5 Mb/s. Range and reliability are improved at the expense of a slower data rate. The maximum available range at 11 Mb/s is about ⅓ of that supported by the 1 or 2 Mb/s data rates, and the range at 5.5 Mb/s is about ⅔ of the low data rates. These four data rates and their modulation methods are part of the 802.11b specification and are shown in Figure 4-6.

Figure 4-6
Modulation
details for
802.11b
signaling

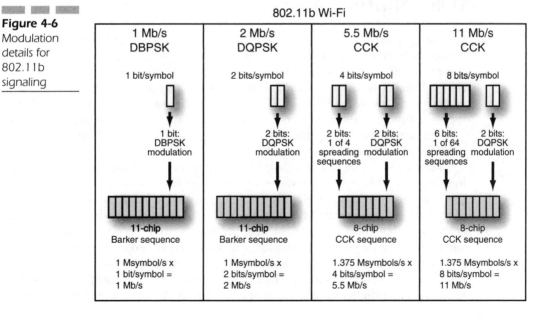

From a coexistence perspective, the 1 and 2 Mb/s low-rate Wi-Fi signals are more robust because of a greater bit energy per symbol. For the Wi-Fi receiver to operate well at the higher data rates using CCK, the SNR must be higher to offset the lower energy per data bit, enabling the receiver to select the correct CCK sequence. As we'll discover in Chapter 6, "Medium Access Control," a 802.11b Wi-Fi packet header is always sent at the lower data rate, and within the header is a field specifying the data rate of the information to follow. If the sending node's retransmission rate is too high, the node will automatically reduce the data rate for improved throughput under the given channel conditions.

Packet Binary Convolutional Coding (PBCC)   *Packet binary convolutional coding* (PBCC) is an optional 802.11b high-rate encoding procedure, in which the transmitted QPSK signal is created through a three-step procedure. The first step takes the data stream and processes it with a rate ½, constraint length 6 convolutional encoder. (Operation of such an encoder was discussed in the previous chapter.) Next, the convolutional encoder's output is mapped onto QPSK for 11 Mb/s or BPSK for 5.5 Mb/s. Finally, a scrambler removes any short-term periodicity from the transmission. This sequence was designed to improve performance in multipath with long delay spreads and in the presence of interference. At practical error rates, the coding gain is about 5 dB or about 3 dB better than CCK [Hee00]. Much of this gain is from the use of coherent (nondifferential) modulation and demodulation. PBCC can also be extended in a rather complex manner to support data rates of 22 and 33 Mb/s within approximately the same transmitted bandwidth as the 11 Mb/s PBCC signal.

DSSS Multiple Access

Spreading sequence correlation properties make it possible for multiple DSSS signals to be transmitted simultaneously on the same carrier frequency, and yet several receivers may be able to extract their own respective data streams and reject others. This is called *multiple access*. Of course, as the number of simultaneous users increases, the BER experienced by each of the users will also increase from the *multiple access interference* (MAI). Multiple access analysis gives an indication of how the system operates over an interference-prone communication channel.

Although multiple access is possible when every user shares the same spreading sequence, this can wreak havoc at a receiver, because its correlator will trigger with every incoming signal, not just the desired one. Alternatively, each participating network could be assigned its own spreading code. If this is done, each spreading code should have good autocorrelation properties, as discussed earlier, so their respective correlators can synchronize properly onto it. However, the codes should also possess good cross-correlation properties; that is, an undesired spreading code shouldn't trigger a receiver's correlator for any possible shift value. As the number of codes increases, it becomes extremely difficult to meet cross-correlation requirements. One way around this, at least from an analysis perspective, is to assume each user is assigned a different spreading code that appears random. That is, each spreading code is generated by flipping an unbiased coin. The resulting BER figures can be accepted as an upper bound, and carefully selected codes will (hopefully) perform better.

To quantify coexistence capabilities, suppose that a total of K users is present, all transmitting DSSS signals using the same carrier, but having different spreading sequences at N chips per data bit. One of these users sends the desired signal, and the others produce MAI. As you might imagine, the mathematics associated with calculating BER in the presence of MAI can be quite intimidating, but we can simplify the situation markedly if we make the following assumptions:

- All users employ BPSK modulation.
- All spreading sequences consist of random $+1$ and -1 values that change from data bit to data bit.
- Every DSSS signal at the desired receiver arrives with equal power.
- The desired receiver can synchronize perfectly on the desired signal.
- The MAI power is high enough that AWGN can be neglected.

Under these conditions the BER at the desired receiver is closely approximated by a weighted average of three Q-function terms given by the following [Hol92], [Mor98]:

$$P_e = \frac{2}{3}Q\left(\sqrt{\frac{N^2}{\mu}}\right) + \frac{1}{6}Q\left(\sqrt{\frac{N^2}{\mu + \sqrt{3\sigma^2}}}\right) + \frac{1}{6}Q\left(\sqrt{\frac{N^2}{\mu - \sqrt{3\sigma^2}}}\right) \quad (4.3)$$

where the Q-function was previously defined in Equation 3.6. The quantities μ and σ^2 are given by

$$\mu = \frac{(K-1)N}{3} \quad (4.4)$$

and

$$\sigma^2 = \frac{23(K-1)N^2}{360} \quad (4.5)$$

Thus, the BER in a DSSS MAI coexistence situation can be expressed in terms of the number of chips per data bit and the number of simultaneous transmissions. Figure 4-7 plots BER values as a function of the number of simultaneous transmissions for some common spreading sequence lengths [Mor98]. As expected, longer spreading sequences allow more users to coexist for a given BER.

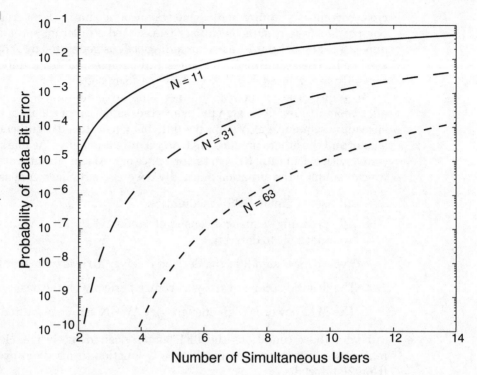

Figure 4-7
BER versus the
number of
simultaneous
DSSS trans-
missions for
random
spreading
sequences and
different values
of N chips per
data bit

By assuming that all MAI is Gaussian distributed, Equation 4.3 simplifies to

$$P_e = Q\left(\sqrt{\frac{3N}{K-1}}\right) \tag{4.6}$$

which is simply the unweighted first term in Equation 4.3. Equation 4.6 can be somewhat optimistic under the previous assumptions of equal power interferers on a nonfaded channel, especially when the number of interfering users is small. The accuracy of a Gaussian MAI assumption improves for larger numbers of interfering users or when the channel experiences Rayleigh fading [Che02].

So how does all this relate to the capability of different Wi-Fi DSSS networks to coexist with each other? The number of chips per data bit for 1 and 2 Mb/s Wi-Fi transmissions is 11, and the BER due to MAI for N = 11 is shown in Figure 4-7. Some differences exist, however, between the assumptions made for generating this plot and the actual 802.11b Wi-Fi operating conditions:

■ All Wi-Fi DSSS transmissions use the same spreading sequence or select one from a small collection of sequences, but Figure 4-7 applies to random spreading sequences for all users.

■ Simultaneous Wi-Fi transmissions from different sources will usually have different incoming power levels at any given Wi-Fi receiver, but Figure 4-7 is accurate only when all incoming signals have equal power at the desired receiver.

■ AWGN and interference other than DSSS MAI exist on a typical Wi-Fi channel, but Figure 4-7 plots BER values from MAI alone.

In spite of these differences, though, it's still possible to gain some important insights into the potential multiple access capability of Wi-Fi transmissions. For reasons that will become apparent in a moment, we'll assume that the desired receiver is already synchronized on the desired packet before interference begins. If the receiver must contend with one interfering Wi-Fi DSSS transmission ($K = 2$) with equal power, its BER is slightly above 10^{-5} from MAI alone, and two such interfering users ($K = 3$) cause the BER to rise to almost 10^{-3}. In other words, the presence of other co-channel Wi-Fi signals can have a significant detrimental effect on the error rate, even if perfect synchronization is achieved on the desired signal.

Unfortunately, from an MAI viewpoint, because all Wi-Fi signals at 1 or 2 Mb/s use the same spreading sequence, interfering transmissions will still cause a correlation peak at the desired receiver when they are chip-aligned with the desired sequence, making efficient multiple access even more difficult. Also, as we'll discover in Chapter 6, a Wi-Fi transmitter using contention-based channel access won't activate when it senses that another signal is present on its operating channel, preventing collisions at the expense of reduced throughput. All these characteristics mean that separate contention-based Wi-Fi networks operating on the same channel and within range of each other essentially share a single network's throughput among them. This same argument applies to 802.15.4 ZigBee, because all users have the same spreading code, or collection of spreading codes. Simultaneous transmissions from more than one ZigBee node operating on the same channel could easily disrupt throughput.

As an aside, a general rule for DSSS systems is that a common channel can support up to $N/10$ simultaneous transmissions while maintaining a reasonable BER of 10^{-6} or better from MAI, assuming that each receiver can synchronize on its respective desired packet and interfering spreading sequences are random. For example, when $N = 31$ the channel can support up to three simultaneous transmissions, and $N = 63$ allows up to six simultaneous transmissions to exist. By this rule, Wi-Fi DSSS using an 11-chip spreading sequence can support only one simultaneous transmission; that is, it has essentially no multiple-access capability. (We discovered this in the previous paragraph using different reasoning.) For ZigBee operating in the 868 and 900 MHz bands, a 15-chip spreading sequence is used, so it has no multiple access capability either. In the 2.4 GHz band, however, the ZigBee spreading sequence is 32 chips long, so according to our rule, perhaps up to three ZigBee simultaneous co-channel transmissions can take place without significant mutual interference. In reality, though, the $N/10$

rule doesn't apply here either because the collection of sixteen 32-chip ZigBee spreading sequences isn't random but instead the sequences are related to one another through cyclic shifts and/or conjugation (the inversion of odd-indexed chips). Consequently, MAI from another ZigBee transmission has a significant potential for causing a symbol error at the desired receiver, so ZigBee also has essentially no multiple-access capability.

DSSS Processing Gain and Narrowband Signal Rejection

The capability of a DSSS receiver to extract a desired signal in the presence of MAI is based on the capability of the receiver's correlator to synchronize on, and then despread, this desired signal. The despreading process takes place because the receiver effectively cancels the transmitter's spreading action through multiplying the data a second time by an identical, synchronized copy of the spreading sequence. Other unrelated signals, regardless of their construction, are actually spread by this processing at the receiver. For example, a coexisting narrowband signal within the passband of the DSSS receiver is spread at the same time the desired DSSS signal is despread. This is depicted in Figure 4-8. The receiver can then use a narrowband filter to eliminate most of the influence of the spread narrowband interference.

As before, the capability of the DSSS receiver to reject such interference depends upon the value of N, as well as other factors such as the relative strengths of the desired and interfering signals and the spreading sequences themselves. Larger values of N equate to greater spreading factors and hence usually a greater capability to reject interference. This capability can be quantified through a factor called *processing gain* (PG), which is simply the value of N

Figure 4-8
An incoming narrowband signal within the passband of a DSSS receiver is spread by the same process that is used to despread the desired signal.

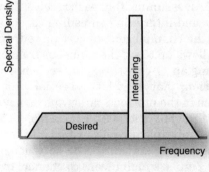

Signals at receiver antenna

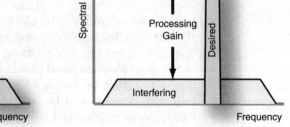

Signals after despreading

expressed in dB. For example, N equals 11 chips per channel symbol in 1 and 2 Mb/s Wi-Fi, and this corresponds to a processing gain of 10.4 dB. The benefits of PG are realized when no special circumstances exist, such as high cross-correlation values, between desired and interfering signals.

Processing gain allows us to quickly determine whether, for example, a Bluetooth transmission that occurs within the Wi-Fi DSSS receiver's passband will disrupt WLAN operation. Interfering Bluetooth signals will automatically be reduced by about 10 dB during Wi-Fi DSSS despreading and filtering. Whether or not the remaining energy is still strong enough to cause disruption is a topic we'll address in subsequent chapters.

Multipath Mitigation Using DSSS

As we've already learned, multipath components exist for nearly all indoor transmissions, and they are manifested in the arrival at the receiver of the transmitted signal's multiple copies, having different strengths, delays, and phases. Another benefit of DSSS is that it includes built-in multipath mitigation. Look again at Figure 4-4. When the incoming Barker sequence in a Wi-Fi transmission is offset by one chip or more, the receiver's correlator won't trigger. Now suppose the correlator is already synchronized on the LOS or OLOS component, and another copy of this same signal arrives with a delay of one or more chips from a multipath reflection. This multipath component isn't despread, and therefore most of its energy is rejected in the receiver's post-despreading filter. Indeed, even multipath components that are delayed by somewhat less than one chip duration will experience some attenuation within the receiver.

The chip rate for DSSS Wi-Fi is 11 Mchips/s, so the chip duration is 91 ns. Therefore, multipath components delayed by 91 ns or more are attenuated by an average of about 10 dB compared to the component on which the receiver is synchronized. Components delayed less than 91 ns will be attenuated less, with the most troublesome being components that arrive at the receiver only a few nanoseconds after the one being despread. These early arriving components have the potential to cause some fading, but very little *intersymbol interference* (ISI), when data rates of 1 or 2 Mb/s are used. We conclude, then, that the processing gain inherent in the Wi-Fi DSSS receiver can also be used as an antimultipath tool that limits ISI and, in many cases, fading. For 802.15.4, the news isn't quite so good, because the chip rate is considerably slower. The chip duration ranges from 3.3 μs at a data rate of 20 kb/s to 500 ns at a data rate of 250 kb/s, and even the latter isn't fast enough to separate multipath components in a typical indoor environment.

For even better performance at the expense of additional hardware, multiple receivers can be employed to synchronize on several multipath components. These are then combined and detected using a RAKE configuration such that the SNR is increased with a corresponding decrease in BER. Taking the concept one step further, additional receivers could even synchronize on any existing interfering DSSS signals, and their influence can then be subtracted from the desired transmission for an even greater performance improvement through multiuser detection.

Frequency Hop Spread Spectrum (FHSS)

Application: 802.15.1

The designers of the Bluetooth specification took a different approach to interference suppression than the designers of the Wi-Fi specification. In almost all Wi-Fi deployments, the carrier frequency remains fixed and the DSSS processing gain is used to improve performance. As pointed out earlier, processing gain depends on the value of N, the number of chips per data bit. Achieving high values of N can be difficult, though, due to the fast chipping rate required for a high data throughput. Furthermore, the 5.5 and 11 Mb/s renditions of Wi-Fi have lower processing gain and lower bit energy; they are thus more susceptible to performance degradation from interference. Fortunately, Wi-Fi nodes can fall back to lower-rate signaling with its built-in 10.4 dB of processing gain.

Bluetooth was meant to be an extremely inexpensive cable replacement technology, so DSSS was ruled out. As the 2.4 GHz band becomes more populated, many users will try to coexist in this band within range of each other. Choosing a single channel and then remaining there for an entire session without any ability to suppress unwanted signals runs the risk of a sudden, catastrophic increase in interference such that the Bluetooth participants may not even be able to coordinate a channel change, and communication will then be lost. To avoid this situation, Bluetooth uses FHSS as an interference-avoidance technique.

FHSS Operation

A Bluetooth FHSS communication system is shown in Figure 4-9. Binary baseband data is *Gaussian-filtered frequency shift keying* (GFSK) modulated and transmitted using a carrier determined by the frequency synthesizer. Instead of producing only a single carrier frequency, the synthesizer is controlled by a hop code generator that causes it to change carrier frequency at a nominal rate of 1,600 hops per second. One Bluetooth data packet is sent per hop.

This hop pattern itself appears to be random, but is actually created by a pseudorandom algorithm in the hop code generator. The generator is duplicated at the receiver and creates the same hopping pattern that the transmitter uses. While communicating, the transmitter and receiver hop together from channel to

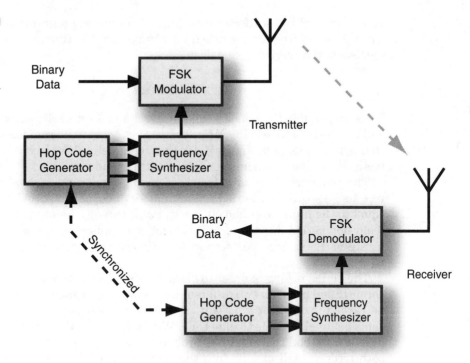

Figure 4-9
An FHSS communication system

channel. Furthermore, the two devices have agreed upon the hop sequence ahead of time, so even if a hop channel contains catastrophic interference, the piconet will survive because all members will soon hop together out of that channel.

FHSS Synchronization

For two devices to communicate using FHSS, they must be properly synchronized so that they hop together from channel to channel. This means that the devices must

- use the same channel set.
- use the same hopping sequence within that channel set.
- be time-synchronized within the hopping sequence.
- ensure that one transmits while the other receives, and vice versa.

All these synchronization items are determined by the piconet master. The master passes the FHSS synchronization parameters to a slave during piconet setup.

Hop Channel Set and Period   An FHSS radio is programmed to operate on a certain set of frequencies called the *channel set*. For Bluetooth, the channel set consists of the carrier frequencies

$$f_c = 2402 + k \text{ MHz}; k = 0, 1, \ldots, 78 \tag{4.7}$$

Thus, 79 possible frequencies exist in the channel set, each spaced 1 MHz apart, and covering 2,402 to 2,480 MHz. The piconet hops pseudorandomly within these channels as shown in Figure 4-10. Each channel is 1 MHz wide, and the Bluetooth GFSK data transmission occupies this bandwidth.

The sequence itself is determined by a pseudorandom hop generator that repeats its sequence after a certain number of hops. The period must be at least equal to the number of hop channels, but it can also be much longer. The Bluetooth hop period during normal piconet communications is 2^{27} hops long, so at a rate of 1,600 hops per second the pattern will begin to repeat after about 23.3 hours.

The period is long in order to greatly reduce the chance of accidental hop synchronization between two different piconets within range of each other. This can be viewed in the following way. Suppose two piconets happen to hop into the same channel at a particular time. With a long pseudorandom sequence that's different for both piconets, the probability is only 1 in 79 that the next hop by both piconets will also be into the same channel. Therefore, both piconet hop sequences, because they are long, can be treated as completely random.

FHSS Multiple Access

One of the strengths of FHSS in general, and Bluetooth FHSS in particular, is its multiple access capability. As long as the number of hop channels is sufficiently

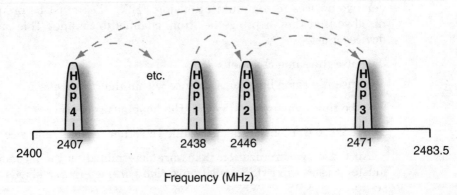

Figure 4-10
Radios in a Bluetooth piconet hop from frequency to frequency in a pseudo-random pattern. Depicted bandwidths are not to scale.

large, two or more FHSS networks using independent hop patterns will transmit together on the same channel relatively rarely. Bluetooth, with its 79 hop channels, can support several piconets within a single room, and we'll define just how many in later chapters. At this point, though, it will be useful to lay some of the groundwork for Bluetooth-on-Bluetooth coexistence analysis.

Because devices in a piconet transmit one packet per hop, we will assume that if two or more piconets within range of each other transmit on the same channel then all affected packets are lost. Also, even though actual piconets are not time-synchronized with each other, for now we will assume that all of them hop at the same time instant, but each uses a sequence that is randomly determined. Thus, the probability of packet error P_E from interference produced by another Bluetooth piconet becomes the probability that two or more transmissions occur within one channel. This is given by the formula

$$P_E = 1 - \left(1 - \frac{1}{M}\right)^{K-1} \tag{4.8}$$

where M is the number of hop channels and K is the total number of piconets operating simultaneously. For most Bluetooth operations, M equals 79, but M may be smaller for use in countries that don't permit the full 79-channel sequence due to restrictions on the 2.4 GHz band or during adaptive frequency hopping. Figure 4-11 shows the packet error probability given by Equation 4.8 as a function of the number of piconets, all of whom can interfere with each other, for the Bluetooth 79-channel hop set. A more detailed analysis of Bluetooth piconet mutual interference will be presented in later chapters, where we will discover that the error probabilities given in Figure 4-11 are optimistic.

As with many formulas involving probabilities, Equation 4.8 looks strange with all the "1 −" parts to it. Its interpretation, though, is quite straightforward. The quantity $1/M$ is the probability that two piconets will hop into the same channel at any given time, so $1 - 1/M$ is the probability that the two will hop into different channels.

Now let's focus on one of the piconets, which we call the *desired piconet*. The quantity $1 - 1/M$ taken to the $K - 1$ power is the chance that none of the other piconets will hop into the desired piconet's channel, so that becomes the probability of packet success. Finally, one minus this quantity is the probability of packet error.

If lots of hop channels exist compared to the number of piconets within range of each other, we can assume that the probability is low that three or more piconets will hop into any particular channel. In this case, Equation 4.8 simplifies to

$$P_E \approx \frac{K-1}{M} \tag{4.9}$$

Figure 4-11
Packet error
probability as a
function of
total piconets
for the
79-channel
Bluetooth
hop set

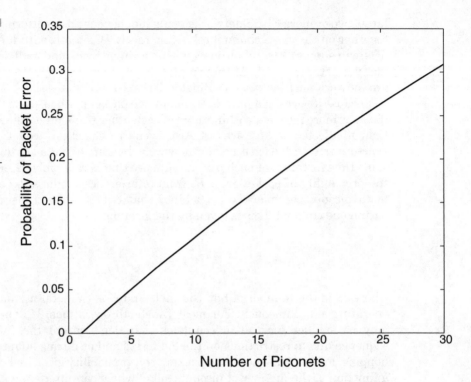

Figure 4-11 Packet error probability as a function of total piconets for the 79-channel Bluetooth hop set

Orthogonal Frequency Division Multiplexing (OFDM)

Application: 802.11a, 802.11g

Look once again at Figure 4-3, the spectrum of the 802.11b transmitted signal. This spectrum has the characteristic rounded shape of *single-carrier modulation* (SCM), in which power is concentrated near the carrier frequency; then the *power spectral density* (PSD) drops for frequencies offset from the carrier. This is a variation of the classic $\text{sinc}^2(x) = [\sin(\pi x)/(\pi x)]^2$ PSD rendition. From an information theory point of view, a nonuniform PSD over the entire bandspread equates to reduced data rates.

By using *multicarrier transmission* (MCT), better use of the available spectrum can be achieved. Incoming data is multiplexed among several relatively slow-rate modulated channels, but the aggregate data rate can be much higher. By spacing the different channel carrier frequencies as close together as possible without mutually interfering with each other, OFDM results. This method is used for 802.11a in the 5.x GHz band and 802.11g in the 2.4 GHz band. OFDM requires significant signal processing, including high-speed implementations of the *fast Fourier transform* (FFT), *inverse FFT* (IFFT), and convolutional coding to be feasible.

OFDM Operation

In the time domain, orthogonality among a set of sinusoids having different frequencies is accomplished by ensuring that an integral number of cycles occurs for each within a particular time window. In the frequency domain, the same set of carriers is spaced such that the peak of any one of these is coincident with the nulls of all the others. Information can be placed onto each carrier using PSK or *quadrature amplitude modulation* (QAM) techniques while maintaining orthogonality among them. Mathematically, the OFDM signal $s(t)$ can be represented as [Wag01]

$$s(t) = \begin{cases} \sum_{k=0}^{n-1} d_k \exp\left[j2\pi\frac{k}{T}(t - t_s) \right]; & t_s \le t < t_s + T \\ 0; & \text{otherwise} \end{cases}$$

(4.10)

where n is the total number of subcarriers, the ith of which is modulated by the complex value d_k, representing the particular amplitude and phase of the QAM symbol associated with that carrier. Each symbol is T seconds in duration.

Demodulating the i-th subcarrier during the time interval $(t_s, t_s + T)$ is accomplished by multiplying $s(t)$ by a sinusoid having frequency i/T and integrating the result over the desired symbol interval; thus

$$\int_{t_s}^{t_s+T} \exp\left[-j2\pi\frac{i}{T}(t - t_s) \right] \left(\sum_{k=0}^{n-1} d_k \exp\left[j2\pi\frac{k}{T}(t - t_s) \right] \right) dt$$

$$= \sum_{k=0}^{n-1} d_k \int_{t_s}^{t_s+T} \exp\left[-j2\pi\frac{k-i}{T}(t - t_s) \right] dt$$

$$= d_i T$$

(4.11)

The QAM symbol associated with carrier i is extracted without any influence from the other carriers due to the orthogonality of the OFDM set.

A useful characteristic of the OFDM signal in Equation 4.10 is that it looks similar to a discrete Fourier transform. This feature can be exploited during signal generation at the transmitter by first working in the frequency domain to select all the constellation points d_k corresponding to all the data contained in a particular OFDM symbol. Then perform an IFFT to obtain the time domain representation for transmission.

An OFDM transmitter and receiver are shown in Figure 4-12 [Wag01]. The transmitter begins its signal construction by operating digitally in the frequency domain. Bits are interleaved and coded, and then mapped into the desired QAM

Figure 4-12
The OFDM
transmitter and
receiver

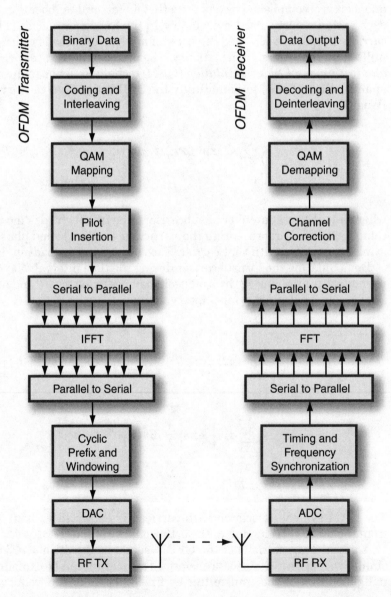

symbols (in which BPSK and QPSK are considered special cases of QAM). Next, the pilot carriers are inserted to assist the receiver in channel estimation and synchronization. After the signal is divided into its respective carriers, an IFFT converts each carrier signal into the time domain. These are combined and windowed. Finally, the digital signal is converted to analog, up-converted, amplified, and transmitted. At the receiver, each operation is reversed until the binary data results.

Wi-Fi OFDM Signal Composition

The organization of 802.11a and 802.11g Wi-Fi OFDM subcarriers is depicted in Figure 4-13. A total of 52 subcarriers is defined, spaced at 312.5 kHz. The total occupied bandwidth is approximately 16.25 MHz (52×0.3125), not including the secondary lobes of the carriers on either end of the spectrum. The 802.11a specification lists the bandwidth as 16.6 MHz, rounding it up slightly, and others sometimes round it down to 16 MHz for convenience. A total of 48 subcarriers are modulated with data using BPSK, QPSK, 16-QAM, or (optionally) 64-QAM, and at convolutional coding rates of ½, ⅔, or ¾, for aggregate data rates ranging from 6 to 54 Mb/s. If each carrier modulates k bits/symbol, where k is 1 (BPSK), 2 (QPSK), 4 (16-QAM) or 6 (64-QAM), then the data rate is $12kr$ Mb/s, where r is the coding rate.

Four pilot subcarriers are spaced within the OFDM signal to be used for channel evaluation and as a phase reference by the receiver for demodulating data in the other carriers. The pilot carriers are BPSK modulated with a pseudorandom code to prevent spectral spurs at the pilot frequencies. Symbol duration in all cases is 4 μs, including a special 800 ns cyclic prefix guard time for antimultipath. The symbol duration minus the guard time is 3.2 μs, and the reciprocal of this produces the correct carrier spacing of 312.5 kHz for orthogonality.

The advantages of using OFDM over other modulation methods include spectral efficiency, better performance in multipath-induced fading and ISI, and the

Figure 4-13
802.11a and 802.11g OFDM carrier set

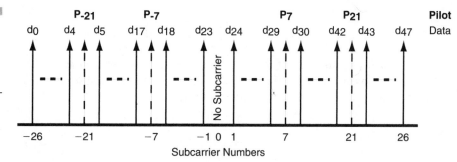

capability to operate when narrowband CCI is present. Each one of these is significant, and together they help motivate the great interest in developing OFDM in a wide variety of communications applications. The high spectral efficiency of OFDM springs from the following characteristics:

- Spacing between subcarriers is the minimum possible while maintaining orthogonality.
- No guard band is needed between subcarriers.
- By dedicating only a few subcarriers to pilot duty, the remainder can be efficiently encoded using QAM.
- The multiple carriers result in an aggregate PSD that is essentially uniform throughout its occupied bandwidth.

Using OFDM, of course, has some disadvantages. One of the greatest is the large amount of signal processing that must be accomplished for generating and demodulating each symbol. Complexity, power consumption, and cost are all adversely affected by this requirement. Also, amplifiers in the transmitter and receiver must have exceptionally good linearity to handle the significant amplitude variations of the composite OFDM signal without introducing unacceptable distortion.

OFDM Performance in AWGN

Unlike simpler modulation schemes like single-carrier BPSK, the OFDM signal can be constructed in many different ways, so a generalization of performance requires accounting for several parameters such as the number of subcarriers, coding rate, and modulation method. For simplicity, we will concentrate on the performance of OFDM as used in 802.11a and 802.11g.

The 802.11 OFDM symbol is 4 μs in duration, of which 800 ns is a guard time. Also, 4 of the 52 subcarriers contain only the pilot signal. As such, the SNR/symbol must be adjusted by the fraction of the symbol that actually contains information. This fraction is

$$F = \frac{48}{52} \times \frac{3.2}{4} = 0.74 \tag{4.12}$$

which is equivalent to an implementation loss of 1.3 dB.

In terms of SNR/bit = E_b/N_0, the bit error probability for each uncoded OFDM carrier modulation method is [Mil03]

$$P_b^{(\text{BPSK})} = P_b^{(\text{QPSK})} = Q\left(\sqrt{F \times 2 \times \frac{E_b}{N_0}}\right) \tag{4.13}$$

$$P_b^{(16\text{-QAM})} = \frac{3}{4}Q\left(\sqrt{F \times \frac{4}{5} \times \frac{E_b}{N_0}}\right) \tag{4.14}$$

$$P_b^{(64\text{-QAM})} = \frac{7}{12}Q\left(\sqrt{F \times \frac{6}{21} \times \frac{E_b}{N_0}}\right) \tag{4.15}$$

These equations include the assumption that a symbol error results in a single bit error within that symbol. The BER versus SNR/bit values for each of these uncoded OFDM modulation methods are given in Figure 4-14. The penalty for 16-QAM and 64-QAM is about 3.8 dB and 8.1 dB, respectively, against BPSK/QPSK.

Simulation results for coded OFDM as used in 802.11a/g, compared to similar uncoded performance, are shown in Figure 4-15 [Sol03]. The BER performance is greatly improved when rate ½, ⅔, or ¾ convolutional coding is included, unless the SNR/bit is low. This explains why uncoded OFDM isn't used in any of the 802.11a/g renditions. As mentioned in the preceding chapter, when the channel BER is high, convolutional *forward error correction* (FEC) codes that are

Figure 4-14
BER versus SNR/bit for uncoded OFDM with 802.11-like structure [Mil03]

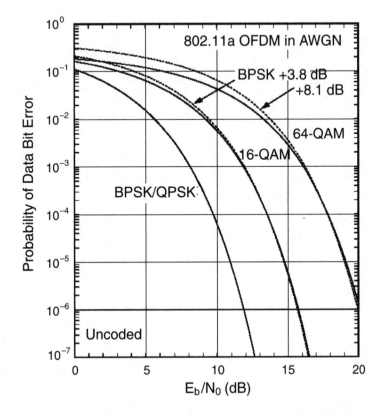

Figure 4-15
BER versus
SNR/bit for
coded OFDM
as used in
802.11a and
802.11g,
compared to
their uncoded
counterparts
[Sol03]

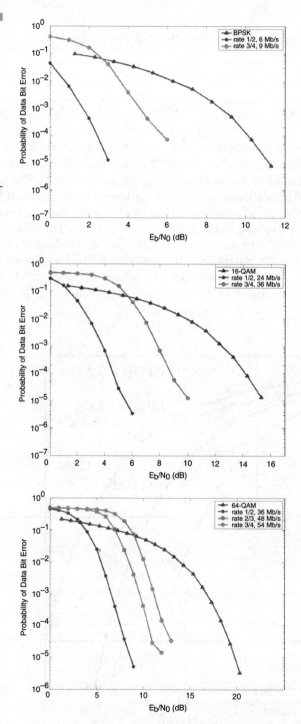

processed with a Viterbi decoder can produce an even higher output BER, which explains the poor performance of coded OFDM at low SNR/bit values.

Multipath Mitigation Using OFDM

From Chapter 2, "Indoor RF Propagation and Diversity Techniques," we learned that multipath can cause fading from Doppler spread and ISI from delay spread. OFDM can mitigate both of these, the former indirectly and the latter directly. The wide bandwidth of the composite OFDM signal means it will probably acquire some frequency-selective fading across the channel. Instead of distorting the entire signal, as with SCM, frequency-selective fading only affects some of the carriers in OFDM. The affected carriers might even be attenuated to the point of loss at the receiver, but the following methods can be employed to recover from this:

- Employ bit interleaving to distribute bit errors within the received data stream such that an FEC can recover those errors (coded OFDM).
- Establish a special guard time to combat ISI from delay spread.
- Use multiple antennas or receivers in a diversity-combining configuration.

Both 802.11a and 802.11g employ all three of these techniques for reducing the effect of fading and ISI from multipath. Performance can be enhanced further by selecting only those carriers that are faded for data rate reduction, or even stopping them temporarily from carrying any data at all, but this isn't done in 802.11a/g. Instead, Wi-Fi will automatically reduce the number of bits per symbol over all OFDM carriers when the frame error rate exceeds a threshold.

Bit Interleaving   The 802.11a or 802.11g OFDM channel is typically slow-faded, so the channel statistics remain constant during transmission of a single OFDM symbol. However, the channel often experiences frequency-selective fading, so a few subcarriers may be unusable at the receiver for the duration of the symbol.

The performance of OFDM can be enhanced through a process called *bit interleaving,* where the data bits in the symbol are interleaved (shuffled) before being assigned to each subcarrier, and then deinterleaved at the receiver. This improves reliability by assigning adjacent data bits to different OFDM subcarriers, greatly reducing the chance that a burst error event from a faded subcarrier will enter the convolutional decoder at the receiver. Remember from our discussion of convolutional codes that a burst error sequence into the decoder may result in an even longer error sequence at the decoder's output. The goal of interleaving is to surround each error-prone bit with several good bits, increasing the chance that all bit errors will be corrected.

Interleaving is performed on a block of data with block size determined by multiplying the number of data subcarriers (48) by the number of bits encoded on each subcarrier for one OFDM symbol. The interleavers for BPSK, QPSK, 16-QAM, and 64-QAM operate with blocks of 48, 96, 192, and 288 bits, respectively. Coding gains through bit interleaving typically range between 2 and 5 dB on a Rayleigh-faded channel withh an rms delay spread of 75 ns [Hei02].

Cyclic Prefix   Delay spread is also handled by OFDM in a direct manner. Although the aggregate data rate using OFDM can be quite high (54 Mb/s maximum in 802.11a/g), the symbol rate within each carrier is much lower. Thus, OFDM has built-in protection from ISI. By insuring that the symbol duration is significantly longer than the maximum anticipated delay spread, ISI from one symbol only affects a small portion of the subsequent symbol and never crosses into multiple symbols within a particular subcarrier. Unfortunately, though, the influence of ISI can cause a loss of carrier orthogonality, resulting in intrasymbol interference. This is avoided by placing a special prefix, called the *cyclic prefix* (CP), at the beginning of each channel symbol. The duration of the CP guard time should be about four times longer than the longest expected delay spread [Wag01]. The transmitter then time-windows the OFDM symbol such that out-of-band spurious transmission is suppressed, but carrier orthogonality is maintained.

Diversity   The reliability of OFDM in a multipath environment is also improved by using antenna diversity, and this is an area of major research interest. In one study, a simulation was performed using 802.11a signaling at 54 Mb/s, with 1,000-byte packets. These were sent over a channel with an rms delay spread of 100 ns. For two receivers using maximal ratio combining, improvements of 6 to 8 dB were achieved for a given PER, corresponding to a range improvement of about 50 to 80 percent. Furthermore, two receivers yielded significant additional performance improvements over a single receiver when synchronization and timing errors were present [Moo03].

UWB Modulation and Coding

According to FCC rules, a UWB signal must have a -10 dB fractional bandwidth greater than 0.20, or a -10 dB bandwidth of at least 500 MHz regardless of fractional bandwidth. In a manner reminiscent of early spark-gap communication, a short pulse produced by the transmitter of a so-called *impulse radio* fits the UWB rules nicely. Alternatively, single-carrier DSSS using a sufficiently high chip rate, or a multicarrier transmission such as OFDM with a large number of carriers, can achieve UWB status. Each of these has its strengths and weaknesses, not only from a purely technical point of view, but also from a regulatory aspect.

For use as an alternate PHY for 802.15.3a, Motorola and Xtreme Spectrum have proposed a DSSS solution, whereas the MultiBand-OFDM Alliance promotes using OFDM instead. Both of these are technically feasible, and their respective promoters claim that they meet FCC regulations, but it's likely that one or the other will have prevailed by the time this work is published. As such, our concern will be demonstrating the differences between these two UWB structures and comparing them to the impulse radio, rather than focusing on specifics in any great detail.

UWB Signaling Methods

Figure 4-16 shows an impulse radio-transmitted pulse of 500 ns and its respective spectral plot. This particular pulse is a member of the Gaussian monocycle class, with a center frequency and bandwidth determined by the pulse width [TDC00]. In the time domain, the pulse is represented by

$$V(t) = \frac{t}{\tau} \exp\left[-\left(\frac{t}{\tau} \right)^2 \right] \tag{4.16}$$

which is similar to the first derivative of the Gaussian function, which is an odd function in time. An even function in time can be obtained through the second derivative of the Gaussian. (The Gaussian pulse itself is usually not used due to its large DC component.) The factor τ determines the monocycle duration, and its center frequency and bandwidth are roughly proportional to the inverse of the pulse duration. For example, the monopulse in Figure 4-16 has a bandwidth of about 2 GHz centered on a frequency of 2 GHz. Other monopulse classes have been studied based upon the Raleigh, Laplacian, and cubic functions [Zha02].

Figure 4-16
Time and frequency domain representation of a UWB Gaussian monopulse [TDC00]

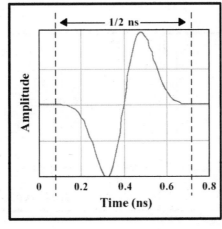 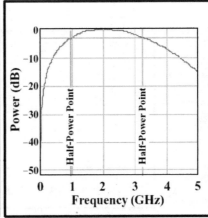

Placing data onto such a signal can be easily accomplished using *pulse position modulation* (PPM), where information is encoded by the position the pulse occupies within a window of time. For example, in a 10 *million pulses per second* (Mp/s) system, one pulse is sent every 100 ns, which is the *pulse repetition interval* (PRI). Figure 4-17 plots this in the time and frequency domains [TDC00]. Within this window, a binary 0 may be represented by transmitting a pulse 100 *picoseconds* (ps) early, and a 1 by sending the pulse 100 ps late. Because pulse transmission and reception are accomplished at baseband, the impulse radio is conceptually simple. Design challenges tend to arise from the need for a broadband antenna and fast-switching circuits rather than from oscillators and filters. Using PPM on a single pulse for data encoding is called *time-modulated UWB* (TM-UWB).

Two technical difficulties arise from the TM-UWB sample given by Figure 4-16. The first is that the spectrum shows spikes at various harmonics associated with the regular pulse shifts of ±100 ps. The second is that coexistence with other similarly modulated UWB signals is poor, because the desired receiver could easily lock onto an interfering signal that happens to have approximately the same pulse transmission times. Both these problems can be solved by placing a pseudorandom dither onto the pulse position, causing it to vary across most of the available window, as shown in Figure 4-18. Now the PSD is much more uniform across its bandwidth. Because the receiver also knows the PN dither code, it can narrow its pulse search window significantly, rejecting MAI from other UWB transmitters and narrowband users. However, the PSD isn't completely uniform across the signal's bandwidth, allowing for some concern that UWB could interfere with some services that require highly sensitive conventional receivers, such as a *global positioning system* (GPS).

As with other modulation methods, TM-UWB data rates can be increased by sending more than one bit per channel symbol. The pulse itself can contain an additional bit of information by inverting it, and the pulse window can be divided into M subwindows containing $\log_2 M$ bits of additional information.

Figure 4-17
Time and frequency domain representation of a UWB signal modulated with random data using PPM [TDC00]

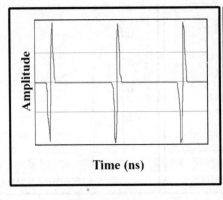

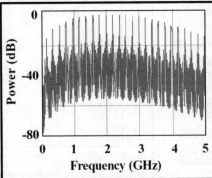

Figure 4-18
Time and
frequency
domain repre-
sentation of a
UWB signal
modulated
with random
data using PPM
with random
dithering of
the time
window
[TDC00]

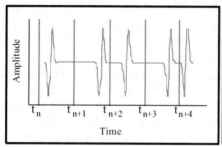

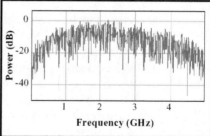

FCC rules require that the UWB signal be concentrated between 3.1 and 10.6 GHz, with very little emissions allowed outside this band. If a Gaussian monopulse is used, the pulse width would need to be about 140 ps in duration for a 7 GHz center frequency and a 7 GHz bandwidth. Further filtering may also be required to meet out-of-band limits. Alternatively, DSSS can be used with an extremely high chip rate to generate a PSD similar to that from a pulse train with random timing dither. A DSSS system meeting UWB bandwidth require- ments has the potential for providing a high processing gain, even at data rates in excess of 100 Mb/s. Furthermore, resolution at these high bandwidths means that a RAKE receiver with a large number of fingers has the potential to improve received SNR significantly. We'll call this subset of DSSS *direct sequence UWB* (DS-UWB). The only significant conceptual difference between TM-UWB and DS-UWB is that DS-UWB typically sends several pulses per data bit for higher bit energy at the expense of either a higher duty cycle or slower data rate.

Instead of using a single, short PPM pulse or DSSS to obtain the bandwidths needed, a multicarrier signal can be developed with the necessary bandwidth for UWB. IEEE 802.11a and 802.11g OFDM transmissions, described earlier in this chapter, use 48 subcarriers and 4 pilot carriers for an occupied bandwidth of 16.6 MHz. For UWB operation, the bandwidth must be increased to at least 500 MHz to comply with FCC rules. At the expense of a more complex transmitter and receiver, multicarrier techniques can exercise significant control over the PSD of the transmission by independently selecting its center frequency, occupied band- width, and shape of its power spectrum. We call this method *multicarrier UWB* (MC-UWB).

UWB Performance in Fading and Interference

Narrowband interference and channel imperfections are mitigated in the tradi- tional way with DS-UWB or MC-UWB. That is, DS-UWB has a processing gain proportional to the spreading factor, and MC-UWB can suffer a loss of some of its carriers, either through fading or interference, and use FEC to recover the missing data. Delay spread is mitigated with a RAKE receiver in the DS-UWB receiver, which can combine several independent paths for improved received SNR. Delay spread mitigation is inherently included in the proposed MC-UWB receiver up to an rms value of about 61 ns, which should be adequate for most LOS situations over distances of less than 10 m. The bandwidth occupied by a DS-UWB signal is determined primarily by its chip duration, which is a function of circuit switching speed, but in the case of MC-UWB, its maximum occupied bandwidth is a function of its signal-processing capability.

In a multipath-prone environment, the performance of the receiver in a TM-UWB radio is closely aligned to that of the DS-UWB receiver, because both are able to exploit their higher bandwidths to independently resolve various multipath components. The processing gain for the DS-UWB receiver is the ratio of the output *signal-to-interference ratio* (SIR) to the input SIR at the desired UWB receiver. It can be roughly determined by converting both the duty cycle and pulse integration factor to their respective dB values and then summing the two [TDC00]. (The TM-UWB receiver's PG is the dB equivalent of the duty cycle if only one pulse is used per data bit.) An example shows how this is done.

Example 3-1:

A DS-UWB has a 1 percent duty cycle and sends 10 Mp/s. For a data rate of 8 kb/s, what is the processing gain?

Solution:

At 10 Mb/s, each data bit is represented by 1,250 pulses. Therefore, PG = 10 log(100) + 10 log(1250) = 54 dB.

Of course, the PG of high-data-rate DS-UWB transmissions will be lower due to the lower pulse integration factor, but PG values between 20 and 30 dB should be readily achievable. The PG can be used to find the UWB system's jam resistance, which is the margin that the processing gain provides above the minimum SIR, called SIR_D, required to meet some system performance specification. When using dB values,

$$JR = PG - SIR_D \qquad (4.17)$$

For example, if the required SIR_D is 16 dB to obtain a BER at the desired UWB receiver of 10^{-6}, the jamming resistance will be 16 dB below the processing gain. In the presence of AWGN, the jam resistance can be calculated against SNR_D instead.

A more accurate representation of UWB PG must take into account not only the UWB pulse shape, but also the nature of the interfering (jamming) signal [Zha02]. Suppose a desired TM-UWB signal uses rectangular pulses with duration T_p and pulse repetition time T_f. Suppose further that an in-band jamming signal exists using a carrier frequency f_J with a bandwidth that is small compared to that of the UWB signal. The processing gain is then given by [Zha02]

$$PG = \frac{\beta(\pi\gamma^2)}{(1 - \cos 2\pi\gamma)^2} \qquad (4.18)$$

where $\beta = T_f/T_p$ is the spreading ratio of the UWB signal, which is typically greater than 100, and $\gamma = f_J T_p$ is the number of jammer carrier cycles during the UWB pulse, which has a value between 0 and 1 for in-band jamming. Figure 4-19 plots the PG as a function of γ for the rectangular pulse with $\beta = 100$. PG is at its minimum when $f_J = 0.4/T_p$. When the jamming signal's bandwidth becomes a significant fraction of the UWB signal's bandwidth, the curve in Figure 4-19 becomes shallower, with PG values between about 7 and 10 dB for an in-band jammer. Rayleigh and certain Gaussian monopulses have about 20 to 40 dB greater PG for small values of γ and about 10 dB less PG for large values of γ, with the Gaussian monopulses performing best [Zha02].

Comparison of TM-UWB and DS-UWB   As we mentioned earlier, the PSD of TM-UWB and DS-UWB signals are similar because both achieve their respective spreading using fast transitions in the time domain. The first major difference between these two signaling methods is that the TM-UWB waveform is baseband (carrierless), whereas DS-UWB is carrier-modulated. The second is that the interference suppression in TM-UWB is performed through time

Figure 4-19
Processing gain for a UWB signal with narrowband jamming within the UWB receiver's passband

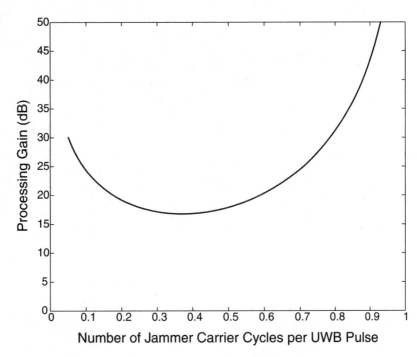

windowing and cross-correlation on the short pulse, whereas DS-UWB reduces interference through despreading and subsequent filtering at the receiver. TM-UWB interference suppression can be clarified by noting that the time window at the receiver for checking the arriving monopulse is extremely short, only $t \pm T_p$, so even at these wide bandwidths aggregate noise and interference power is very low. Furthermore, the high correlation between the interference present at times t and $t \pm T_p$ results in the significant reduction of its effect.

To quantify the PG of DS-UWB for comparison with TM-UWB, we assume a 1 ns Gaussian chip waveform for the former and a 1 ns Gaussian monopulse with even symmetry for the latter, both with unit energy [Zha02]. We define for TM-UWB the quantity $\alpha = W_J T_p$ as the ratio of interference and UWB bandwidths, and for DS-UWB we have, for chip duration T_c, $\nu = (f_J - f_c)T_c$, which is the ratio of the interference carrier frequency offset to system bandwidth. For the same occupied bandwidth, the DS-UWB carrier frequency isn't necessarily the same as the UWB monopulse center frequency. When Gaussian pulses are used, the two offsets γ and ν are related by

$$\gamma = \nu + 0.65 \tag{4.19}$$

A comparison of *jam resistance* (JR) versus ν for TM-UWB and DS-UWB is shown in Figure 4-20 for wideband jamming where α, the ratio of jammer to UWB bandwidths, equals 0.5. The assumption is that a BER of 10^{-6} requires SNR_D to be 10 dB, so the PG is 10 dB greater than the JR shown in Figure 4-19. The bandwidth spreading ratio is 100 for both TM-UWB and DS-UWB. Of

Figure 4-20
JR for TM-UWB and DS-UWB against wideband interference [Zha02] © IEEE

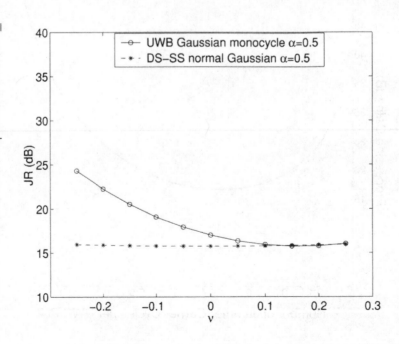

course, the energy per data bit is 100 times greater with DS-UWB than with TM-UWB, so we can expect that an SNR_D of 10 dB will occur over a greater TX-*receive* (RX) separation using the spread spectrum. For DS-UWB, the JR is fairly constant at about 16 dB, but TM-UWB exceeds this value by as much as 18 dB for large, negative ν. For narrowband interference ($\alpha = 0.1$ and $\alpha = 0.001$), the difference in JR between TM-UWB and DS-UWB is even more profound, as depicted in Figure 4-21.

Proposed 802.15.3a UWB Signaling

The MultiBand-OFDM Alliance, which we mentioned in Chapter 1, "Introduction," was formed to promote using MC-UWB for the 802.15.3a PHY. A DS-UWB method has been proposed by Xtreme Spectrum and Motorola. Both of these are outlined in Table 4-1 [Man03].

One of the disadvantages of the proposed MC-UWB solution is the relatively narrow 528 MHz occupied signal bandwidth, which would limit average TX power to −14.1 dBm (−41.3 + 10log(528)), or about 39 μW. A method similar to frequency hopping has been proposed to distribute the average signal power over a larger bandwidth comprised of several channels to satisfy FCC rules for higher TX power. The existence of the two competing UWB signaling methods makes it

Figure 4-21
JR for TM-UWB
and DS-UWB
against
narrowband
interference
[Zha02] © IEEE

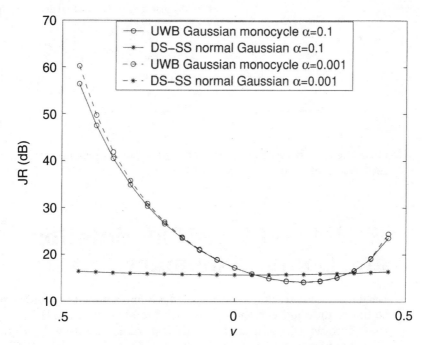

Table 4-1

Proposed 802.15.3a PHY with DS-UWB and MC-UWB

	DS-UWB	MC-UWB
Number of bands	2	3 mandatory 10 optional
Bandwidth	1.368 GHz 2.736 GHz	528 MHz
Frequency range	3.2–5.15 GHz 5.825–10.6 GHz	Group A: 3.168–4.752 GHz Group B: 4.752–6.072 GHz Group C: 6.072–8.184 GHz Group D: 8.184–10.296 GHz
Modulation	DSSS (BPSK)	OFDM
WLAN coexistence method	5 GHz null	5 GHz null
Multiple access method	CDMA	Time-frequency interleaving
Number of simultaneous piconets	8	4
FEC method	Convolutional code Reed-Solomon code	Convolutional code
FEC rates	1/2 @ 110 Mb/s RS(255,223) @ 200 Mb/s RS(255,223) @ 480 Mb/s	11/32 @ 110 Mb/s 5/8 @ 200 Mb/s 3/4 @ 480 Mb/s
Link margin	6.7 dB @ 10 m @ 110 Mb/s 11.9 dB @ 4 m @ 200 Mb/s 1.7 dB @ 2 m @ 480 Mb/s	5.3 dB @ 10 m @ 110 Mb/s 10.0 dB @ 4 m @ 200 Mb/s 11.5 dB @ 2 m @ 480 Mb/s
Chip/symbol duration	731 ps (low band) 365.5 ps (high band)	312.5 ns
Multipath mitigation method	Equalizer and RAKE receiver	Inherent up to 60.6 ns delay spread

clear that UWB technology is in its infancy and progress will continue in both technical and regulatory arenas.

WLAN and WPAN Modulation and Coding Summary

Each of the WLAN and WPAN specifications is aimed at a different application, so the modulation and coding for each are also different. Here is a summary of these techniques, along with the BER formulas used by the IEEE 802.15 TG2 coexistence task group, along with those appearing in the IEEE 802.15.3 and

802.15.4 standards. In these formulas, the term *signal-to-interference-and-noise ratio* (SINR) refers to the ratio of the signal power to the sum of the powers in the interference and in any AWGN that may be present.

IEEE 802.11b Wi-Fi

The modulation and coding used by 802.11b Wi-Fi are given in Table 4-2. Wi-Fi in its 802.11b rendition requires supporting low-rate data at 1 and 2 Mb/s, and high-rate data at 5.5 and 11 Mb/s. Low-rate transmission is accomplished using traditional DSSS with an 11-chip Barker spreading sequence, and high rate requires supporting CCK. Optionally, the higher data rates can employ PBCC. The 802.11b specification refers to 1 or 2 Mb/s as the basic rate set, 5.5 and 11 Mb/s CCK as the *high-rate DSSS* (HR/DSSS), and 5.5 and 11 Mb/s PBCC as HR/DSSS/PBCC.

$$\text{BER at 1 Mb/s} \;=\; Q(\sqrt{11 \times \text{SINR}}) \tag{4.20}$$

$$\text{BER at 2 Mb/s} \;=\; Q\!\left(\sqrt{5.5 \times \frac{\text{SINR}}{2}}\right) \tag{4.21}$$

$$\text{BER at 5.5 Mb/s} = \frac{8}{15}[14Q(\sqrt{8 \times \text{SINR}}) + Q(\sqrt{16 \times \text{SINR}})] \tag{4.22}$$

$$\text{BER at 11 Mb/s} = \frac{128}{255}[24Q(\sqrt{4 \times \text{SINR}} + 16Q(\sqrt{6 \times \text{SINR}}) \tag{4.23}$$
$$+ 174Q(\sqrt{8 \times \text{SINR}}) + 16Q(\sqrt{10 \times \text{SINR}})$$
$$+ 24Q(\sqrt{12 \times \text{SINR}}) + Q(\sqrt{16 \times \text{SINR}})]$$

Table 4-2

802.11b modulation methods (2.4 GHz ISM band)

Data rate (Mb/s)	Mandatory modulation	Optional modulation
1	DBPSK/DSSS	
2	DQPSK/DSSS	
5.5	DQPSK/CCK	PBCC
11	DQPSK/CCK	PBCC

In each case, the SINR is taken over a single chip; that is, $\text{SINR} = E_C/N_C$ where E_C is the energy contained in a single chip and N_C is the total noise and interference energy within the chip.

IEEE 802.11a/g Wi-Fi

Table 4-3 lists the different data rates and how they're created within the OFDM signal set for both 802.11a and 802.11g. The SNR increases needed to support 16-QAM and 64-QAM are given in Table 4-3.

The data rates supported by 802.11a range from 6 to 54 Mb/s, depending on channel conditions and throughput requirements. As shown in Table 4-4, only three data rates are mandatory, and the remaining five are optional, all of which use OFDM with various carrier modulation and coding methods. The manufacturers of 802.11a chipsets usually support all the mandatory and optional data rates.

The OFDM signal structures for 802.11g are identical to those used by 802.11a and are shown in Table 4-3. The modulation methods for 802.11g are listed in Table 4-5. The 802.11g devices include the mandatory and optional data rates depicted in Table 4-4, the low- and high-rate data of 802.11b (as shown in Table 4-2), and optional support for high-rate PBCC. The modulation methods shown in Table 4-5 that differ from those in Table 4-2 are called *extended rate PHY* (ERP). These modes include ERP-DSSS/CCK (802.11b HR/DSSS with some minor differences), ERP-OFDM (OFDM at 6, 9, 12, 18, 24, 36, 48, or 54 Mb/s),

Table 4-3

OFDM symbol structure for 802.11a (5.x GHz U-NII band) and 802.11g (2.4 GHz ISM band)

Data rate (Mb/s)	Modulation	Coding rate	Coded bits per subcarrier	Coded bits per OFDM symbol	Data bits per OFDM symbol
6	BPSK	1/2	1	48	24
9	BPSK	3/4	1	48	36
12	QPSK	1/2	2	96	48
18	QPSK	3/4	2	96	72
24	16-QAM	1/2	4	192	96
36	16-QAM	3/4	4	192	144
48	64-QAM	2/3	6	288	192
54	64-QAM	3/4	6	288	216

Table 4-4

802.11a modulation methods (5.x GHz U-NII band)

Data rate (Mb/s)	Mandatory modulation	Optional modulation
6	OFDM	
9		OFDM
12	OFDM	
18		OFDM
24	OFDM	
36		OFDM
48		OFDM
54		OFDM

Table 4-5

802.11g modulation methods (2.4 GHz ISM band)

Data rate (Mb/s)	Mandatory modulation	Optional modulation
1	DSSS	
2	DSSS	
5.5	CCK	PBCC
6	OFDM	DSSS-OFDM
9		OFDM, DSSS-OFDM
11	CCK	PBCC
12	OFDM	DSSS-OFDM
18		OFDM, DSSS-OFDM
22		PBCC
24	OFDM	DSSS-OFDM
33		PBCC
36		OFDM, DSSS-OFDM
48		OFDM, DSSS-OFDM
54		OFDM, DSSS-OFDM

ERP-PBCC (PBCC at 22 or 33 Mb/s), and DSSS-OFDM (described in the following paragraph).

Because nodes conforming to 802.11g operate in the 2.4 GHz band, it's quite possible that they will be required to coexist with 802.11b units. Therefore, 802.11g is required to be backward compatible with 802.11b, supporting 1, 2, 5.5, and 11 Mb/s data rates. Optionally, 802.11g devices can transmit packet headers using DSSS so that these can be decoded by 802.11b nodes for improved network traffic handling in mixed 802.11g/802.11b environments. This DSSS-OFDM mode will be examined in greater detail in Chapter 6.

IEEE 802.15.1 Bluetooth

Bluetooth is unique among the WLAN and WPAN networks discussed in this book in that modulation is done using GFSK and interference is mitigated through FHSS. A summary of Bluetooth operation is given in Table 4-6. The BER for GFSK signaling involves a Bessel function and a correlation coefficient based on the modulation index [IEEE802.15.2]. The BER given here is an approximation based on standard noncoherent FSK detection [IEEE802.15.4].

$$\text{BER at 1 Mb/s} \approx \frac{1}{2}\exp\left[\frac{-\text{SINR}}{2}\right] \tag{4.24}$$

IEEE 802.15.3 WiMedia

WiMedia was designed to carry high-data-rate video and audio streams over short distances. Because RX SNR values are expected to be high and TX powers low, TCM is used for most renditions to achieve the required data rates. A summary of WiMedia modulation is given in Table 4-7. Note that the symbol rate is always 11 Mbaud, but the data rate varies with the number of bits encoded per symbol. The 11 Mb/s data rate is available to maintain the communication link under poor channel conditions.

$$\text{BER at 22 Mb/s} = Q(\sqrt{\text{SINR}}) \tag{4.25}$$

Table 4-6

802.15.1 Bluetooth modulation method (2.4 GHz band)

Data rate (Mb/s)	Modulation	Number of FHSS channels	Nominal hop rate
1	GFSK/FHSS	79	1600/s

Table 4-7

802.15.3 Wi-
Media modula-
tion methods
(2.4 GHz band)

Data rate (Mb/s)	Modulation	Coding	Symbol rate (Mbaud)	Bits per symbol
11	QPSK	8-state TCM	11	1
22	DQPSK	None	11	2
33	16-QAM	8-state TCM	11	3
44	32-QAM	8-state TCM	11	4
55	64-QAM	8-state TCM	11	5

IEEE 802.15.4 ZigBee

ZigBee was developed as an extremely low cost network for use in sensor and other applications requiring low data rates and highly asynchronous channel access. The data rates and modulation methods vary with each band and are given in Table 4-8.

$$\text{BER at 250 kb/s} = \frac{8}{15} \times \frac{1}{16} \times \sum_{k=2}^{16} -1^k \binom{16}{k} \exp\left[20\left(\frac{1}{k} - 1\right)\text{SINR}\right] \quad (4.26)$$

BER Comparisons

BER values for all these formulas are plotted in Figure 4-22 [IEEE802.15.4]. The best BER performance for SNR values above about −2 dB is achieved by 802.15.4 ZigBee due to the relatively slow data rate of 250 kb/s, which provides higher energy per data bit. The four 802.11b DSSS and CCK signals follow, with an SNR penalty of 1 to 2 dB for each step up in speed. Finally, 802.15.3 WiMedia and 802.15.1 Bluetooth require the highest SNR for a given BER due to different reasons. Bluetooth uses a noncoherent FSK demodulation method, which, despite the fairly low data rate, still exhibits a significant SNR penalty. Both 802.11b Wi-Fi at 2 Mb/s and WiMedia at 22 Mb/s use DQPSK, which requires a SNR 3 dB higher to match the BER of QPSK. The much lower energy per data bit means that 22 Mb/s WiMedia has an additional SNR penalty compared to 2 Mb/s Wi-Fi. Within the 802.11b Wi-Fi ensemble, higher data rates produce a higher BER for a given SNR, as expected.

Data rate (kb/s)	Band	Modulation	Spreading sequence length (chips)	Chip rate (kc/s)
20	868 MHz	DBPSK	15	300
40	900 MHz	DBPSK	15	600
250	2.4 GHz	OQPSK	32	2000

Figure 4-22

BER versus SNR
for the four
data rates of
802.11b Wi-Fi,
along with
values for
802.15.1
Bluetooth,
802.15.3
WiMedia, and
802.15.4
ZigBee, all
operating in
the 2.4 GHz
band
[IEEE802.15.4]
© IEEE

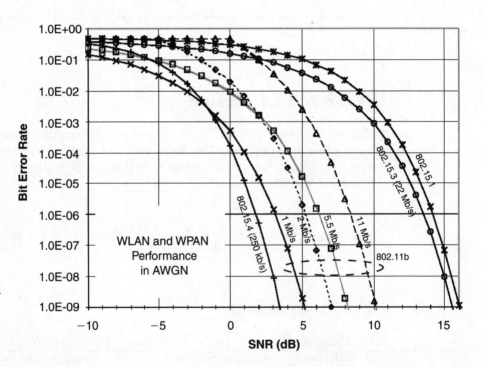

Summary

DSSS is used by 802.11b devices to spread signal energy over a larger bandwidth to allow higher TX power and still meet government regulations. The processing gain for low-rate communications is about 10 dB, providing some reduction of interference such as that from Bluetooth, but essentially no multiple-access capability exists. DSSS does help mitigate multipath from delays beyond that of about one chip period. CCK is used for faster data rates while producing TX PSD values equivalent to those from DSSS. CCK operates by encoding information into the spreading sequence, but the corresponding reduction in bit energy dictates shorter operating ranges for a given BER. ZigBee is aimed at low-speed data transfers over longer ranges, so DSSS is employed for all three data rates of 20, 40, and 250 kb/s, and all three bands of operation (868, 900, and 2,400 MHz). Four bits per symbol are sent at 250 kb/s by selecting one of 16 nearly orthogonal spreading sequences.

FHSS is used to reduce the detrimental effect of interference for Bluetooth systems. A total of 79 channels, each 1 MHz wide, is employed in a pseudorandom hop pattern. This allows several Bluetooth piconets, and other interfering signals as well, to coexist without a catastrophic disruption to Bluetooth communication.

OFDM is used by 802.11a in the 5.x GHz band and by 802.11g in the 2.4 GHz band to enable data rates up to 54 Mb/s within a relatively small bandwidth of about 16 MHz. This is done by modulating 48 subcarriers with relatively long symbol durations, along with using four pilot subcarriers for a channel estimation and synchronization aid. OFDM can combat multipath fading through interleaving and FEC. It can also prevent multipath-induced ISI from its slow symbol period and use its cyclic prefix to help maintain subcarrier orthogonality.

UWB signaling promises to accommodate very high data rates within an extremely large bandwidth, such that the transmitted signal's PSD per unit bandwidth is extremely low. Indeed, the UWB signal operates below the noise floor in most narrowband and conventional DSSS, FHSS, and OFDM receivers. For high-speed communication over a few meters, UWB promises to be an excellent choice given its capability to combat multipath and coexist with many other services.

References

[Che02] Cheng, J., and Beaulieu, N., "Accurate DS-CDMA Bit-Error Probability Calculation in Rayleigh Fading," *IEEE Transactions on Wireless Communications*, January 2002.

[Dix94] Dixon, R., *Spread Spectrum Systems with Commercial Applications*, New York: John Wiley & Sons, Inc., 1994.

[Hee00] Heegard, C., "High Performance Wireless Local Area Networks (WLAN) in the ISM, 2.4-GHz, Band," presented at the Wireless Symposium, Fall 2000.

[Hei02] Heiskala, J., and Terry, J., *OFDM Wireless LANs: A Theoretical and Practical Guide*. Indianapolis: SAMS Publishing, 2002.

[Hol92] Holtzman, J., "A Simple, Accurate Method to Calculate Spread-Spectrum Multiple-Access Error Probabilities," *IEEE Transactions on Communications*, March 1992.

[IEEE802.15.2] IEEE 802.15.2-2003, "Coexistence of Wireless Personal Area Networks with Other Wireless Devices Operating in Unlicensed Frequency Bands," August 28, 2003.

[IEEE802.15.4] IEEE 802.15.4-2003, "Wireless Medium Access Control (MAC) and Physical Layer (PHY) Specifications for Low-Rate Wireless Personal Area Networks (LR-PANS)," October 1, 2003.

[Lit01] Litwin, L., and Pugel, M., "The Principles of OFDM," *RF Design*, January 2001.

[Man03] Mandke, K., et al., "The Evolution of Ultra Wide Band Radio for Wireless Personal Area Networks," *High Frequency Electronics*, September 2003.

[Mil03] Miller, L., "Validation of 802.11a/UWB Coexistence Simulation," National Institute of Standards and Technology white paper, October 2003.

[Moo03] Moon, J., and Kim, Y., "Dual RX Boosts WLAN OFDM," *Electronic Engineering Times*, September 29, 2003.

[Mor98] Morrow, R., "Accurate CDMA BER Calculations with Low Computational Complexity," *IEEE Transactions on Communications*, November 1998.

[Pea00] Pearson, B., "Complementary Code Keying Made Simple," Intersil Application Note AN9850.1, May 2000.

[Sho01] Shoemake, M., "Wi-Fi (IEEE 802.11b) and Bluetooth: Coexistence Issues and Solutions for the 2.4 GHz ISM Band," Texas Instruments white paper, Version 1.1, February 2001.

[Skl88] Sklar, B., *Digital Communications: Fundamentals and Applications*, New Jersey: Prentice Hall, 1988.

[Sol03] Soltanian, A., "Coexistence of Ultra-Wideband System and IEEE 802.11a WLAN," National Institute of Standards and Technology white paper, revised by L. Miller, October 2003.

[TDC00] "Time Modulated Ultra-Wideband for Wireless Applications," Time Domain Corporation white paper, revision 2, May 2000.

[Wag01] Wagner, K., "Use OFDM to Compress Large Data Rates into Small Bandwidths," *Portable Design*, March 2001.

[Zha02] Zhao, L., and Haimovich, A., "Performance of Ultra-Wideband Communications in the Presence of Interference," *IEEE Journal on Selected Areas in Communications*, December 2002.

Radio Performance

The radio, which comprises a major part of the *physical* (PHY) layer of the *Open Systems Interconnection* (OSI) model, is a marvel of high technology. Not only does it extract data at a rate of several megabits per second from an extremely weak incoming signal, but it must work in a hostile electromagnetic environment in which noise and interference abound. When the first Wi-Fi products were released in the 1990s, meeting radio performance criteria dictated the high component count, and thus the high cost, of these products. With the development of *complementary metal oxide semiconductor* (CMOS) technology that is fast enough to operate in the 2.4 and 5.x GHz bands, the radio can be integrated into the same substrate in which digital processing takes place. Inexpensive, single-chip solutions for *wireless local area network* (WLAN) and *wireless personal area network* (WPAN) nodes are now achievable.

The radio consists of two major units, the transmitter and the receiver. These were designed as two separate entities in early wireless systems such as analog cellular, because full-duplex operation required both the transmitter and receiver to be operating at the same time. With the advent of fast *transmit-receive* (TX-RX) switching times, or *turnaround*, TX and RX units can now share items such as the frequency synthesizer between them, reducing cost and space.

Despite the importance of the radio as an essential component in wireless communication, WLAN and WPAN specifications are surprisingly sparse when discussing this part of the system. The specifications dictate *what* the radio must do, but is silent on *how* those requirements are to be implemented. Many of these requirements are actually quite difficult to accomplish, so manufacturers have developed a substantial amount of intellectual property associated with the design of an inexpensive radio.

The radio equipment is also expected to meet government regulations on TX power, RX and TX spurious outputs, and other criteria. Because these rules can be different in different regulatory domains, we mention them only if they are listed in the respective specification. Although the corresponding specifications try to conform to as many regulatory domains as possible, many of the rules are conflicting and don't directly affect the capability of the wireless device to meet a user's expectations. Conformance is still required, even though the criteria may not appear in the corresponding specification.

While designing a short-range wireless system, it's tempting to build a transmitter with the highest power possible. This can be detrimental to coexistence, especially if the other nodes competing for channel access also have high TX power, but this is only one of many associated costs:

- In moderate clutter, increasing the range by a factor of 4 requires increasing transmit power by a factor of about 100.
- Both transmitters on the link must have higher power before the range can increase (assuming identical receiver performance).

- Interference potential to other users increases rapidly with increased transmit power.

- Components in the transmit circuitry become more complex and expensive.

- Transmit power control and its added complexity is required for levels higher than +20 decibels relative to 1 *milliwatt* (mW), or dBm (802.11b), +4 dBm (802.15.1), or 0 dBm (802.15.3).

- Battery life can be significantly shortened.

- *Radio frequency* (RF) power levels higher than a few milliwatts can severely interfere with the operation of digital and analog circuits in the near field.

In this chapter we present some of the requirements that the WLAN and WPAN radios must meet to operate properly. We won't necessarily cover all the TX and RX requirements for radio certification, but only those that have relevance to coexistence issues. Wi-Fi was conceived as a WLAN for the office environment, so cost considerations often took a secondary role to performance. Bluetooth was envisioned to be an inexpensive cable replacement technology for ubiquitous use, so cost became a primary concern. Unlike the Wi-Fi specification, which defers many performance criteria to the applicable government regulatory domain or manufacturer, the Bluetooth specification contains considerably more detailed requirements for the radio. Perhaps this is a result of the extreme cost pressures anticipated for the Bluetooth cable replacement applications, so the *Bluetooth Special Interest Group* (BSIG) may have been concerned about the potential for compromises in the radio design.

It's a bit ironic to note that extreme market pressures on both technologies have resulted in low-cost Wi-Fi nodes that still perform very well, and Bluetooth devices perform much better than the specification dictates but still cost very little. We can expect to see the same cost and performance measures to be addressed as 802.15.3 WiMedia and 802.15.4 ZigBee radio devices begin to proliferate.

The 802.11 Wi-Fi Radio

As a WLAN, 802.11 is usually expected to operate over a longer distance than WPAN, so the devices use more TX power. Since the *access point* (AP) is expected to always be present to serve the *station* (STA) units as they come and go, the AP is almost always powered by the mains rather than by a battery. Traditionally, the radio in a WLAN node is designed with performance ahead of cost.

802.11 Channel Sets

IEEE 802.11b and 802.11g occupy the 2.4 GHz band, with a channel set as shown in Table 5-1 for the United States and the countries of Europe that fall under *European Telecommunications Standards Institute* (ETSI) rules. Thirteen of these channels are spaced 5 MHz apart, with an occupied bandwidth of approximately 22 MHz for 802.11b and 16.6 MHz for 802.11g. The carrier frequency assigned to channel 14 for use in Japan is 12 MHz, not 5 MHz, higher than channel 13. Different subsets of these channels are available for use in other countries.

Nodes operating on adjacent channels have significant bandwidth overlap. Aside from the obvious interference potential, network traffic on adjacent channels may also activate a receiver's carrier-sense mechanism and reduce throughput considerably. Because of this, multiple Wi-Fi deployments in the same area typically operate on nonoverlapping channels, designated as channels 1, 6, and 11 in the United States, and channels 1, 7, and 13 in Europe. Figure 5-1 shows how three 802.11b transmissions fit into the U.S. version. By restricting network

Table 5-1

802.11b/g channel set in the United States and most of Europe

Channel number	Center frequency (MHz)	USA	ETSI	Nonoverlapping channel set
1	2,412	x	x	USA/ETSI
2	2,417	x	x	
3	2,422	x	x	
4	2,427	x	x	
5	2,432	x	x	
6	2,437	x	x	USA
7	2,442	x	x	ETSI
8	2,447	x	x	
9	2,452	x	x	
10	2,457	x	x	
11	2,462	x	x	USA
12	2,467		x	
13	2,472		x	ETSI
14	2,484			

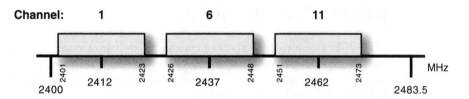

Figure 5-1
Nonoverlapping Wi-Fi
channel set in
the U.S. 2.4
GHz ISM band.
TX bandwidth
is approximately 22 MHz
for 802.11b
(shown) and
16.6 MHz for
802.11g.

operation to the nonoverlapping channel set, APs that are close to each other can operate without mutual interference, and those with greater physical separation can be assigned the same channel for frequency reuse.

For 802.11a, the 5.x GHz *industrial, scientific, and medical* (ISM) and *Unlicensed National Information Infrastructure* (U-NII) bands can accommodate a total of 24 nonoverlapping channels in the United States (see Figure 5-2). Channel numbers in these bands are assigned using the formula $5,000 + 5n$ MHz. Channel 165, operating within the 5.x GHz ISM band, isn't listed in the 802.11a specification, but it often appears in the channel set provided by manufacturers of 802.11a devices for sale within the United States. In Europe, the operation of approved devices is allowed in the lower three 5.x GHz bands for a total of 19 nonoverlapping channels.

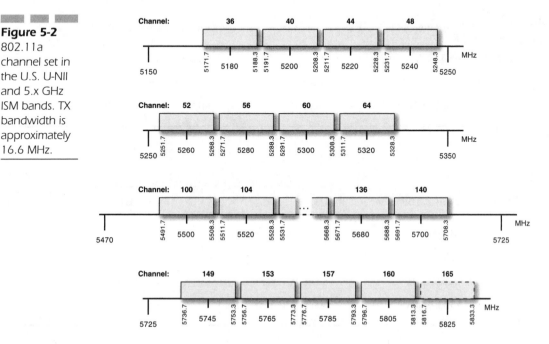

Figure 5-2
802.11a
channel set in
the U.S. U-NII
and 5.x GHz
ISM bands. TX
bandwidth is
approximately
16.6 MHz.

802.11 TX Performance

The Wi-Fi transmitter exists, of course, as an intentional radiator of electromagnetic signals within the 2.4 and 5.x GHz bands. The transmitter must produce a signal that conforms to the modulation requirements in the band of interest while limiting spurious output products both within and outside this band.

Power Levels The maximum TX power level depends on the regulatory domain in which the Wi-Fi device is operating, and these power levels are given in Table 5-2 for the 2.4 GHz band. The minimum TX power is 1 mW (0 dBm), and power control is required for levels greater than 100 mW. Wi-Fi power control is meant to be implemented easily, with a maximum of only four power levels required. As a minimum, a transmitter capable of operating with a power level above +20 dBm need only switch its power level down to 100 mW or less to conform to this requirement.

For the 5.x GHz bands, maximum TX power levels allowed by the FCC and by the European Conference of Postal and Telecommunications Administration are given in Table 5-3. In the United States, these power levels are based on a 16 MHz *orthogonal frequency division multiplexing* (OFDM) bandwidth, and *equivalent isotropic radiated power* (EIRP) can be further increased by using an antenna with up to 6 *decibels relative to an isotropic source* (dBi) of gain. An

Table 5-2

Maximum TX power levels in the 2.4 GHz band

TX power level	Location	Compliance document
1000 mW output	USA	FCC 15.247
100 mW EIRP	Europe	ETS 300-328
10 mW/MHz	Japan	MPT (Radio) article 49-20

Table 5-3

Maximum TX power levels in the 5.x GHz bands

Band (GHz)	Maximum TX output power	
	USA	**ETSI**
5.15–5.25	40 mW (2.5 mW/MHz)	200 mW
5.25–5.35	200 mW (12.5 mW/MHz)	200 mW
5.470–5.725	200 mW (12.5 mW/MHz)	1 W
5.725–5.825	800 mW (50 mW/MHz)	----

802.11a device must operate within at least one of these bands to be certified. For ISM channel 165, maximum TX power in the United States is 1 W (+30 dBm).

Spectrum Mask The TX spectrum mask is provided in the Wi-Fi specification to serve as an absolute limit on the transmitted signal's *power spectral density* (PSD). For testing, a spectrum analyzer can be attached to the antenna port and the plot compared to the mask. This mask for *direct sequence spread spectrum* (DSSS) and *complimentary code keying* (CCK) Wi-Fi is shown in Figure 5-3. The TX spectral products must be less than −30 *dB relative to the center peak* (dBr) for the second sinc(*x*) lobe between 11 and 22 MHz on either side of the carrier, and −50 dBr beyond 22 MHz on either side of the carrier frequency.

For OFDM, the spectrum mask has a different appearance, as shown in Figure 5-4. The mask is built around the highest amplitude that each carrier will

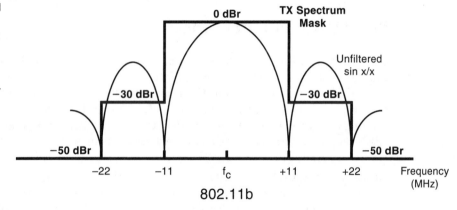

Figure 5-3
Spectrum mask for Wi-Fi DSSS and CCK signaling

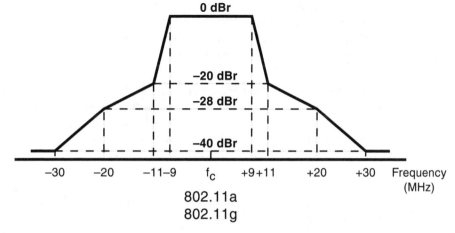

Figure 5-4
Spectrum mask for Wi-Fi OFDM signaling

achieve, which will be the case when they are *binary phase shift keying* (BPSK) or *quadrature phase shift keying* (QPSK) modulated. The −20 dBr bandwidth is 22 MHz, eventually dropping to −40 dBr at 30 MHz on either side of the center frequency.

Other TX Requirements   Modern WLAN transmitters must have the capability of turning on and off very rapidly to enable the fast duplexing of data or audio information. For DSSS and CCK operation in the 2.4 GHz band, the 10 to 90 percent TX power-on ramp and the 90 to 10 percent power-down ramp must each be no greater than 2 *microseconds* (μs). Ramp times for 802.11a/g OFDM are implementation dependent as long as required turnaround times and emissions regulations are met. There is a limit, though, to how fast the TX signal can ramp up and down because these transients can produce unacceptable in-band and out-of-band spurious emissions.

802.11 RX Performance

The receiver is the most difficult to design component in the PHY layer, and it must be able to extract and demodulate the desired signal within the required error limits even when it is in the midst of interference, some of which may be several orders of magnitude more powerful. The bottom-line performance of any digital receiver is its *bit error rate* (BER), also manifested in a *packet error rate* (PER) or *frame error rate* (FER) for packets or frames of a specified size.

Standard Frame Error Rate   For many of the measurements associated with 802.11b receiver performance, a standard FER of not more than 8×10^{-2} has been specified for a *medium access control* (MAC) frame length of 1,024 bytes. For low-rate data, the frame is sent at 2 Mb/s using *differential QPSK* (DQPSK), and for high-rate data the frame is sent at 11 Mb/s using CCK. (See Chapter 6, "Medium Access Control," for how MAC frames are structured.) The standard FER for OFDM in the 2.4 and 5.x GHz band is 10^{-1} for a frame length of 1,000 bytes sent at the speed being measured.

Sensitivity   The receiver sensitivity has both high and low required levels listed in the Wi-Fi specification. The minimum sensitivity is the weakest signal that the receiver must be able to demodulate without exceeding the standard FER. This limit is dictated in part by the noise figure of the receiver, which is the *signal-to-noise ratio* (SNR) degradation that occurs as the signal passes through the receiver. Although we saw in earlier chapters that the theoretical SNR limits of detection in *additive white Gaussian noise* (AWGN) are quite low, actual receivers, especially when cost is a major consideration, are less capable due to their relatively high noise figure.

The required maximum RX signal power sensitivity is based on the capability of the receiver to respond to large input signal levels without component saturation. Receivers have an *automatic gain control* (AGC) to vary the gain of the early stages in the RX amplifier chain to prevent overload later in the chain, but the AGC has its own operating limits. Required lower and upper sensitivity limits are given in Table 5-4 for low-rate and high-rate 802.11b. The minimum sensitivity exhibits the usual rise as the data rate (and hence modulation complexity) increases. A signal exceeding the maximum sensitivity level is extremely powerful, usually occurring when the TX antenna is only a few centimeters from the RX antenna.

Receiver sensitivity limits for 802.11a/g OFDM operation are given in Table 5-5. Signal levels are at the antenna connector, and a receiver noise figure of 10 dB and an implementation margin of 5 dB are assumed. Again, as the data rate increases, the standard FER is allowed to be reached with a stronger signal at the receiver. The effect is much more noticeable for 802.11a/g due to the high precision needed to properly demodulate OFDM. The minimum *received signal strength indication* (RSSI) for a properly decoded 54 Mb/s signal can be about

Table 5-4

802.11b receiver sensitivity limits

Modulation	Required minimum sensitivity	Required maximum sensitivity
2 Mb/s DSSS	−80 dBm	−4 dBm
11 Mb/s CCK	−76 dBm	−10 dBm

Table 5-5

802.11a/g receiver sensitivity limits

Modulation	Required minimum sensitivity	Required maximum sensitivity 802.11a	802.11g
6 Mb/s OFDM	−82 dBm	−30 dBm	−20 dBm
9 Mb/s OFDM	−81 dBm	−30 dBm	−20 dBm
12 Mb/s OFDM	−79 dBm	−30 dBm	−20 dBm
18 Mb/s OFDM	−77 dBm	−30 dBm	−20 dBm
24 Mb/s OFDM	−74 dBm	−30 dBm	−20 dBm
36 Mb/s OFDM	−70 dBm	−30 dBm	−20 dBm
48 Mb/s OFDM	−66 dBm	−30 dBm	−20 dBm
54 Mb/s OFDM	−65 dBm	−30 dBm	−20 dBm

50 times stronger than the minimum RSSI needed for the accurate decoding of data sent at 6 Mb/s. The maximum sensitivity limit for 802.11a/g is considerably lower than that for 802.11b due to the detrimental effect of receiver nonlinearities on OFDM detection.

Interference Immunity   As interference in the 2.4 GHz band becomes more prevalent, the capability of a receiver to selectively detect the desired signal and reject the others will become more and more critical to proper operation of the network. Interference performance generally falls into the following categories: *carrier-to-interference ratio* (C/I), blocking, image rejection, intermodulation distortion, and spurious responses. Of these, only the first is directly addressed in the Wi-Fi specification. The others are indirectly addressed in the C/I performance figures.

C/I is the desired (carrier) power in dBm minus the unwanted (interfering) power in dBm when both are present at the receiver's antenna input. The interfering signal can be at the same carrier frequency as the desired signal, which is termed *co-channel C/I*, or in an adjacent channel, either overlapping or nonoverlapping. The latter is called *adjacent channel C/I* at some specified carrier offset. The Wi-Fi specification includes performance requirements for nonoverlapping adjacent channel C/I.

To measure low-rate adjacent channel C/I, for example, a −74 dBm, 2 Mb/s DQPSK signal is combined at the receiver's antenna with an interfering signal offset at least 30 MHz at a power level of −39 dBm and also using 2 Mb/s DQPSK. With the adjacent channel C/I thus set at −35 dB, the desired FER cannot exceed the standard. For CCK signals, the adjacent channel is offset 25 MHz instead. No co-channel C/I values are given in the Wi-Fi specification because it's assumed that any nearby co-channel signal will interfere with WLAN operations. As we'll see in the next section and in following chapters, this assumption may or may not be valid, and it depends partly upon how the receiver assesses whether or not the channel is clear.

The adjacent channel C/I requirements for 802.11b are given in Table 5-6. The numbers are expressed in negative dB to indicate that the interfering signal is stronger than the desired signal, but because the interference is in an adjacent channel the receiver is required to reject it sufficiently to maintain the standard FER or better. The Wi-Fi specification lists these numbers as positive dB, so they

Table 5-6

802.11b receiver C/I limits

Input signal	25 MHz adjacent channel C/I limit	30 MHz adjacent channel C/I limit
2 Mb/s DSSS	—	−35 dB
11 Mb/s CCK	−35 dB	—

represent adjacent channel rejection values instead. However, we'll be consistent with cellular and Bluetooth terminology and change the sign to represent C/I. The low-rate (2 Mb/s) adjacent channel is offset by 30 MHz, which conforms to the spacing of the ETSI nonoverlapping channel set, and the high-rate (11 Mb/s) adjacent channel is offset by 25 MHz, corresponding to the spacing of the U.S. nonoverlapping channel set.

As was the case for minimum RX sensitivity values, 802.11a OFDM in the 5.x GHz band and 802.11g OFDM in the 2.4 GHz band have adjacent channel C/I values that gradually loosen as the data rate increases, as shown in Table 5-7. The 802.11a specification defines *adjacent channel C/I* as an interfering OFDM signal in the disjoint channel 20 MHz away from the desired channel. The specification also defines *alternate adjacent channel C/I* as adjacent channel C/I with the interfering signal 40 MHz or more away. The required C/I values for 6 and 54 Mb/s are depicted in Figure 5-5. (Only one of the interfering signals needs to be present for testing FER performance.) For 802.11g, the adjacent channel C/I limits are the same, but the adjacent channel center frequency is offset 25 MHz from the desired channel to conform to the U.S. nonoverlapping channel set. The alternate adjacent channel C/I limits in Table 5-7 don't apply to 802.11g.

Blocking Immunity Although Table 5-7 and Figure 5-5 imply a higher allowable interfering signal power as its carrier frequency moves further away from the desired carrier frequency, common sense tells us that eventually the interfering signal will be powerful enough to cause a general detrimental effect on desired receiver performance regardless of the interfering signal's carrier frequency. The front-end in most receivers contains a wideband amplifier that will saturate when exceptionally strong signals appear within its passband.

Table 5-7

802.11a/g receiver C/I limits

Input signal	20 MHz (802.11a) or 25 MHz (802.11g) adjacent channel C/I limit	≥40 MHz alternate adjacent channel C/I limit (802.11a only)
6 Mb/s OFDM	−16 dB	−32 dB
9 Mb/s OFDM	−15 dB	−31 dB
12 Mb/s OFDM	−13 dB	−29 dB
18 Mb/s OFDM	−11 dB	−27 dB
24 Mb/s OFDM	−8 dB	−24 dB
36 Mb/s OFDM	−4 dB	−20 dB
48 Mb/s OFDM	0 dB	−16 dB
54 Mb/s OFDM	+1 dB	−15 dB

Figure 5-5
Required
802.11a
interference
rejection for
data rates of 6
and 54 Mb/s

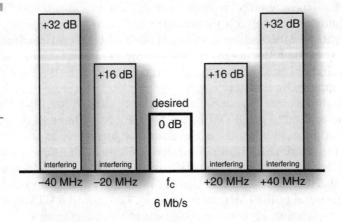

6 Mb/s

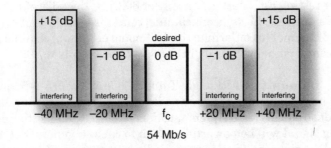

54 Mb/s

Table 5-8

Strengths and
weaknesses of
802.11b CCA
modes

Mode	Strength	Weakness
1	Responds to all nearby wireless signals	May falsely indicate channel is busy
2	Responds to other 802.11b low-rate signals	May cause interference to non-802.11b users
		May be too sensitive to other distant 802.11b networks
3	Responds to other 802.11b low-rate signals	May cause interference to non-802.11b users
4	Prevents interfering with longest high-rate frame	Adds unnecessary delay for shorter frames
		May be too sensitive to other distant 802.11b networks
5	Prevents interfering with high-rate signals	May cause interference to non-802.11b users

Amplifier or filter stages beyond the front end may saturate as well. When this happens, the receiver becomes overwhelmed and can't properly amplify and detect a weak desired signal. This effect is called *blocking*.

The Wi-Fi specification doesn't directly address the blocking issue beyond listing a maximum required sensitivity level, so we'll need to make some assumptions to decide under what conditions the receiver is blocked. It is perhaps reasonable to consider the receiver blocked when the C/I reaches the highest negative value in its applicable specification, because the standard FER will probably not be met under these conditions. Also, the receiver will most likely be blocked if the interference level is at or above the maximum receiver sensitivity figure.

Clear Channel Assessment (CCA)   The capability of the Wi-Fi receiver to detect the presence or absence of a signal is vital to efficient network operation using *carrier sense multiple access* (CSMA). The purpose of the *clear channel assessment* (CCA) is to determine whether the channel is busy or idle. For example, CCA might be programmed to react to any signal above an *energy detect* (ED) threshold. CCA can be considered a Boolean variable; thus, when CCA equals *true,* the channel is idle, and when CCA is *false* it indicates a busy channel. The definition of a busy channel is, however, not as simple as it might seem. As we'll discover in Chapter 9, "Coexisting with Other Services," a microwave oven emits energy in the 2.4 GHz band, but the 802.11b/g network shouldn't consider these emissions as an indication of a busy channel.

The 802.11b specification defines several modes of CCA operation, any or all of which can be incorporated into the node, depending on the type of signal being considered. These are as follows:

- **Mode 1: Energy above threshold (low- and high-rate data)**   CCA reports a busy medium upon detection of any signal energy above the ED threshold.

- **Mode 2: Carrier sense only (low-rate data)**   CCA reports a busy medium only upon detection of a DSSS signal, either above or below the ED threshold.

- **Mode 3: Carrier sense with energy above threshold (low-rate data)**   CCA reports a busy medium upon detection of a DSSS signal above the ED threshold.

- **Mode 4: Carrier sense with timer (high-rate data)**   CCA starts a 3.65 ms timer upon detection of a high-rate data signal. Upon timer expiration, CCA reports an idle medium if no high-rate signal is still present. This timer accommodates the longest packet sent at 5.5 Mb/s.

- **Mode 5: Carrier sense with energy above threshold (high-rate data)**   CCA reports a busy medium upon detection of a high-rate data signal with energy above the ED threshold.

Modes 1, 2, or 3 can be used when communicating at 1 or 2 Mb/s, whereas data exchanged at 5.5 or 11 Mb/s requires using modes 1, 4, or 5. Each of the five CCA modes has strengths and weaknesses, some of which are listed in Table 5-8. Most 802.11b implementations use mode 3 at 1 or 2 Mb/s and mode 5 at 5.5 and 11 Mb/s. In other words, the channel is considered busy when a valid 802.11b signal with energy above the ED threshold is present.

As we'll discover in Chapter 6, Wi-Fi data frames begin with a preamble sent at 1 Mb/s containing a code that designates the frame as either low rate or high rate. CCA modes 1, 2, and 3 respond to the low-rate data preamble, whereas modes 1, 4, and 5 will respond to the high-rate data preamble. Furthermore, both low-rate and high-rate 802.11b frames contain a length field in their headers, which indicates the time needed to transmit that particular frame. If this field is correctly received, the CCA is required to assert that the channel is busy for the expected frame duration (the virtual channel sense), even if the actual carrier sense is lost before the time has expired. Generally, the CCA mechanism for 802.11b devices must respond within 15 μs to the presence of a signal under any of the different modes of CCA operation. If the radio has just completed a transmission, CCA is allowed up to 10 μs for *transmit-to-receive* (T/R) turnaround and up to 15 μs additional time for energy detection. Incidentally, these response times are fast enough to sense the presence of the shortest Bluetooth packet, which requires 68 μs to transmit, provided that the hop frequency is within the Wi-Fi receiver's passband, the RSSI is above the ED threshold, and CCA Mode 1 is used.

The ED threshold varies depending on the data rate and node's corresponding TX power level according to Table 5-9. It's interesting that the required ED level decreases as the TX power level increases at the same node. The reason is that the node's higher TX power has the potential to interfere with other networks over a greater distance, so the node must in turn be willing to sense that the channel is busy when a weaker signal is present.

For 802.11a, the preamble portion of a frame is transmitted at the minimum rate of 6 Mb/s, and the CCA must indicate a busy channel within 4 μs with a probability of 90 percent for an RSSI of −82 dBm or higher. This corresponds to the minimum sensitivity required for the 802.11a receiver in the presence of a 6 Mb/s OFDM signal. If the preamble is missed by the receiver, the CCA must indicate that the channel is busy if any signal appears with an RSSI of −62 dBm or higher.

Table 5-9

802.11b ED threshold values

TX power	ED for low-rate data	ED for high-rate data
Above 100 mW	−80 dBm or less	−76 dBm or less
>50 mW to 100 mW	−76 dBm	−73 dBm
50 mW or less	−70 dBm	−70 dBm

For 802.11g, the CCA situation is somewhat complex because these devices must not only communicate among themselves using DSSS, CCK, or OFDM, but they also must be able to communicate with 802.11b devices and coexist with them without causing excessive disruption. Therefore, the 802.11g radio must have the capability to assert a busy channel condition whenever any low-rate, high-rate, or OFDM preamble is present on the operating channel and the ED threshold of -76 dBm is exceeded. When 802.11b transmissions are present on the network, the 802.11g node CCA response time must not exceed 15 μs, and the CCA must indicate a busy channel with 99 percent probability. If only 802.11g OFDM signals are present, the CCA response time must not exceed 4 μs, and the CCA must indicate a busy channel with 90 percent probability. As with 802.11b devices, if the frame length field is successfully decoded, the CCA must indicate that the channel is busy for the expected frame duration, even if the actual carrier sense is lost before the time has expired.

Turnaround Times

Successful communication between nodes in a WLAN requires careful timing coordination between the transmitting node and receiving node, and other nodes must remain silent during the exchange. The *receive-to-transmit* (R/T) and T/R turnaround times are functions of the radio that set a limit on the minimum time a sending node waits for an *acknowledgment* (ACK) to appear, for example, or the speed at which synchronous packet exchange can occur. Unlike Bluetooth FHSS, 802.11a/b/g doesn't change its carrier frequency each time a packet is sent or received, so both R/T and T/R switching can be quite fast.

For 802.11b, the R/T turnaround time is 5 μs or less, including any TX power-up ramp. The T/R turnaround time must be less than 10 μs, including the TX power-down ramp. If the channel agility option is enabled, the time to change from one operating frequency to another can't exceed 224 μs. For 802.11a, the R/T turnaround time is more stringent at 2 μs maximum. No specific value for the T/R turnaround time is given in the specification, but the receiver must be ready to provide a channel assessment within 4 μs of the appearance of an incoming signal after the transmitter has switched off. For 802.11g, the RX-TX turnaround time is 5 μs or less, and the T/R turnaround time is 10 μs or less.

The 802.15.1 Bluetooth Radio

In some ways, the Bluetooth radio presents the greatest design challenge of all the WLAN and WPAN radios discussed here for a number of reasons. First, the Bluetooth implementation of FHSS requires that the radio change its channel a nominal 1,600 times per second (3,200 times per second during inquiry and page functions), placing significant demands upon the RF synthesizer. Second, co-channel and adjacent channel C/I figures for the Bluetooth radio are somewhat stringent to provide increased coexistence capabilities with other users, including other Bluetooth piconets, in the 2.4 GHz band. Finally, pressure is intense to keep costs low, and many manufacturers have responded by placing both the radio and baseband protocols onto a single CMOS chip. Making an analog radio work with a digital baseband on the same substrate is itself a significant coexistence challenge!

802.15.1 Channel Set

Bluetooth devices in most regulatory domains use 79 hop channels within the 2.4 GHz ISM band conforming to the formula

$$f_k = 2402 + k \text{ MHz} \tag{5.1}$$

where k is the channel number as an integer from 0 to 78. This set is depicted graphically in Figure 5-6. Some countries, such as France, have required a reduced 23-channel hop set under certain conditions, but the BSIG is working hard to standardize all Bluetooth communication to 79 hop channels. If they are successful, one rendition will be legal in any regulatory domain.

802.15.1 TX Performance

The Bluetooth transmitter is, of course, also an intentional radiator of electromagnetic signals within the 2.4 GHz ISM band. The Bluetooth specification devotes a few pages to TX performance values, which are selected to meet government regulations for coexistence while allowing for low manufacturing costs.

Figure 5-6
Bluetooth FHSS 79-channel set. The TX bandwidth in each channel is approximately 1 MHz.

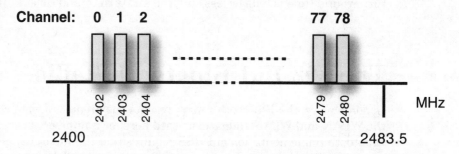

Power Classes Bluetooth transmitters fall into three basic classes determined by their maximum power output. The class 1 transmitter is the most powerful at 100 mW (+20 dBm) maximum power, the class 2 transmitter has a maximum power of 2.5 mW (+4 dBm), and the class 3 transmitter produces 1 mW (0 dBm). To conserve battery power, most Bluetooth devices contain class 2 or class 3 transmitters. Class 1 units are better suited for clients such as laptop and desktop computers, and in fixed servers such as printers and access ports. Class 1 transmitters must have a power control feature to reduce the power to a level adequate for communication in order to prevent excessive interference to other users in the band. The other two classes don't require power control, but it can be implemented as an optional feature.

The class 2 Bluetooth device is allowed to have a maximum power output of +4 dBm, with a "nominal" power output of 0 dBm and a minimum power (at the maximum power setting, if power control is implemented) of −6 dBm. The maximum power output for the class 3 Bluetooth transmitter is 0 dBm. Power control is optional but will probably not find its way into very many class 2 and class 3 Bluetooth devices.

The maximum transmit power output for class 1 Bluetooth devices is up to +20 dBm, and power control is required down to +4 dBm, or lower if desired. The lower power limit is suggested to be −30 dBm (1 μW), but TX power control is not mandatory for levels below +4 dBm. (According to our analysis in Chapter 2, "Indoor RF Propagation and Diversity Techniques," a transmit power of −30 dBm will result in a nominal range of about 1 meter.) All these power levels are summarized in Table 5-10.

Power control is implemented using a feedback mechanism between the master and a slave in the piconet. In describing its operation, let's assume that the master's transmitter power is being adjusted by a slave. For power control to be possible, the master's transmitter must have the capability to change its power level automatically, and the slave's receiver must have a calibrated RSSI. Furthermore, there must be a means for the slave to direct the master to adjust

Table 5-10

802.15.1 TX power class summary

Power class	Maximum TX power P_{max}	Maximum TX power at least	Suggested minimum TX power P_{min}	TX power control
1	100 mW +20 dBm	1 mW 0 dBm	1 μW −30 dBm	+4 dBm to P_{max} Optional from P_{min}
2	2.5 mW +4 dBm	0.25 mW −6 dBm	1 μW −30 dBm	Optional P_{min} to P_{max}
3	1 mW 0 dBm	n/a	1 μW −30 dBm	Optional P_{min} to P_{max}

its power, either up or down. This is done using *link manager protocol* (LMP) packets.

A Bluetooth power-controlled transmitter must have the capability to adjust its output level in steps that range in size between 2 and 8 dB. The adjustment range should be between +4 dBm (or, optionally, lower) and the maximum power level (up to +20 dBm). As an example, suppose a transmitter's maximum power output is +20 dBm. The power control requirement could be met by implementing a step size of 8 dB, in which case minimum performance would be met by providing only three power levels: +20 dBm, +12 dBm, and +4 dBm. If a step size of 2 dB is used instead, then nine power levels would be needed to cover the same span.

A receiver participating in the power control process attempts to place the incoming signal's power level within the *Golden Receive Power Range*. The lower range is selected to be a specific value between −56 dBm and 6 dB above the actual sensitivity of the receiver. For example, if a receiver has a sensitivity of −80 dBm, the lower threshold of the range is between −74 and −56 dBm. The upper threshold is 20 dB above the lower threshold to an accuracy of ±6 dB. See Figure 5-7 for a graphical display of the Golden Receive Power Range.

Continuing with our example, suppose the slave's Golden Receive Power Range is between −60 and −40 dBm, and the master's transmit power step size is 8 dB. Now assume that the master sends a packet to the slave, and the slave's RSSI shows −65 dBm. The slave will then return an LMP packet to the master, asking it to raise its transmit power by one step. Assuming no changes occur in the propagation path, the slave receiver's RSSI will now read −53 dBm upon arrival of the next packet from the master, which is within the desired range. Of course, power control can also compensate for the changes in path loss as the master and slave change positions relative to each other.

Figure 5-7
The Golden
Receive Power
Range

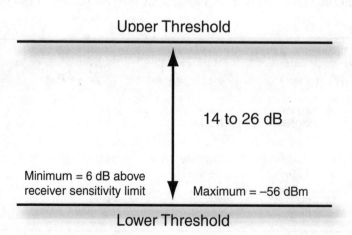

Upper Threshold

14 to 26 dB

Minimum = 6 dB above
receiver sensitivity limit

Maximum = −56 dBm

Lower Threshold

If a class 1 Bluetooth device is communicating with a device without RSSI, then power control cannot be accomplished. When the class 1 device discovers the lack of RSSI capability at the other end of the link, it must reduce its transmitter power to class 2 or 3 levels to reduce interference to users outside the piconet. This discovery is made during the early stages of the link setup through a special LMP "supported features" packet. A device without RSSI probably has a class 2 or 3 transmitter, so if the class 1 device can hear the signal then higher power levels are usually not needed for reliable communication.

Spurious Emissions   As mentioned earlier, an intentional radiator such as a Bluetooth transmitter also produces unintentional radiation at various frequencies, and these spurious emissions are restricted by various governments to prevent interference to other users. In an attempt to satisfy the rules in as many regulatory domains as possible, the Bluetooth specification offers detailed spurious output requirements. Furthermore, unlike most Wi-Fi implementations, Bluetooth devices will often be collocated with other wireless devices such as cell phones and GPS receivers, both of which have sensitive receivers, so tight control of spurious emissions is a necessity.

Spurious emissions can occur either within the 2.4 GHz band or outside it. Within adjacent channels, power levels in any 100 kHz band segment must be -20 *decibels relative to the carrier* (dBc) compared to the 100 kHz band segment within the transmitter's channel that contains the highest power level when the signal is modulated with pseudorandom data. For example, if f_c equals 2,410 MHz, then the transmitter is intentionally radiating between 2,409.5 and 2,410.5 MHz. Suppose the highest-power 100 kHz band segment within that channel contains 5 dBm of power measured at the antenna connector. Then power levels in any 100 kHz segment within the two adjacent channels (2408.5– 2409.5 MHz and 2410.5–2411.5 MHz) are required to be below -15 dBm.

Adjacent-channel spurious emissions must total less than -20 dBm within a channel that is 2 MHz away from the tested transmitter's channel, and be less than -40 dBm for channels that are 3 or more MHz away. These power levels are the total within the entire 1 MHz bandwidth of the channel. An exception is made for up to three channels that are 3 or more MHz away, and these power levels can total up to -20 dBm in each channel. In all cases, applicable government regulations must also be met.

For out-of-band spurious emissions, measurements are made using a 100 kHz bandwidth throughout a wide range of frequencies, and the resulting power levels cannot exceed those in Table 5-11. The last two entries are additional restrictions placed upon some of the frequencies covered in the second entry of the table. The *Personal Communications Service* (PCS) operates in the 1.8 GHz band, and the *Mobile Satellite Services* (MSS), which are satellite phones, use the 5.15 GHz band. Bluetooth-enabled, hands-free headsets are useful in phones that employ either of these services, so spurious emissions from the Bluetooth transmitter must be extremely low to avoid interference.

Table 5-11

802.15.1 transmitter out-of-band spurious emission limits

Frequency band	Operating mode	Idle mode
30 MHz–1 GHz	−36 dBm	−57 dBm
1 GHz–12.75 GHz	−30 dBm	−47 dBm
1.8 GHz–1.9 GHz	−47 dBm	−47 dBm
5.15 GHz–5.3 GHz	−47 dBm	−47 dBm

These spurious emissions are absolute (dBm) and not relative (dBc). This factor places an additional burden on class 1 transmitters, because their relative spurious emissions limits are 20 dB (100 times) less than the limits from a class 3 transmitter.

802.15.1 RX Performance

The circuitry in the receiver contains the greatest amount of intellectual property within the Bluetooth hardware set provided by most manufacturers. Unlike Wi-Fi, where FER is used as the RX performance measure, the Bluetooth receiver's performance is based on the raw BER. One reason for this is that Bluetooth packets have optional error correction coding, and the packet structure is different depending on whether data or real-time, two-way voice is being communicated. Using BER instead of PER reduces the number of variables to be tested.

Reference Sensitivity Level The Bluetooth specification uses −70 dBm as the *reference sensitivity level*. The actual sensitivity level for a particular receiver is the weakest signal at which a BER of 0.1 percent, or 10^{-3}, is obtained. Several performance criteria are supposed to be met without exceeding the 10^{-3} BER out of the receiver's detector circuitry. Of course, a BER that averages 1 in every 1,000 received bits is unacceptable for file transfers and most other types of data exchange between two Bluetooth users. Both *forward error correction* (FEC) and *automatic repeat request* (ARQ) can be used to reduce the actual error rate to negligible levels.

Sensitivity For Bluetooth certification, the receiver must at least meet the reference sensitivity level of −70 dBm for a raw BER of 10^{-3}. This sensitivity level is actually quite poor compared to state-of-the-art receivers, which can have sensitivities of −100 dBm or better, or even compared to the low-rate Wi-Fi receiver with a required sensitivity of at least −80 dBm. Bluetooth, though, is intended for inexpensive, short-range links, so the receiver's sensitivity can be comparatively low. Bluetooth chipset manufacturers are producing receivers with typical sensitivities of −80 to −85 dBm. For Bluetooth certification, the receiver must be able to detect signal levels as high as −20 dBm without exceeding a BER of 10^{-3}.

Interference Immunity   The C/I values listed in the specification for the Bluetooth receiver must be tolerated without exceeding a BER of 10^{-3}. As with Wi-Fi, Bluetooth is a channelized radio system, so these values are given for co-channel and adjacent-channel interfering signals that are Bluetooth modulated. A summary of these values is given in Table 5-12 and graphically in Figure 5-8. Because the Bluetooth signal in each hop channel is only 1 MHz wide, C/I values cover a wider range of channel offsets than they do with Wi-Fi. Also, performance is measured relative to an interfering Bluetooth signal, so the values don't strictly apply to other interference such as Wi-Fi. However, we can make a reasonable assumption that a Wi-Fi signal of equivalent energy within the Bluetooth's receiver passband will also disrupt operations equivalently.

Table 5-12

802.15.1 receiver C/I limits

Requirement	Value
Co-channel C/I	11 dB
Adjacent-channel (1 MHz) C/I	0 dB
Adjacent-channel (2 MHz) C/I	−30 dB
Adjacent-channel (≥3 MHz) C/I	−40 dB

Figure 5-8
Required interference performance for a Bluetooth receiver

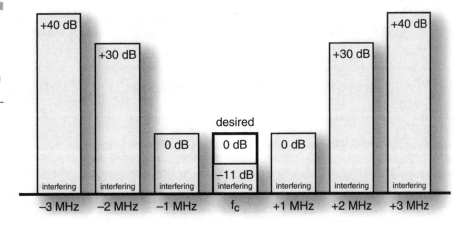

Unlike Wi-Fi, the Bluetooth specification includes a co-channel C/I value that the receiver must meet or exceed. This is important because the FHSS nature of Bluetooth means that occasional collisions are expected during normal operation when the Bluetooth piconet hops into an occupied channel. The *Gaussian-filtered frequency shift keying* (GFSK) modulation scheme includes a capture characteristic, in which the stronger co-channel signal completely dominates a receiver's detector. As long as the desired signal's RSSI is at least 11 dB stronger than that from a competing piconet, the Bluetooth receiver should be able to decode the desired data with a raw BER no higher than 10^{-3}.

The required C/I value drops as the carrier frequency of the interfering signal moves further away from the desired signal. In other words, the further the separation in frequency becomes between the desired and interfering carriers, the higher the allowable interfering power level before it causes the desired BER to reach 0.1 percent. For example, an interfering Bluetooth transmission 1 MHz away from the desired signal can have the same power level as the desired signal before the desired BER reaches 10^{-3} (C/I = 0 dB), but when the interferer is 2 MHz away its power level can be 1,000 times greater (C/I = -30 dB) before the BER can be exceeded. Incidentally, receiver designers have discovered that the adjacent channel C/I values are often more difficult to achieve than the co-channel C/I.

Blocking Immunity   The out-of-band blocking signals are measured with the desired signal at -67 dBm, which is 3 dB above the -70 dBm reference signal. Unlike the C/I measurements, the blocking signal is assumed to be *continuous wave* (CW). The allowable blocking signal power within each associated band is given in Table 5-13. These represent the out-of-band power levels under which the Bluetooth receiver's BER is not to exceed 10^{-3}. For the range of 2.0 to 3.0 GHz (except for the 2.4 GHz ISM band), the CW blocking signal's power level is equivalent to a C/I of -40 dB, which is the same as that required for Bluetooth-modulated interfering signals that exist 3 MHz or more away from the desired carrier frequency within the 2.4 GHz ISM band.

Many receivers, especially those controlled digitally (as Bluetooth receivers are), can be more sensitive to blocking signals at certain out-of-band frequencies.

Table 5-13

802.15.1 receiver out-of-band blocking power levels

Interfering carrier frequency	Power level	Equivalent C/I
30–2,000 MHz	-10 dBm	-57 dB
2,000–2,399 MHz	-27 dBm	-40 dB
2,498–3,000 MHz	-27 dBm	-40 dB
3,000–12,750 MHz	-10 dBm	-57 dB

This is because of various harmonic effects from component selection and their orientation on the circuit board, as well as from crystal frequency harmonics in both the analog and digital parts of the circuit. As a result, the Bluetooth specification allows exceptions to the blocking levels in Table 5-13 for 24 frequencies at integer multiples of 1 MHz. At 19 of these frequencies, a -50 dBm blocking signal may be sufficient to bring the BER to 10^{-3}, and at the remaining 5 frequencies the receiver is allowed to be blocked at any power level. The latter has significant coexistence ramifications if a transmission from another service resides at one of these frequencies.

Image Rejection   In a superheterodyne receiver, conversion from the RF f_{RF} to the *intermediate frequency* (IF) f_{IF} is performed by the *mixer*, as shown in Figure 5-9. The output of the mixer contains two new frequencies equal to $f_{RF} + f_{LO}$ and $f_{RF} - f_{LO}$. The higher (sum) component is filtered out and the lower (difference) component becomes the IF; thus, $f_{IF} = f_{RF} - f_{LO}$. This implies, of course, that $f_{RF} > f_{LO}$, in which case the receiver uses *low injection*. Because f_{IF} is fixed, the receiver is tuned to different carrier frequencies by changing the *local oscillator* (LO) frequency.

Unfortunately, another frequency also exists, called the *image frequency*, f_{Image} $= f_{RF} - 2f_{LO}$, such that $f_{IF} = f_{LO} - f_{Image}$. This means that another transmitter using f_{Image} as its carrier frequency could severely interfere with the desired signal with carrier frequency f_{RF}. To make matters even more complex, it's also possible for a receiver to have $f_{IF} = f_{LO} - f_{RF}$ (high injection) and in this case $f_{Image} =$ $f_{RF} + 2f_{LO}$. Both of these situations are depicted in Figure 5-10.

If a Bluetooth receiver has a desired signal power level of -67 dBm, an interfering signal at the image frequency is required to be stronger than -58 dBm before the BER is driven above 10^{-3}, corresponding to a C/I$_{Image}$ of -9 dB. The image C/I performance figures sometimes conflict with the adjacent channel C/I requirements given in Table 5-12. In these cases, the more relaxed specification applies.

Figure 5-9
Conversion
from RF to IF
in a super-
heterodyne
receiver

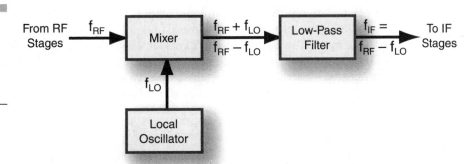

Figure 5-10
Location of
image
frequencies for
high-injection
and low-
injection
superhetero-
dyne receivers

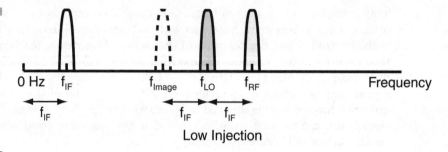

Low Injection

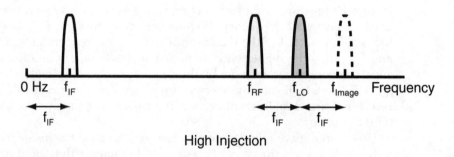

High Injection

A lot of interest has arisen recently in designing a receiver using a method called *direct conversion*, also known as *zero-IF* (ZIF). In this design, the data stream is extracted using a single mixer operation in which the LO is tuned to the same frequency as the RF carrier. Images are nonexistent, but the receiver has the potential to jam itself if some of the LO energy enters the mixer port to which the antenna is connected. Such a situation is alleviated while maintaining reasonable image immunity by designing a receiver with *near-zero-IF* (NZIF) circuitry.

Intermodulation Distortion   All amplifiers have some nonlinearity associated with them, and, of course, the amplifiers used in Bluetooth receivers are no exception. An amplifier with nonlinearities produces new frequencies at its output that weren't present at its input. If one of the new frequencies happens to be the same as the desired signal's carrier, then interference could occur.

Often the most troublesome new frequency, from an interference standpoint, is generated as a third-order product from two other frequencies, which we call f_1 and f_2. A third-order product f_3 has the relationship

$$f_3 = 2f_1 - f_2 \tag{5.2}$$

Two interfering signals at f_1 and f_2 can combine to form a third-order product at the desired signal's carrier frequency f_c in many different ways. For example, suppose the desired receiver hops to $f_c = 2,410$ MHz. Then interfering signals at $f_1 = 2,413$ MHz and $f_2 = 2,416$ MHz can produce a third-order product at 2,410 MHz. In other words, if three Bluetooth piconets are within range of each other, and if they hop to the previous frequencies, the desired piconet may experience interference from the other two. This type of interference is called third-order *intermodulation distortion* (IMD). It could be argued that the probability of such an event happening is exceedingly small, but third-order IMD could also occur under many different hop-frequency combinations, one or another of which could happen quite often.

The Bluetooth specification requires that third-order IMD be tested under the following conditions:

- The desired signal f_c has a power level of -64 dBm.
- The interfering signal f_1 is CW with a power of -39 dBm.
- The interfering signal f_2 is reference Bluetooth modulated at a power of -39 dBm.

The desired receiver's BER cannot exceed 10^{-3} when the relationship given in Equation 5.2 holds and the difference between f_2 and f_1 is selected to be *one* of the following: 3, 4, or 5 MHz.

Spurious Responses and Emissions Bluetooth receivers (and most other digital receivers, for that matter) contain oscillators, timers, and other circuitry that can produce their own RF emissions. The result is that a receiver may attempt to detect its own generated signals, and it may transmit signals as well. The former are called *spurious response frequencies*, and the latter are *spurious emissions*.

A receiver will probably fail to meet the C/I requirements in Table 5-12 as well as the requirements in the "Image Rejection" section on its spurious response frequencies. Up to five of these are allowed at frequencies at least 2 MHz away from the frequency to which the receiver is tuned. A relaxed C/I value of -17 dB is permitted for these frequencies.

Spurious emissions are listed in Table 5-14 for two different band segments. The maximum power requirement must not be exceeded for any 100 kHz band segment within the ranges listed.

Table 5-14

802.15.1 receiver spurious emission limits

Frequency band	Limit
30 MHz–1 GHz	−57 dBm
1–12.75 GHz	−47 dBm

The 802.15.3 WiMedia Radio

The emphasis of WiMedia is on the transfer of fast isonchronous data over ranges as short as 1 m. As such, the modulation scheme has been set to exploit the high SNR available at these short ranges by sending up to 5 data bits per symbol. The symbol rate is 11 Ms/s, with data rates up to 55 Mb/s. This symbol rate is identical to the chip rate in 802.11b Wi-Fi.

802.15.3 Channel Set

The 802.15.3 radio occupies the 2.4 GHz band, with the channel set depicted in Figure 5-11. A total of five channels are available for use by an 802.15.3 *device* (DEV), divided into two groups. The high-density group may be employed when no 802.11b networks are in range, allowing up to four WiMedia networks to operate in the same area without channel overlap. When Wi-Fi networks are nearby, however, WiMedia operation on the high-density channel set could interfere with Wi-Fi more than necessary. This is because WiMedia channel 2 partially overlaps Wi-Fi channels 1 and 6, and WiMedia channel 4 partially overlaps Wi-Fi channels 6 and 11. Therefore, a single WiMedia network has the potential to interfere with two nonoverlapping Wi-Fi channels. The center frequencies of the three available channels in the WiMedia coexistence channel set correspond to the center frequencies of Wi-Fi channels 1, 6, and 11, so a single WiMedia piconet operating on WiMedia channels 1, 3, or 5 can disrupt at most one Wi-Fi channel.

Figure 5-11
IEEE 802.15.3
WiMedia
channel set for
high density
and Wi-Fi
coexistence. TX
bandwidth is
approximately
15 MHz.

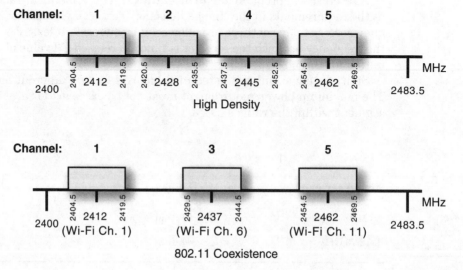

802.15.3 TX Performance

The WiMedia transmitter must support modulation ranging from simple QPSK to 64-QAM if it is capable of sending data at the maximum rate of 55 Mb/s. The TX spectrum mask required for all data rates is shown in Figure 5-12. The −30 dBr bandwidth is 15 MHz, which is less than both 802.11b and 802.11g. The transmitter may have one spurious output within the 2.4 GHz band (called an *in-band image* in the 802.15.3 specification) with a PSD of −40 dBr over a 15 MHz bandwidth.

The maximum allowable TX power in the United States is about +10.5 dBm under FCC Section 15.259, and the typical 802.15.3 TX powers are expected to be about +8 dBm. If the transmit power can exceed 0 dBm, then TPC must be implemented. TPC should be able to reduce TX power to less than 0 dBm in monotonic steps between 3 and 5 dB. The 10 to 90 percent TX power-on ramp and the 90 to 10 percent power-down ramp must each be no greater than 2 μs. To avoid DC offset problems at the receiver, the RF carrier suppression at the channel center frequency must be at least 15 dB below the peak sinx/x power spectrum using DQPSK modulation, while sending a repetitive 0101 data sequence, using a 100 kHz resolution bandwidth.

802.15.3 RX Performance

For the 802.15.3 receiver, a standard FER of not more than 8×10^{-2} has been specified for a MAC frame length of 1,024 bytes of pseudorandom data. This FER is the same as that used for 802.11b performance measurements, but because 802.15.3 uses *trellis coded modulation* (TCM), a degree of error correction is built

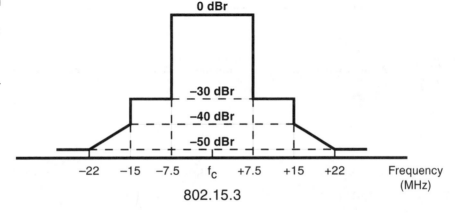

Figure 5-12
Spectrum mask for 802.15.3 WiMedia signaling

into the demodulation process and cannot be separately evaluated. Therefore, the 802.15.3 standard FER includes any benefit gained by TCM.

Sensitivity The receiver sensitivity has both high and low required levels dictated by the 802.15.3 specification. The required minimum sensitivity is the weakest signal that the receiver must be able to demodulate without exceeding the standard FER, whereas the required maximum sensitivity is the strongest signal that the receiver must be able to demodulate without exceeding the standard FER. These values are given in Table 5-15. As usual, the required minimum sensitivity is less stringent as the data rate increases.

Interference Immunity Table 5-16 lists the interference immunity (the jamming resistance) required for the 802.15.3 radio. The high-density channel set is used for testing the receiver; that is, channel 3 is not used in the test. The adjacent channel is the channel nearest to the desired channel, and the alternate channel is one removed from the adjacent channel.

Interference immunity is measured by placing a desired 802.15.3 signal appropriately modulated into the desired receiver at a level of 6 dB above the sensitivity limit in Table 5-15. The interfering 802.15.3 signal is then placed into

Table 5-15

802.15.3 receiver sensitivity limits

Modulation	Required minimum sensitivity	Required maximum sensitivity
QPSK-TCM	−82 dBm	−10 dBm
DQPSK	−75 dBm	−10 dBm
16-QAM-TCM	−74 dBm	−10 dBm
32-QAM-TCM	−71 dBm	−10 dBm
64-QAM-TCM	−68 dBm	−10 dBm

Table 5-16

Required 802.15.3 receiver C/I limits

Modulation	Adjacent channel C/I limit	Alternate channel C/I limit
QPSK-TCM	−33 dB	−49 dB
DQPSK	−26 dB	−41 dB
16-QAM-TCM	−25 dB	−40 dB
32-QAM-TCM	−22 dB	−37 dB
64-QAM-TCM	−19 dB	−34 dB

the corresponding adjacent or alternate channel at the C/I limit given in Table 5-16, and the FER performance checked.

Clear Channel Assessment (CCA) The CCA for the 802.15.3 receiver is accomplished using ED in the received signal bandwidth. The CCA should indicate that the channel is busy with a probability of 90 percent or higher when it detects a valid 802.15.3 preamble sequence at an RSSI level at or above the required minimum sensitivity level for DQPSK (-75 dBm). The CCA response time should be less than 7.3 μs, and the channel should remain indicated as busy until the end of the detected frame. For enhanced coexistence with non-WiMedia transmissions, the receiver should report the channel as busy when any signal is detected at an RSSI of -55 dBm, which is 20 dB above the minimum DQPSK sensitivity level.

RSSI and SNR Reporting Each 802.15.3 receiver should be able to report RSSI relative to the required maximum sensitivity of -10 dBm in 8 steps of 8 dB with $\pm$dB step-size accuracy and over a minimum range of 40 dB. The receiver must also provide a channel SNR estimate at the receiver's decision point. Noise shall include all signal impairments such as thermal noise, distortion, and interference. The SNR is represented as a 5-bit number covering equal steps between 6 and 21.5 dB. Both RSSI and SNR provide important information for power control implementation.

Turnaround Times

For 802.15.3, the R/T and T/R turnaround times are each 10 μs or less, including any TX power-up or power-down ramps. If a channel switch is required, the time to change from one operating frequency to another shouldn't exceed 500 μs.

The 802.15.4 ZigBee Radio

ZigBee emphasizes transferring asynchronous data over ranges of a few meters with a particular focus on extremely low cost. The modulation scheme is DSSS, and data rates are relatively low at 20, 40, and 250 kb/s, depending on the band of operation.

802.15.4 Channel Set

The 802.15.4 device must operate in at least one of three bands, with a channel set as shown in Figure 5-13. Channel 0 was intended for use in Europe, channels 1 through 10 for use in the United States and other countries that allow operation

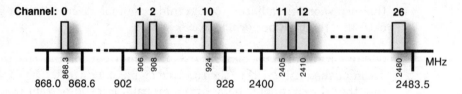

Figure 5-13
802.15.4 ZigBee channel set. TX bandwidths are approximately 1.2 MHz in the 900 MHz band and 3.5 MHz in the 2.4 GHz band. TX bandwidth for 868 MHz is not specified.

in the 900 MHz ISM band,* and channels 11 through 26 for worldwide operation. Channel separation in the 900 MHz band is 2 MHz to accommodate a chip rate of 600 kchips/s (1.2 MHz nominal bandwidth), and separation in the 2.4 GHz band is 5 MHz to allow a faster chip rate of 2 Mchips/s (3.5 MHz nominal bandwidth) without overlap.

802.15.4 TX Performance

The transmit PSD masks for the 900 MHz and 2.4 GHz ISM bands are given in Table 5-17. The average spectral power is measured using a 100 kHz resolution bandwidth. The relative limit for the 900 MHz band is measured within 600 kHz of the carrier frequency, and for the 2.4 GHz band the relative limit is measured within 1 MHz of the carrier frequency. No equivalent PSD mask is provided in the specification for operation at 868.6 MHz, so local regulations apply.

Transmit power and spurious emissions in each of the bands of operation shall conform to local regulations. For example, in the 2.4 GHz ISM band TX power is limited to the values given in Table 5.3, and in the United States the same power limits also apply to the 900 MHz ISM band. Transmitters shall be capable of a TX power of at least −3 dBm, but should transmit at a lower power when possible to reduce interference.

Table 5-17

802.15.4 transmit PSD mask

Band	Offset from carrier frequency	Relative limit	Absolute limit
900 MHz ISM	> 1.2 MHz	−20 dB	−20 dBm
2.4 GHz ISM	> 3.5 MHz	−20 dB	−30 dBm

*The 802.15.4 specification calls this the 915 MHz band.

802.15.4 RX Performance

A standard PER of not more than 10^{-2} has been specified for a packet length of 20 bytes of data. Although this PER is lower than that in other WLAN and WPAN specifications, it is achieved with a much smaller packet length. As such, the BER can be considerably higher while still meeting this criteria. For example, attaining a PER of 8×10^{-2} over 1,024 bytes with no FEC requires a BER of about 10^{-6}, whereas a lower PER of 10^{-2} over 20 bytes with no FEC requires a BER of only about 6×10^{-5}.

Sensitivity The receiver sensitivity has both high and low required levels dictated by the 802.15.4 specification. The required minimum sensitivity is the weakest signal the receiver must be able to demodulate without exceeding the standard PER. The required maximum sensitivity is the strongest signal the receiver must be able to demodulate without exceeding the standard PER. These values are given in Table 5-18.

Interference Immunity Table 5-19 lists the interference immunity (the jamming resistance) required for the 802.15.4 radio. The adjacent channel is the channel nearest the desired channel, and the alternate channel is one removed from the adjacent channel. No jamming resistance is specified for channel 0 because no adjacent channels exist.

Interference immunity in both bands is measured by placing a desired 802.15.4 signal appropriately modulated into the desired receiver at a level of 3 dB above the sensitivity limit in Table 5-18. The interfering 802.15.4 signal is

Table 5-18

802.15.4 receiver sensitivity limits

Band	Required minimum sensitivity	Required maximum sensitivity
868 MHz	−92 dBm	−20 dBm
900 MHz	−92 dBm	−20 dBm
2.4 GHz	−85 dBm	−20 dBm

Table 5-19

802.15.4 receiver C/I limits

Band	Adjacent channel C/I limit	Alternate channel C/I limit
900 MHz	0 dB	−30 dB
2.4 GHz	0 dB	−30 dB

then placed into the corresponding adjacent or alternate channel at the C/I limit given in Table 5-19, and the PER performance is checked.

Clear Channel Assessment (CCA)   The 802.15.4 specification defines three modes of CCA operation, at least one of which can be incorporated into the node. These are similar to the first three modes of 802.11b CCA:

- **Mode 1: Energy above threshold**   The CCA reports a busy medium upon detection of any signal energy above the ED threshold.
- **Mode 2: Carrier sense only**   The CCA reports a busy medium only upon detection of a signal with modulation and spreading characteristics of 802.15.4, either above or below the ED threshold.
- **Mode 3: Carrier sense with energy above threshold**   The CCA reports a busy medium upon detection of a signal with modulation and spreading characteristics of 802.15.4 and with energy above the ED threshold.

The strengths and weaknesses of each of these three modes given in Table 5-8 apply here when 802.15.4 is substituted for 802.11b. The CCA detection time should be 8 symbol periods, which is 400 μs in the 868 MHz band, 200 μs in the 900 MHz band, and 128 μs in the 2.4 GHz band. The ED threshold for use by the CCA should be at most 10 dB above the specified RX minimum sensitivity in Table 5-18. Modes 2 and 3 are not required to respond to the presence of non-ZigBee signals.

Energy Detection (ED) and Link Quality Indication (LQI)   The 802.15.4 receiver must have ED capability in the received signal bandwidth without any attempt to identify the type of signal being transmitted. ED is reported as an 8-bit integer RSSI value, with 0x00 representing an RSSI of less than 10 dB above the minimum sensitivity level given in Table 5-18, and the RSSI range shall be at least 40 dB with a linear mapping having ±6 dB step-size accuracy. The ED response time should be less than 8 symbol periods, which is 400 μs in the 868 MHz band, 200 μs in the 900 MHz band, and 128 μs in the 2.4 GHz band.

The receiver must also provide a channel *link quality indication* (LQI) estimate at the receiver's decision point. The value can be based upon ED, SNR, or a combination of these. The actual use of LQI is not specified in the 802.15.4 standard. LQI is represented as a 8-bit number covering uniformly distributed steps from maximum to minimum, and at least eight unique LQI values shall be used.

Turnaround Times

For 802.15.4, the R/T and T/R turnaround times are each 12 symbol periods or less, including any TX power-up or power-down ramps. The maximum turnaround time equates to 600 μs in the 868 MHz band, 300 μs in the 900 MHz band, and 192 μs in the 2.4 GHz band.

Summary

The radio makes up the lowest protocol in the OSI model. The transmitter is the last unit to process data at the sending node, and the receiver is the first unit to process data at the receiving node. The various WLAN and WPAN specifications list specific performance requirements of the TX and RX units that comprise a node, but it's important that the radio meets government regulations as well.

For acceptable coexistence with other networks, the transmitter should use the lowest power possible for a reliable communication link. The receiver should be sensitive enough to provide a reasonably good maximum range but be selective enough to reject adjacent and alternate channel interference. Unlike the other WLAN and WPAN implementations, the 802.15.1 Bluetooth receiver has co-channel C/I performance requirements listed in the specification to improve FHSS performance.

All the WLAN/WPAN implementations except Bluetooth use a form of *listen before talk* to reduce the chance of packet collisions on the channel. Some nodes will defer transmission if they sense any signal on the channel, whereas others defer transmission only if they sense another signal having the same characteristics.

Fast T/R and R/T turnaround times are also needed for high-speed communication to prevent excessive delays while waiting for an ACK frame and for efficient synchronous communication. When the turnaround times are standardized, minimum timing at the MAC layer for packet exchanges can be specified.

References

[IEEE802.11] IEEE 802.11-1999, "Wireless LAN Medium Access Control (MAC) and Physical Layer (PHY) Specifications," August 20, 1999.

[IEEE802.11a] IEEE 802.11b-1999, "High-Speed Physical Layer in the 5 GHz Band," September 16, 1999.

[IEEE802.11b] IEEE 802.11b-1999, "Higher-Speed Physical Layer Extension in the 2.4 GHz Band," September 16, 1999.

[IEEE802.11g] IEEE 802.11g-2003, "Amendment 4: Further Higher Data Rate Extension in the 2.4 GHz Band," June 27, 2003.

[IEEE802.11h] IEEE 802.11h-2003, "Amendment 5: Spectrum and Transmit Power Management Extensions in the 5 GHz Band in Europe," October 14, 2003.

[IEEE802.15.1] IEEE 802.15.1-2002, "Wireless Medium Access Control (MAC) and Physical Layer (PHY) Specification for Wireless Personal Area Networks (WPAN)," June 14, 2002.

[IEEE802.15.2] IEEE 802.15.2-2003, "Coexistence of Wireless Personal Area Networks with Other Wireless Devices Operating in Unlicensed Frequency Bands," August 28, 2003.

[IEEE802.15.3] IEEE 802.15.3/D17, "Wireless Medium Access Control (MAC) and Physical Layer (PHY) Specifications for High Rate Wireless Personal Area Networks (WPAN)," draft standard, February 2003.

[IEEE802.15.4] IEEE 802.15.4-2003, "Wireless Medium Access Control (MAC) and Physical Layer (PHY) Specifications for Low-Rate Wireless Personal Area Networks (LR-PANS)," October 1, 2003.

Medium
Access Control

The *medium access control* (MAC) is, as a minimum, the method by which the nodes in a network access the communication channel. Sometimes the MAC layer performs additional functions. For example, the 802.11 MAC controls access to the medium, provides reliable data delivery to users, and protects data. Because the latter is implemented through security measures and is not directly applicable to coexistence, we will concentrate on the first two.

If the network is unidirectional point to point, and if no consideration is given to coexisting with other networks, the MAC is trivial: Simply transmit whenever data is available. Most networks, however, support two-way traffic, so even a point-to-point link must employ a MAC to at least ensure that one node is receiving while the other is transmitting. The situation becomes more complex when the network is point to multipoint, with several nodes competing for channel access, especially when other independent networks are also present on the same channel.*

Access to the channel can be accomplished using contention-free techniques, in which transmissions are scheduled in such a way that *collisions*, or simultaneous transmissions, cannot occur. This type of MAC works well when traffic is mostly synchronous and data is always available when a transmission is scheduled. Alternatively, when traffic is predominately asynchronous, contention-based channel access can be more efficient, allowing a node the opportunity to send its data shortly after it becomes available from a higher protocol layer without having to await its scheduled transmission time.

In practical networks, the MAC may be structured to provide coexistence at three possible levels: 1) only within the desired network itself, 2) within the desired network and across to other networks of the same type, or 3) within the desired network and across to other networks of all types. The first level was of primary importance during the early days of wireless networking, because the odds were high that only one network existed in any one area. The MAC was developed to enhance communication among users in that particular network. As interference became more prevalent, MAC designers have paid more attention to the third level, so that all users in different, independent networks can access the shared communication channel without causing catastrophic interference to each other.

We begin this chapter by laying the foundation of network analysis through examining the various measures of network throughput, followed by the channel states and access summary. Next, both contention-free and contention-based

---

*To harmonize with network design terminology, we consider the term "channel" to be synonymous with "communication medium," which is related to, but not necessarily the same as, the carrier frequency channel. For example, the *frequency hopping spread spectrum* (FHSS) accesses the channel (medium) to send data, but this "channel" consists of several carrier frequency channels accessed in a pseudorandom hop pattern.

access methods are defined and analyzed. Finally, actual MAC methods used by the various *wireless local area network* (WLAN) and *wireless personal area network* (WPAN) systems are presented in their basic renditions. Attention is devoted mostly to Wi-Fi and Bluetooth not only because they're the most mature, but also because both 802.15.3 WiMedia and 802.15.4 ZigBee MAC on the 2.4 GHz *physical layer* (PHY) have many similarities to the Wi-Fi MAC. All the systems enable various modifications to the MAC layer to enhance coexistence, but we will postpone discussion of these until later chapters.

MAC Performance Metrics and Analysis Background

Determining how the performance of the MAC should be measured can be somewhat difficult, due in part to the fact that network goals can be different depending upon which perspective is being considered. For example, the network engineer wants the aggregate network throughput to be as high as possible, but the individual user wants to maximize his or her own throughput.

Measuring Network Throughput

The means by which a MAC can be evaluated usually involves some form of network throughput, given in units of packets/second or bits/second. The former is most often used in the various theoretical models, because these models rarely look any deeper than the packet level. If a network has variable data rate options, such as 802.11 and 802.15.3, evaluating throughput in bits/second may also be necessary to provide an accurate assessment of whether or not the network meets user expectations.

Various ways to measure throughput exist, both at the bit and packet levels, requiring caution when comparing the performance of different networks to ensure that the comparison is fair. Some of these performance categories are as follows:

- **Raw data rate** The rate at which the transmitter sends data over the PHY. The rate can be fixed (Bluetooth at 1 Mb/s) or variable (802.11b at 1, 2, 5.5, or 11 Mb/s). Using a raw data rate as an indication of throughput can be highly misleading because neither packet duty cycle nor overhead is considered.

- **Packet throughput** The fraction of time a packet is successfully transmitted on average. This is often the performance metric used by network analysis in which the actual contents of the packet are considered to be irrelevant. For a channel that can convey only one packet at a time, this value S is in the range 0 to 1.

■ **Channel capacity**   The maximum packet throughput allowed by the selected MAC protocol over the given communication channel. This is often used to compare the performance of different MAC protocols. This value S_{max} is between 0 and 1 if the medium can convey only one packet at a time.

■ **User data throughput**   The rate at which the MAC layer transfers packet payload data. This value, usually given in bits/second, is adjusted to remove any overhead such as headers and error control, and it also accounts for the packet duty cycle.

■ **Higher-layer throughput**   The rate at which higher layers can exchange data. *Transport Control Protocol/Internet Protocol* (TCP/IP) throughput or the speed at which a file can be transferred in bits/second are examples of higher-layer throughput. This can be surprisingly low compared to the raw data rate if any inefficiencies exist.

Channel States

Most basic network performance models assume that the communication channel can carry at most a single transmission successfully. That is, if two or more transmitters are active at the same time, a collision occurs and all affected packets are lost. This approach has the twofold benefit of not only simplifying the analysis to make it tractable, but also in providing a slightly pessimistic set of results due to the lack of capture in the model.

Capture is manifested by a receiving node's capability to decode one of multiple signals that may be present, in spite of the interference. In other words, the existence of a reasonable co-channel *carrier-to-interference ratio* (C/I) at a receiver provides capture capabilities for desired signals that are stronger than interfering signals beyond some minimum value. Capture can also include the capability of a receiver to successfully decode a desired signal that arrives prior to a relatively strong interfering signal, because such interference occurs only after the receiver has synchronized on the desired transmission. Traditional network analysis usually ignores capture because either 1) all transmissions are assumed to have equal RSSI, making capture difficult, or 2) no assumptions are made about whether the strongest signal is sent by a desired or interfering transmitter. Another assumption typically made by network engineers is that packets are lost only during the collisions. If only one transmitter is active, then that packet is always received correctly. In other words, *additive white Gaussian noise* (AWGN) is ignored.

Under the assumptions outlined previously, a communication channel can be in one of three possible states. These are

- **Idle**   No transmitter is active and the channel is available.
- **Communicating**   Exactly one transmitter is active and data is being transferred.
- **Collision**   Two or more transmitters are active and the channel is useless.

Of the three states, only the communicating state results in data being transferred from sender to receiver. No communication takes place during the idle or collision states.

It is tempting to assume that keeping the network in the communicating state as much as possible results in the best performance. That's true from a network engineer's perspective, because a network that is always communicating has the highest possible throughput. However, a user would prefer that the network is idle whenever access is needed, because an idle network is available for immediate use, minimizing the time required for the data to be transferred from the sending to receiving node. If any particular user almost always finds the network in the communicating state, then delays for everyone can be extremely long, even though the network appears to be performing well. Both network engineer and user can agree, though, that the collision state should be avoided whenever possible.

Poisson Arrival Process

Another important aspect to modeling network performance is determining the length and arrival times of various packets to be sent over the channel. The most common approach involves an assumption that all users in a network independently generate packets for transmission over the network. New packets originate in the upper layers of the *Open System Interconnection* (OSI) model and are handed down to the MAC layer for processing. These packets have exponentially distributed arrival times, and their lengths may either be identical or exponentially distributed as well. When a large number of users operates independently, the time between the arrivals of packets for transmission over the network also has an exponential distribution. Both these situations lead to a network performance analysis that is greatly simplified, and yet the resulting throughput and delay calculations are very close to real-world performance, even for nonexponential distributions of packet lengths and arrival times.

Exponentially distributed packet interarrival times imply a Poisson distribution of the probability $P_k(t)$ that k arrivals occur during a particular time

Figure 6-1
Poisson
distribution
when $\lambda t = 3$

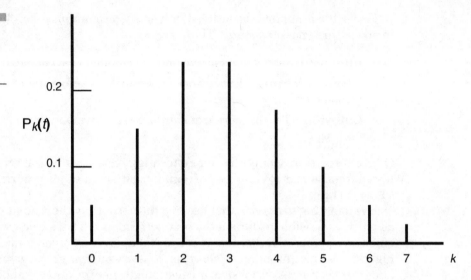

Figure 6-1
Poisson distribution when $\lambda t = 3$

interval of t seconds. This distribution, named after the French mathematician Siméon Denis Poisson (1781-1840), is given by

$$P_k(t) = \frac{(\lambda t)^k}{k!} \, e^{-\lambda t} \tag{6.1}$$

where λ is the mean (average) arrival rate in packets per second. Figure 6-1 portrays the arrival probabilities for the Poisson distribution when $\lambda t = 3$. For example, if the mean arrival rate is 3 packets/second and the time interval being examined is 1 second, then the distribution of Figure 6-1 holds. Under these conditions, the probability that one arrival occurs within 1 second is about 0.15, and the probability that no arrivals occur during the same time period is about 0.05.

Contention-Free Channel Access

Channel access is broadly divided into two categories: *contention-free* and *contention-based*. The contention-free access method creates a structure that allows all members of a network, or within multiple networks that agree among themselves to share a particular contention-free MAC, to experience essentially uncontested access to the channel. In this way the channel can be either idle or communicating, but the collision state is avoided under normal conditions. Contention-free access can be accomplished by queuing the packets for transmission one at a time through a first-in, first-out buffer, which works well when only one central node is performing all transmissions within a particular network. When transmitting nodes are distributed throughout the network and act independently, which is usually the case for practical networks, other methods are more suitable. These include scheduled and demand access techniques, both of which involve either a controlling supervisor or some sort of previous agreement among the nodes concerning access details.

Queuing

If a central controller is available to queue packets as they arrive, and it transmits them using a first-come, first-served protocol, then two general observations are immediately apparent. First, the channel is never in the collision state, and second, if the queue always contains at least one packet, then throughput S is at its maximum possible value of 1. In other words, the channel is never in the idle state either and is therefore always communicating. This is why queuing is sometimes called the ideal MAC. The problem, of course, is that most networks don't employ such a centralized transmission scheme. (We'll also discover that if the queue never empties, it is unstable when arrivals are Poisson distributed.) Even with its implementation difficulties, queuing can be used as a standard by which other access techniques can be compared, whether contention-free or contention-based.

This queuing process is shown in Figure 6-2. Arriving packets enter the queue and are processed by a server (transmitter) and sent over the channel. The type of queue being analyzed is given the designation X/Y/Z, where X is the arrival interval distribution, Y is the processing time distribution, and Z is the number of servers available. Sometimes a fourth designator appears, which specifies the maximum length of the queue.

The M/M/1 Queue The M/M/1 queue is the most commonly analyzed, where both the arrival and processing time distributions are Markov (exponential), and one server is used. When the fourth designator is missing, the maximum queue length is assumed to be infinite; that is, an arriving packet is never blocked because the queue is full. Suppose packets arrive in the queue with mean arrival

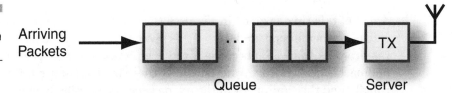

Figure 6-2
The queue as a
MAC protocol

Arriving
Packets

Queue

Server

rate λ, and they are processed (transmitted) at a mean rate μ. It should be obvious that μ must be greater than λ for the queue to be *stable*, that is, for the server to keep up with arrivals so that the queue doesn't grow without bound. The *traffic intensity* is defined as

$$\rho = \frac{\lambda}{\mu} \tag{6.2}$$

which must always be less than 1 for a stable queue. It follows that the probability that the queue is empty is $1 - \rho$, which is, of course, greater than 0 when ρ is less than 1. In other words, for the M/M/1 queue to be stable, the queue must occasionally empty. From a network point of view, stability requires that the channel sometimes be idle.

To determine the mean waiting time in the queue, we first note that the average number of packets in the system (in the queue and the one being serviced) is [Tan81]

$$\overline{N} = \frac{\rho}{1 - \rho} \quad \text{M/M/1 queue} \tag{6.3}$$

The mean waiting time $\overline{T}_s$, which includes service time, can be found by dividing the average number of packets in the system by the mean arrival rate (Little's result). Thus

$$\overline{T}_s = \frac{\overline{N}}{\lambda} = \frac{\rho/\lambda}{1 - \rho} = \frac{1/\mu}{1 - \rho} = \frac{1}{\mu - \lambda} \quad \text{M/M/1 queue} \tag{6.4}$$

The mean queuing time $\overline{T}_w$ that a packet must wait before it begins its transmission is found by subtracting the average transmission time $1/\mu$ from the right side of Equation 6.4, giving [Kle76a]

$$\overline{T}_w = \frac{\rho}{\mu(1 - \rho)} \quad \text{M/M/1 queue} \tag{6.5}$$

A plot of this time is given in Figure 6-3 when the average packet transmission rate is one per second, and thus $\overline{T}_w$ is given in seconds. The plot emphasizes the requirement that $\mu > \lambda$, because if the average rate of new packet arrivals even approaches the average rate at which they are transmitted, the queue grows rapidly and the time spent in the queue becomes extremely long.

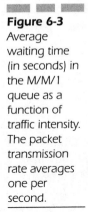

Figure 6-3
Average waiting time (in seconds) in the M/M/1 queue as a function of traffic intensity. The packet transmission rate averages one per second.

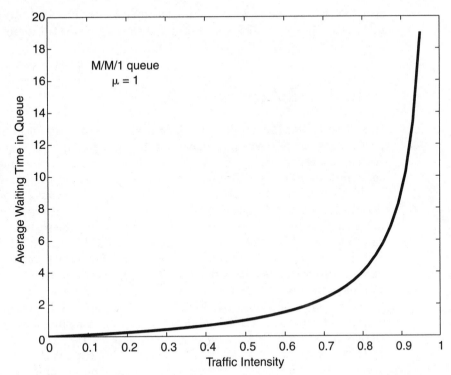

The M/D/1 Queue If every packet has a fixed length instead of being exponentially distributed, the resulting time to send a packet is fixed (assuming a constant bit-rate transmitter). This type of queue is designated M/D/1, where the *D* means *deterministic*. The average number of packets in the system now becomes [Kle76a]

$$\overline{N} = \frac{\rho}{1 - \rho} - \frac{\rho^2}{2(1 - \rho)} \quad \text{M/D/1 queue} \tag{6.6}$$

When packets are fixed in length instead of exponentially distributed, the mean number of packets in the system is reduced by $\rho^2/[2(1 - \rho)]$. For example, when $\rho = 0.9$, the average number of packets in the M/M/1 system is 9, but in the M/D/1 system the average number is 5. As $\rho \to 1$, the average number of packets in the M/D/1 system approaches half the average number in the M/M/1 system.

The average waiting time in the M/D/1 queue is

$$\overline{T}_w = \frac{\rho}{2\mu(1 - \rho)} \quad \text{M/D/1 queue} \tag{6.7}$$

As expected, the mean waiting time for a packet in the M/D/1 queue is half that of a packet in the M/M/1 queue, but the waiting time still increases rapidly as $\rho \rightarrow 1$.

Scheduled Access

In a decentralized network, it is much easier to schedule packets for transmission using some agreed-upon algorithm than it is to implement queuing. When a centralized controller is present, downlink packets can be queued, but uplink communication lends itself more easily to scheduled access. Scheduled access is itself divided into two major categories. *Frequency division multiple access* (FDMA) assigns each communication link to a separate carrier frequency (channel) such that simultaneous transmissions will always be separated in frequency. Entering the collision state is impossible with FDMA, even if all nodes transmit at once. For example, sensors can be deployed throughout a factory, each of which periodically sends its data using a separate carrier frequency. A central receiver can select the desired sensor by tuning to its carrier. A variation of FDMA called *frequency division duplexing* (FDD) is used by the cellular telephone network, which uses separate uplink (handset to cell) and downlink (cell to handset) carriers to allow simultaneous two-way voice conversation.

The disadvantage of using FDMA within a single network is that devices operating on different carrier frequencies cannot readily communicate with each other, nor can a central controller monitor all nodes simultaneously with a single receiver. FDMA is therefore better suited to enhance coexistence between independent networks that don't cross-communicate. Of course, a channel access mechanism may also be needed on each of the FDMA channels. Figure 6-4 shows networks operating on different carrier frequencies, with a guard band between them providing the necessary isolation. This is analogous to three independent 802.11 networks operating in the 2.4 GHz band on the U.S. nonoverlapping channel set 1, 6, and 11.

Figure 6-4
FDMA with guard bands to properly isolate independent networks

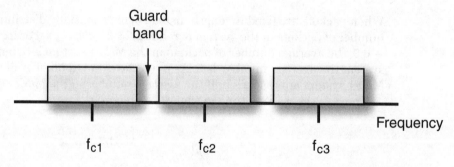

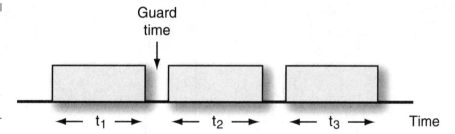

Figure 6-5
TDMA with guard times to protect disjoint time slots from jitter and delays

The other major scheduled access method is *time division multiple access* (TDMA) in which several nodes are assigned nonoverlapping time windows for transmissions, all of which operate on the same carrier frequency. In return for implementing fast TX/RX switching and accurate timing control, all nodes in a network can communicate with each other by selecting the proper time window in which to transmit or receive (see Figure 6-5). Guard times prevent collisions from timing jitter or propagation and processing delays. In practice, TDMA is often used for communication within a network, and FDMA is used to separate independent networks to prevent mutual interference.

Aside from the need for strict timing, the major disadvantage of TDMA is that each node is given a fixed time window for transmitting data, which is not efficient when the data to be sent is asynchronous (bursty) in nature. When nothing is available to send, the time window is wasted, and when the information is lengthy it may require transmission over several time slots, increasing delay even more. TDMA is much better suited for synchronous applications, where fixed-length packets are available at fixed intervals and their transmission must take place with very low latency. If guard times are short compared to packet transmission times, throughput using TDMA can approach 1. A variation of TDMA, called *time division duplexing* (TDD), allows two nodes to rapidly exchange packets in a manner that simulates full duplex communication, even though they share only a single carrier frequency. This is the manner by which Bluetooth implements two-way voice.

Demand Access

The shortcomings of TDMA for asynchronous transmissions can be mitigated significantly by using *demand access* methods, whereby nodes with data to send are given enhanced access to the channel. Demand access can take the form of either *polling* or *reservation*. Polling requires a central controller, and reservation access can take place in either a centralized or decentralized network. Many demand access schemes can be categorized as a combination of contention-free and contention-based methods.

Figure 6-6
Roll call polling
process

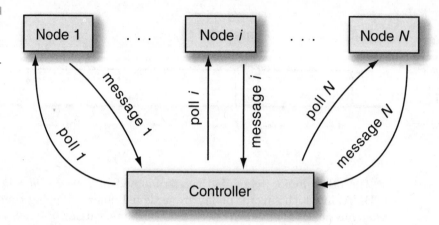

Polling   A basic *roll call polling* method is depicted in Figure 6-6, where a central controller polls nodes in sequence for transmissions. At first, this scheme appears to be no better than TDMA, because nodes with no data still consume polling overhead. However, when a node's message queue is empty, it sends a short "no traffic" frame, and the controller can quickly move on to another node without waiting for the duration of a full time slot. Conversely, if a node has a relatively long message, it can be given a longer transmission window up to some maximum value. Efficiency under asynchronous data conditions is thus increased significantly over that provided by TDMA, at the expense of requiring a controller. A polling mechanism is included in most WLAN and WPAN specifications because a controller is usually available.

To find the average cycle time T_c to poll all participating nodes, the polling server moves from node to node, stopping at node i if it contains any messages. Sending the queued messages requires T_i seconds. The time required for the server to move from node $i - 1$ to i (the *walk time*) is W_i seconds, and there are a total of N nodes. The cycle time is thus [Hay84]

$$T_c = \sum_{i=1}^{N} W_i + \sum_{i=1}^{N} T_i \qquad (6.8)$$

Despite the dependencies in the random variables listed in Equation 6.8, the average cycle time is simply the sum of the average W_i and T_i values in Equation 6.8 due to the linearity of the expectation operator.

The average time to empty a queue at node i is the time required to transmit all the messages that arrived during a cycle. If the arrivals are Poisson and the queue has infinite buffer space, then

$$\overline{T}_i = \frac{\lambda_i \overline{T}_c}{\mu_i} = \rho_i \overline{T}_c \qquad (6.9)$$

If the W_i are equal and the ρ_i are equal, then combining Equations 6.8 and 6.9 gives [Hay84]

$$\overline{T}_c = \frac{N\overline{W}}{1 - N\rho} \qquad (6.10)$$

Equation 6.10 shows a common characteristic of queuing theory results, where the numerator is a function of overhead, which is independent of traffic, and the denominator is a function of traffic information. Also characteristic is the fact that the network becomes unstable as the total load $N\rho \to 1$. Finding the average delay that a message spends in a node's queue is quite involved, but it exhibits the familiar behavior of increasing rapidly as the total load approaches 1.

Reservation Demand access through making a *reservation* has a large number of possible implementations. The reservation mechanism is used by nodes that have data to transmit so they can obtain a time allocation, and then the data transmission itself can proceed efficiently. When used in conjunction with TDMA, there are typically fewer time slots than nodes. Slot reservations are either *explicit* or *implicit*. For explicit reservations, a portion of a time is set aside for the reservation process. The protocol used for making explicit reservations is independent of the protocol used for data transmission. Indeed, the reservation protocol is often implemented using one of the contention-based access methods discussed in the next section, so this protocol is actually a hybrid of contention-based and contention-free channel access. The reservation protocol can be distributed among the network nodes, with a node announcing its intention to use time by transmitting a special packet within the reservation slot. This method is included in the 802.11 MAC. Alternatively, reservation activity can be centralized, with a designated node taking reservation requests and then assigning transmission times. The latter method is used in WLAN and WPAN systems that operate with a central controller.

Implicit reservations during the TDMA cycle are made simply by transmitting in a particular time slot. If the transmission is successful, the same slot during the next TDMA cycle is implicitly reserved for follow-on transmissions. When the node is finished using the sequence of slots, the corresponding slot following its last transmission will be vacant, indicating to the other nodes in the network that the slot will be available during the following TDMA cycle. This method, also called *reservation ALOHA* (for reasons that will become apparent in the next section), works well when the network consists of a few nodes, each of which exchange highly asynchronous data.

Contention-Based Channel Access

Contention-based channel access establishes a protocol by which users compete for access to the channel. This competition invariably results in the occasional collision, so all three channel states (idle, communicating, and collision) are possible when contention-based access is used. In return for the necessity of implementing a collision recovery scheme, contention-based access is conceptually simple, easy to implement, has very little overhead, and lends itself to good performance under lightly loaded channels containing primarily asynchronous transmissions.

Contention-based access can be divided into two broad categories: *ALOHA* and *carrier sense multiple access* (CSMA). The former, developed at the University of Hawaii (hence the name), is characterized by a node transmitting a packet either immediately upon its arrival (pure ALOHA) or at the beginning of the time slot following its arrival (slotted ALOHA). CSMA requires that the channel first be sensed idle before a transmission can commence, so it is more complex, and more polite, than ALOHA. CSMA reduces collision probabilities under heavier channel loads, but in certain conditions ALOHA outperforms CSMA.

Network Model

In order to accommodate the possibility of collisions occurring in a network that allows contention-based access, the model in Figure 6-7 is useful. This model considers only total network performance, and no attempt is made to analyze activity at individual nodes. The source generates new packets that are Poisson distributed at rate S. Some of these are transmitted successfully and others are

Figure 6-7
Network activity model for contention-based access

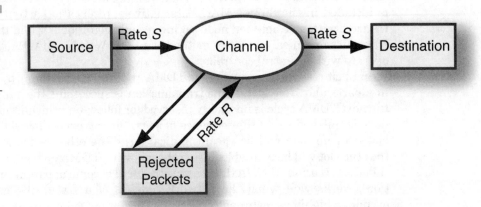

rejected due to collisions. The rejected packets are reintroduced into the channel at rate R, also assumed to be Poisson. Because the sum of two Poisson processes is also Poisson with a rate equal to the sum of the two original rates, the channel experiences an *offered rate* G that is Poisson with

$$G = S + R \qquad\qquad (6.11)$$

In a *stable channel*, defined as one in which the rejected packets bin occasionally empties, all newly generated packets eventually reach their destination. In other words, packets leave the channel and reach their destination at rate S, equal to the rate of newly generated packets leaving the source and entering the channel. The quantity S, then, is the packet *throughput*.

One of the goals of network performance analysis is to discover the relationship between G and S under different loading conditions. It's easy to conclude that, for very light channel-offered rates, $S \approx G$ because collisions are extremely rare. On the other hand, for very high offered rates the throughput $S \to 0$ because the channel is predominately in the collision state and it's rare for a packet to successfully reach its destination. What happens in between these two extremes is what makes network throughput analysis interesting.

Strictly speaking, throughput S is the independent variable and the channel-offered rate G is the dependent variable whose value is adjusted to achieve a required throughput. However, in plots of offered rate versus throughput, we will place the offered rate on the abscissa simply because such plots are easier to read and comprehend.

ALOHA

The ALOHA channel access method is completely distributed among the network participants, with no controller of any kind required. An arriving packet at any node is transmitted immediately for pure ALOHA, or at the beginning of the next time slot for slotted ALOHA, without regard to the channel state. Obviously, if the channel would have been idle otherwise, the transmission would have been successful, but if another transmission were to take place then a collision would occur and the affected packets would enter the rejected packets bin in Figure 6-7, where they would eventually be retransmitted. Channel activity for pure ALOHA and slotted ALOHA is shown in Figure 6-8. In pure ALOHA, partially overlapping transmissions can occur, and the resulting collision is assumed to destroy all affected packets. In slotted ALOHA, collisions are always fully overlapping.

As before, we assume that the new packet generation process is Poisson with rate λ packets per second. If packet length is fixed, then we define the quantity $m = 1/\mu$ as the time required to transmit the packet. Thus, ρ equals $m\lambda$, and this

Figure 6-8
Sample
channel activity
for pure
ALOHA and
slotted ALOHA

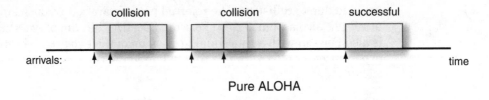

Pure ALOHA

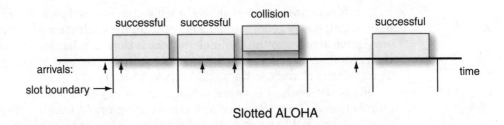

Slotted ALOHA

quantity must be less than 1 for the network to be stable. Equivalently, a stable network must be able to support the eventual transfer of packets from the source to destination at a channel load of ρ, so this loading is equal to the throughput S. When considering total channel activity, if newly generated and retransmitted packets together constitute a Poisson process with rate Λ, then the total offered load on the channel is $G = m\Lambda$. The probability of a collision is always greater than 0 for a positive throughput, so it follows that $G > S$. By normalizing on a packet length of $m = 1$, we can work directly with the quantities G and S for throughput calculations without a loss of generality. The quantity S/G is the probability of a successful transmission and its reciprocal G/S is the average number of times a packet must be transmitted until success.

ALOHA Throughput   To find the throughput in terms of the offered rate for a pure ALOHA transmission scheme, it is necessary to discover the probability that a packet of length $m = 1$ will be sent successfully. That is, what is the probability that an overlapping transmission will not occur? Figure 6-9 depicts a timeline in which a single "desired" packet exists. The region of vulnerability dur-

Figure 6-9
Period of
collision
vulnerability for
a packet on
the pure
ALOHA
channel

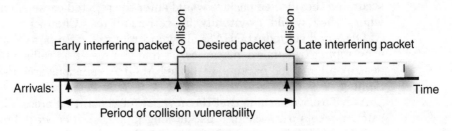

ing which an interfering packet of length 1 will at least partially overlap the first begins one time unit before the desired packet and ends at the end of the desired packet. The probability of a successful transmission is thus the probability that no other packets arrive during an interval that is two time units long. From Equation 6.1, for an offered rate of G this probability is e^{-2G}. The throughput is this probability multiplied by the offered rate. Therefore,

$$S = Ge^{-2G} \quad \text{pure ALOHA} \tag{6.12}$$

Put another way, Equation 6.12 is the probability that exactly one arrival occurs within a two time unit period.

For slotted ALOHA, the region of vulnerability is the length of a slot, which has a duration of one packet length (assuming the guard time is short), so packet throughput is the offered rate times the probability that exactly one arrival occurs during an interval of one time unit, giving

$$S = Ge^{-G} \quad \text{slotted ALOHA} \tag{6.13}$$

Both of these throughput versus offered rate curves are shown in Figure 6-10. Incidentally, simulation studies have shown that Equations 6.12 and 6.13

Figure 6-10
Throughput
versus offered
rate for both
pure and
slotted ALOHA

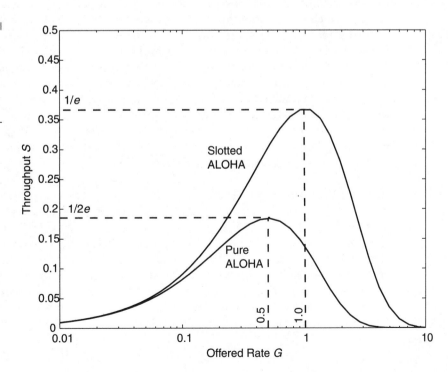

generally hold for most realistic channel traffic flow situations, even if the packet arrivals aren't Poisson [Hay84].

ALOHA Capacity and Stability   Peak throughput (channel capacity) is found by taking the derivative of Equations 6.12 and 6.13 with respect to G, setting the result to 0, and solving for $S = S_{max}$. These values are given in Figure 6-10. Although the channel capacity of slotted ALOHA is twice that of pure ALOHA, neither of the two ALOHA schemes is particularly efficient. In slotted ALOHA, for example, channel capacity is about 0.36, which is reached for an offered rate of 1. Because capacity is roughly ⅓ of the offered rate, it follows that when the channel is operated at capacity about two of every three transmitted packets are involved in a collision.

Suppose an ALOHA network is operating with an offered rate below that which achieves capacity. If the offered rate increases slightly, then throughput increases as well. In other words, if the required throughput increases as manifested by an increase in the number of newly generated packets entering the channel, the collision probability increases, the retransmission rate increases, and (provided capacity has not yet been reached) the channel adjusts to the higher throughput. In other words, the channel is *stable*.

Consider now what happens when the network is operating with an offered rate above that which achieves capacity. If the rate of newly generated packets increases slightly, the corresponding increase in collision probability results in a *decrease* in channel throughput, which increases the retransmission rate, which in turn increases the offered rate, leading to even higher collision probabilities. In other words, the channel is *unstable*, and the throughput rapidly drops toward 0 as the offered rate becomes large. Therefore, it is extremely important to operate the network in the stable region, with an offered rate well below capacity. If capacity is approached even from the region of stability, packet delay rapidly increases, as expected. For example, for pure ALOHA operating at about 80 percent capacity, the average delay is about 30 packet lengths, depending on the retransmission scheme, and in slotted ALOHA running at 80 percent capacity the average delay is about 6 packet lengths [Tob74].

ALOHA Applications   Due to its relatively low capacity, strict ALOHA is rarely used as a MAC by members of a WLAN or WPAN piconet. However, all four networks studied in this book have at least one type of TDMA or TDD in their specifications, none of which includes a listen-before-talk protocol. Furthermore, 802.11, 802.15.3, and (optionally) 802.15.4 all transmit beacons to indicate the presence of the network. These are often sent without regard to whether or not the channel is in use. Hence, many transmissions appear to be ALOHA-like to other networks communicating on the same channel, and these can cause a drop in throughput.

Carrier Sense Multiple Access (CSMA)

CSMA is a listen-before-talk channel access protocol that purports to decrease the probability of entering the collision state. Before making a transmission, the node activates its receiver momentarily to sense channel activity. If the channel is idle, the transmission takes place; otherwise, the node enters a backoff routine, the details of which are implementation dependent. Provided the channel sense (*clear channel assessment* [CCA]) mechanism can respond quickly to changes in the channel status, CSMA has the potential to increase throughput markedly over the ALOHA schemes.

Figure 6-11 gives a sample of channel activity using CSMA. In this example, a busy period begins when a packet arrives, the channel is sensed idle, and the transmission begins. As before, we assume constant-length packets with a normalized time duration of 1. During the time period a, which is the combined propagation and CCA processing delay, the channel is still sensed idle by the other users, so a collision is possible. Upon expiration of this vulnerability period, the channel is sensed busy by all remaining nodes and no collisions can occur. Most indoor wireless networks experience much shorter propagation delays than CCA response times, so it's realistic to use the same value of a for all users within a particular network, or sometimes even across different networks if they all use similar CCA methods. When the transmission ends, an additional time a must expire before the channel is sensed idle due once again to the time required for the CCA mechanism to react. The channel then enters the idle state until the next transmission occurs. The combined busy and idle times constitute a *cycle*.

Provided the vulnerability period a is short compared to the length of a packet, CSMA performs better than ALOHA. This makes intuitive sense because the

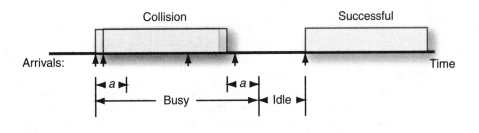

Figure 6-11
General network activity sample for the CSMA protocol. The first and second arrivals collide, the third and fourth arrivals are deferred for later transmission, and the fifth arrival is successfully transmitted.

network effectively operates as ALOHA during the vulnerability period, followed by a time during which any new packet arrivals are deferred for later transmission. CSMA performance depends not only upon the vulnerability period, but also upon the rescheduling algorithm used both for packet arrivals during the busy period, and for transmitted packets that experience a collision. Throughput using CSMA still cannot be greater than 1, but offered rates can be quite high before instability results because all packet arrivals are considered part of the offered load in CSMA, even if they occur during the busy period and are deferred.

A slight performance improvement in CSMA can be realized by slotting the transmission times. Unlike slotted ALOHA, CSMA slot times are based on the CCA time a and are usually much shorter than the packet length. If the channel is sensed idle during a slot, then transmission can commence at the beginning of the following slot. In practical indoor wireless networks, the slot boundaries can be aligned to beacons or controller transmissions, and with negligible propagation delay all users can achieve relatively close slot synchronization. Analyzing slotted CSMA can be quite complex, so most of the equations we show will be for unslotted CSMA. For WLAN and WPAN systems, the slot time is defined as the basic time unit to ensure that consistent timing at the PHY and MAC layers are used by devices from different manufacturers that may be running at different processor speeds.

Nonpersistent CSMA The nonpersistent CSMA protocol operates in the following way. When a packet arrives at a node, it first senses the channel and transmits the packet if it is idle. If the channel is busy, the node waits a random time and checks the channel again. This is a power-efficient algorithm because the receiver can be shut down during the random waiting period, but as we'll discover, delays can be long during high channel traffic. The throughput for nonpersistent CSMA as a function of the offered rate and vulnerability period is [Kle75]

$$S = \frac{Ge^{-aG}}{G(1 + 2a) + e^{-aG}} \tag{6.14}$$

which is the ratio of the probability that the channel successfully transmits a packet to the average cycle time. As $a \to 0$, channel sensing becomes instantaneous, so collisions are impossible and $S \to 1$ as G becomes large. In fact, for both unslotted and slotted nonpersistent CSMA,

$$\lim_{a \to 0} S = \frac{G}{1 + G} \tag{6.15}$$

For practical networks, a is always greater than 0, so $S \to 0$ as G becomes large.

1-Persistent CSMA The purpose of 1-persistent CSMA is to reduce delay by avoiding the idle channel state if a packet is available to send. When a packet arrives at a node, it is transmitted if the channel is sensed idle. If the channel is sensed busy, the node waits until the channel becomes idle and then transmits the packet with probability 1 (hence the name 1-persistent). The disadvantage, of course, is that if more than one node has a packet to send at the end of a busy period, they will always collide. Also, a power penalty is paid because a waiting node's receiver must remain on to sense the end of the current busy period.

The throughput of 1-persistent CSMA is given by [Kle75]

$$S = \frac{G[1 + G + aG(1 + G + aG/2)]e^{-G(1+2a)}}{G(1 + 2a) - (1 - e^{-aG}) + (1 + aG)e^{-G(1+a)}} \tag{6.16}$$

Equation 6.16 is complex because it must account for the number of packets awaiting transmission at the end of a busy period. A packet arriving during a busy period will be successful if and only if no other packets arrive during the same busy period and no packets arrive during the first a seconds of transmission time following the busy period. For both unslotted and slotted 1-persistent CSMA, we have

$$\lim_{a \to 0} S = \frac{Ge^{-G}(1 + G)}{G + e^{-G}} \tag{6.17}$$

Unlike nonpersistent CSMA, the throughput of 1-persistent CSMA when $a = 0$ doesn't approach 1 for large offered rates; instead, throughput peaks at about $S = 0.54$ for an offered rate of about 1 and then drops toward 0 as G increases further [Kle75]. This is due to the increased probability of collision after a busy period has ended for high offered rates, even when a equals 0.

P-persistent CSMA In an attempt to improve the operation of both nonpersistent and 1-persistent CSMA, the p-persistent CSMA scheme was developed. The intent is to randomize the start time of transmissions that take place at the end of a busy period. In this way, only one transmitter usually activates first, and the others again sense a busy channel and defer their transmissions again. This is done by introducing an additional parameter p, which is the probability that a node with a ready packet persists, that is, the probability that the packet is sent immediately. It follows that the probability is $1 - p$ that the transmission is delayed by τ seconds, which is the duration of a slot. The parameter p is chosen such that the collision rate is tolerable while reducing the idle period between two consecutive transmissions. This so-called p-persistent CSMA can employ a fixed p, or it can dynamically change as offered traffic changes.

The protocol works as follows. When a new packet arrives, the node senses the channel. If idle, the node sends the packet with probability p, or it delays by one slot with probability $1 - p$ and senses the channel again, proceeding as before. If the channel is sensed busy, the node waits until the channel is sensed idle and operates as above. The protocol is thus a generalization of the 1-persistent CSMA. Several WLAN and WPAN specifications employ variations of the p-persistent CSMA. The throughput equation for this protocol is quite complex and is not given here.

CSMA Throughput and Delay Comparisons Figure 6-12 shows the plots of channel throughput versus the offered rate for the various CSMA protocols for $a = 0.01$, along with the throughput of both pure and slotted ALOHA. All the CSMA protocols have a higher capacity than either of the ALOHA schemes, at least when a is relatively small. The capacity of 1-persistent CSMA occurs at about the same offered rate as that of slotted ALOHA, but 1-persistent CSMA has a capacity of greater than 0.5, higher than ALOHA. As expected, the throughput of 1-persistent CSMA drops off quickly in the unstable region.

Nonpersistent CSMA requires much higher offered rates before reaching its capacity, and for small values of a, the capacity is close to the maximum possible value of 1.0. However, the capacity is reached at offered rates of about 20, so fewer than 1 in 20 arriving packets are sent successfully during the first attempt. Most arriving packets are deferred due to a busy channel, and some are involved in collisions. Either way, delay can be significantly high. Also, when operating in the stable region below capacity, nonpersistent CSMA has lower

Figure 6-12
Throughput versus the offered rate for the various CSMA protocols, along with the throughput of pure and slotted ALOHA [Kle75] © IEEE

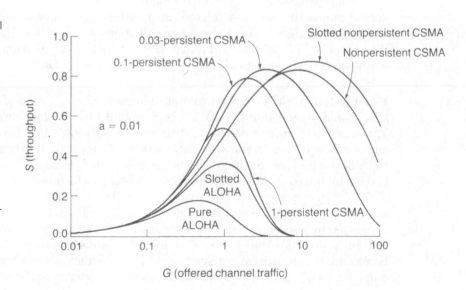

throughput than the other CSMA schemes due to its nonaggressive channel access mechanism.

Throughput performance of p-persistent CSMA falls between 1-persistent and nonpersistent. The capacity is reached for offered rates higher than for 1-persistent and lower than for nonpersistent CSMA, and capacity is slightly lower than for nonpersistent CSMA but significantly higher than for 1-persistent CSMA. Of all the channel access schemes shown, 0.1-persistent CSMA is probably the best compromise between throughput and delay, because its capacity of about 0.8 has its peak at a relatively low offered rate of about 2, and throughput is comparatively high for given offered rates in the stable performance region.

As mentioned earlier, the capacity of the various CSMA methods is affected significantly by the CCA time as manifested through the parameter a. Figure 6-13 depicts how the capacity drops as the CCA time increases due to the increased probability that a collision will occur during the first portion of a busy period. If the CCA time exceeds about 0.2 of the packet length, slotted ALOHA is a better protocol because it doesn't rely on CCA information that is (for large a) often erroneous. This is one reason that maximum allowable CCA times are listed in WLAN and WPAN specifications.

Finally, we present the delays normalized to packet length for the various protocols in Figure 6-14. These curves show the characteristic sudden rise as throughput demands approach channel capacity. Of the CSMA schemes, p-persistent has the best delay performance because the parameter p can be optimized for channel conditions to minimize delay. The standard by which the others are compared, though, is the M/D/1 queue. Delay is small for the queue

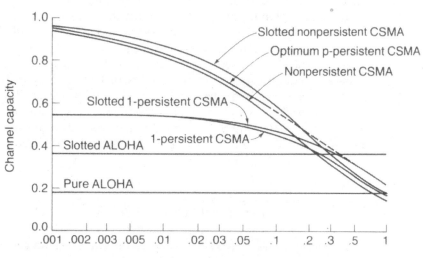

Figure 6-13
Channel capacity for the various CSMA protocols for different vulnerability values a, along with the capacity of pure and slotted ALOHA [Kle75] (© IEEE

Figure 6-14
Normalized
delay versus
channel
throughput for
different
channel access
mechanisms
[Kle75] © IEEE

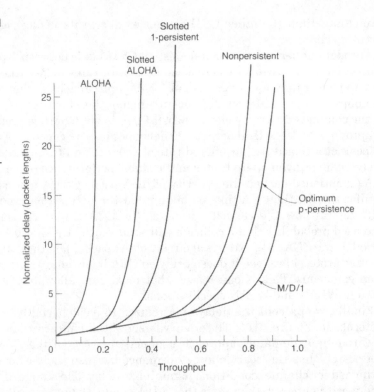

even for throughputs in excess of 0.8 partly due to the complete avoidance of the collision state. Also, for fixed-length packets found in this type of queue, if a packet's position in the queue is known then the delay is also known.

For a practical network, the information in this section leads us to several conclusions. First, when operating at a throughput well below its respective capacity, any of the channel access schemes work well. Second, if CCA is fast enough, a higher channel capacity is achieved with CSMA than with ALOHA. Third, when the capacity is approached from the stable region, the delay increases rapidly regardless of the channel access method used. Finally, the *p*-persistent scheme, or a variation thereof, presents the best compromise between throughput and delay if a queue cannot be implemented for practical reasons.

802.11 Wi-Fi Operations

IEEE 802.11 is the only WLAN discussed in this book, and several renditions of it have become extremely successful. Single-chip solutions are available that are only slightly larger physically than Bluetooth devices and are also only slightly more costly. (The higher nominal TX power of 802.11 leads to higher-energy consumption, though.) Although 802.11 can be operated in the ad hoc peer-to-peer mode, almost all implementations include an *access point* (AP) as the central controller and the means by which the *station* (STA) units access network services. We therefore direct most of our attention toward the infrastructure *basic service set* (BSS) rather than the *independent basic service set* (IBSS).

The basic 802.11 specification, developed before any of the letter-suffix renditions were created, included three PHY entities: *direct sequence spread spectrum* (DSSS), FHSS, and *diffuse infrared* (DFIR). Because 802.11b attained its faster data rates using the DSSS-based derivative CCK, both FHSS and DFIR have become obsolescent. The various renditions of 802.11 all have a similar MAC protocol.

Wi-Fi PHY Frame Structure

During its development, Wi-Fi was given some interesting terminology, sometimes requiring abbreviations of abbreviations. The basic PHY frame is one of these, burdened with the name PPDU. This stands for *PLCP protocol data unit*, and PLCP in turn means *physical layer convergence protocol*. The specific PPDU format is different for DSSS and CCK used by 802.11b, OFDM used by 802.11a/g, and DSSS-OFDM used by 802.11g.

802.11b PPDU   The 802.11b PPDU can have either a long or short format, as shown in Figure 6-15. Following standard wireless practice, the transmitted bits

Figure 6-15
Long and short
802.11b PPDU

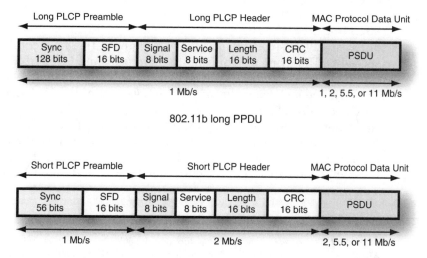

802.11b long PPDU

802.11b short PPDU

are whitened to prevent DC bias at the detector by randomizing the binary 1's and 0's and forcing their occurrence to be about equal in number. In 802.11b devices, support for the long PPDU is mandatory, yet is optional for the short PPDU. The 802.11g device must support both PPDU lengths.

The long PPDU begins with a PLCP preamble of 128 whitened binary 1's in the *sync* field, followed by 0xF3A0 as a *start-of-frame delimiter* (SFD). Next comes the header, consisting of a *signal* field containing the data rate by which the *PLCP service data unit* (PSDU) will be sent, a *service* field listing whether CCK or PBCC is used for a high rate (5.5 or 11 Mb/s) PSDU, and a PSDU *length* field. The header is protected by a 16-bit *cyclic redundancy check* (CRC), and both the preamble and header are sent at 1 Mb/s. The PSDU follows the PPDU header. The header and preamble of the long PPDU require 192 μs to transmit.

The optional short PPDU enhances throughput by reducing overhead in two ways. First, the sync field consists of only 56 bits, sent as (whitened) binary 0's to distinguish it from the long PPDU sync field. Second, the short PPDU header is sent at 2 Mb/s; consequently, the PSDU must be sent at 2 Mb/s or faster. The SFD is 0x05CF, which is the time reversal of the long PPDU SFD. This SFD will not be recognized by nodes without short PPDU support. The header and preamble of the short PPDU require 96 μs to transmit, half that of the long PPDU. The 802.11b specification doesn't say when to use the short or long PPDU.

The PSDU consists of a header field up to 30 bytes long, containing the frame type, addressing, and sequence fields; a frame body between 0 and 2,304 bytes long (with an additional 8 bytes if *wired equivalent privacy* [WEP] encryption is enabled); and a 4-byte *frame check sequence* (FCS). Therefore, the PSDU length can be between 28 and 2,346 bytes. This is the only part of the frame that the MAC layer sees, whether generating or receiving, so the PSDU is also called the *MAC protocol data unit* (MPDU).[*] The MPDU frame body is called the *MAC service data unit* (MSDU) in a nonfragmented frame.

802.11a/g PPDU   The 802.11a and 802.11g OFDM modulation scheme has a different PPDU structure, as shown in Figure 6-16. The *preamble* consists of 12 special OFDM symbols requiring 16 μs to transmit, followed by the first part of the *PLCP header* consisting of a one-symbol (24-bit) *signal* field sent at 6 Mb/s. The remainder *data* portion of the frame is sent at the rate specified in the signal field. This field begins with a two-byte *service* field containing whitener initialization values, followed by the *PSDU* (MPDU) field, and concludes with six *tail bits* and sufficient *pad bits* to complete the last transmitted OFDM symbol. An unusual feature of the OFDM PPDU is that the PLCP header contains both the signal and service fields, which may be sent at different bit rates.

---

[*]Some publications exclude the MAC header, FCS, and security overhead in the PSDU when referring to the MPDU. However, we will follow notation in the 802.11 specification and consider the PSDU and MPDU to be the same collection of bits viewed from, respectively, the PHY or MAC perspective.

Figure 6-16
PPDU for
802.11a and
802.11g
OFDM

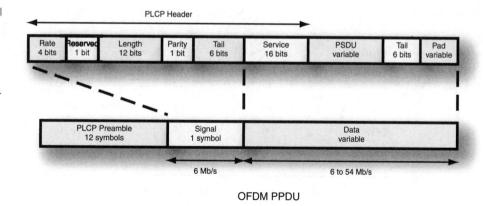

OFDM PPDU

For enhanced coexistence in a mixed network, 802.11g can optionally employ the DSSS-OFDM PPDU, consisting of a DSSS header preceding an OFDM body, making header information available to any nearby 802.11b device. The implementation of DSSS/OFDM is technically challenging because the recipient's receiver must transition smoothly and rapidly from single-carrier to multicarrier demodulation after the header is sent. The transition is facilitated if the sending node can carefully control power, spectrum, frequency, phase, and timing.

Wi-Fi MAC Timing

Channel access methods are essentially the same for all Wi-Fi renditions. To reduce the probability of collision, especially among the members of a particular BSS, several timing entities have been specified:

- **Slot time**   Basic MAC time measurement unit
- **Short interframe space (SIFS)**   Prevents interruption of basic frame exchange sequence
- **Point coordination function interframe space (PIFS)**   Used by the point coordinator to capture a channel, defined as SIFS + 1 slot time
- **Distributed coordination function interframe space (DIFS)**   Waiting time before a new contention window opens, defined as PIFS + 1 slot time
- **Extended interframe space (EIFS)**   Waiting time after receiving a frame with errors, defined as SIFS + (8 × *acknowledgment* [ACK] length) + PLCP (preamble + header)

The definitions of these timing values will make more sense once we describe MAC channel activity in subsequent subsections. The actual values for these

Table 6-1

MAC timing
entities for
802.11a/b

Entity	Value (µs)			
	802.11a	802.11b	Only 802.11g	802.11g with 802.11b
Slot time	9	20	20 or 9*	20
SIFS	16	10	10**	10**
PIFS	25	30	30 or 19*	30
DIFS	34	50	50 or 28*	50

*The shorter times are optional when a BSS contains only 802.11g nodes. The optional PIFS and DIFS times are based upon the 9 µs slot time, so the optional times must be used together.

**For 802.11g transmissions using OFDM, an additional 6 µs of no transmission, called the *signal extension*, follows the frame. This time, combined with the 10 µs SIFS time, allows the convolutional decoder to return to its initial state.

time units (except EIFS, which can vary) are listed in Table 6-1 for 802.11a/b/g. It's interesting to note that an 802.11g WLAN has the potential to perform slightly better than an 802.11a WLAN due to the shorter PIFS and DIFS optionally available to a BSS containing only 802.11g nodes. (Although the SIFS is also shorter in 802.11g, the required signal extension time offsets this advantage.)

Distributed Coordination Function (DCF)

The *distributed coordination function* (DCF), also called the *basic access mechanism*, uses a channel access method that is similar to *p*-persistent CSMA in which the parameter *p* adjusts dynamically to channel loading. This access method is called *carrier sense multiple access with collision avoidance* (CSMA/CA). When a frame is available for transmission, the sending node monitors the channel for a time equal to DIFS and sends the frame if the channel remains idle during this time. If the channel is busy or becomes busy during the monitoring period, the sending node waits until the channel becomes idle, waits an additional DIFS, and then enters a contention window.

Contention window timing is shown in Figure 6-17. If this is the first contention window for that particular frame, the node loads a counter with a random number uniformly distributed on the set {0, 1, ... , 31} and decrements the counter at each slot boundary during the contention window. If the channel becomes busy before the backoff time expires, the backoff counter stops decrementing slots until the next contention window opens, upon which time the backoff counter resumes decrementing slots. In this manner, nodes that have waited

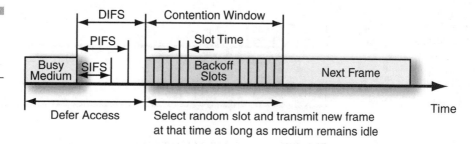

Figure 6-17
DCF timing between busy periods

for the longest number of contention windows obtain a statistical advantage for access to the channel.

Upon expiration of the random backoff time, if the channel is still idle, the data frame is transmitted. The intended recipient waits for SIFS after the end of the data frame, during which it performs a CRC check. An ACK frame is then returned to the sender if all is well; otherwise, no transmission is made for an implied *negative acknowledgment* (NAK). If no ACK is forthcoming, a collision is assumed, and the sending node effectively doubles its average backoff time by selecting a new random number uniformly distributed on the set $\{0, 1, \ldots, 63\}$. In this way the slot transmission probability p is reduced in an attempt to keep network offered loads within the stable region of the throughput curve at the expense of increased average delay. This exponential backoff process is shown in Figure 6-18. With each failure to receive an ACK, the sending node again effectively doubles its average backoff time until reaching a maximum. Each frame during this data/ACK exchange is separated by the short SIFS to prevent other nodes from transmitting until the exchange is complete. For example, a node participating in DCF must wait at least DIFS before it is allowed to transmit on a channel that is sensed idle, and if an exchange is still in progress the node's CCA will indicate that the channel is busy before DIFS expires.

The price of a collision on a long data frame is high, because the entire frame must be retransmitted. Furthermore, the channel is in the collision state for the duration of this long frame, increasing delay for all nodes awaiting channel access. To enhance throughput when long frames are awaiting transmission, a four-way frame exchange can be used, depicted in Figure 6-19. When a node is ready to send a data frame, it monitors the channel for DIFS, with additional backoff as required, and then sends a short *request-to-send* (RTS) frame. The RTS frame contains a duration field from which a *network allocation vector* (NAV) can be determined. The NAV specifies the time needed to complete the entire four-way exchange. Other nodes, even those outside the desired BSS, that successfully decode the RTS frame will defer their own transmissions until the NAV expires.

The recipient node responds to the RTS with a *clear-to-send* (CTS) frame from which a (shorter) NAV can be extracted, indicating when the remaining frames in the four-way exchange will finish. In an infrastructure BSS, the AP will always send either the RTS or CTS, increasing the probability that all nearby Wi-Fi

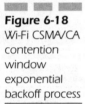

Figure 6-18
Wi-Fi CSMA/CA
contention
window
exponential
backoff process

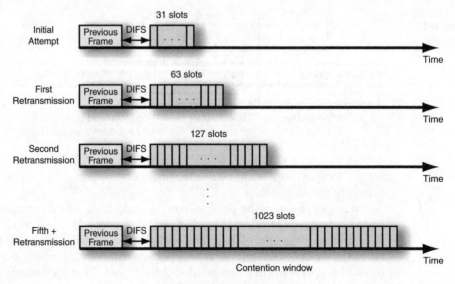

Figure 6-19
Wi-Fi CSMA/CA
channel access
timing using
RTS and CTS to
begin a four-
way exchange

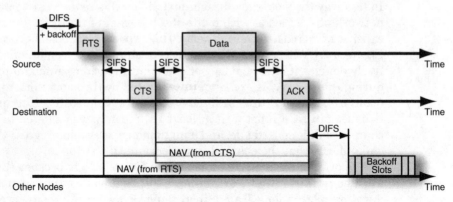

nodes will hear the frame and thus defer any transmissions until the NAV expires. In this way, physical carrier sense via CCA and virtual carrier sense via NAV combine to enhance the possibility that the frame exchange will be successful. The 802.11 specification doesn't state when RTS/CTS should be employed, but many manufacturers allow the user to enable or disable RTS/CTS globally, and/or manually select an MPDU threshold length above which RTS and CTS frames will automatically be sent. RTS/CTS can be used only for *unicast frames* (frames directed at one recipient).

The throughput of a long data frame during high channel loading or intermittent interference can be further enhanced through a method called *frame*

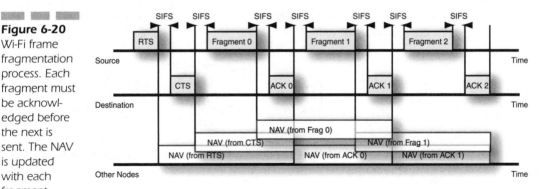

Figure 6-20
Wi-Fi frame fragmentation process. Each fragment must be acknowledged before the next is sent. The NAV is updated with each fragment and ACK.

fragmentation, depicted in Figure 6-20. An MSDU is divided into several pieces, all of which are equal in length except perhaps the last, and an MPDU is formed from each. These in turn are assembled into a set of PPDU frames and sent under one RTS/CTS. Each fragment is separately ACKed, and an updated NAV is included with each fragment and ACK. A particular fragment must be successfully transferred before the next one can be sent. The data frame length threshold for enabling fragmentation is also not given in the specification, but many implementations suggest making the RTS/CTS and fragmentation thresholds identical. Fragmentation is available only for unicast frames.

DCF Frame Transmission Times and User Data Throughput It's obvious that Wi-Fi can exchange frames using many different methods, with many data rate options, but we will examine the performance of only a few of these after deriving some user data throughput generalizations over a perfect channel [Cho01]. The basic transmission cycle consists of the following steps:

1. DIFS deferral
2. Backoff within the contention window (if necessary)
3. Data frame transmission with PSDU length from 28 to 2,332 bytes
4. SIFS deferral
5. ACK frame transmission

The time in μs to send a DSSS/CCK data frame with an L-byte MAC frame body (MSDU) at a PHY rate of m Mb/s using a long PPDU is given by

$$T_{\text{data}}^{m} = t\text{PLCPPreamble} + t\text{PLCPHeader} + \frac{8 \times (28 + L)}{m}$$

$$= 192 + \frac{8 \times (28 + L)}{m} \qquad \text{DSSS/CCK} \qquad (6.18)$$

An RTS frame has a length of 20 bytes, so the time in μs to send this frame at a PHY rate of n Mb/s is

$$T_{\text{RTS}}^n = 192 + \frac{8 \times 20}{n}$$

$$= 192 + \frac{160}{n} \qquad \text{DSSS/CCK} \qquad (6.19)$$

The MPDU in a CTS or ACK frame is 14 bytes long. The time in μs to send the CTS/ACK frame at a PHY rate of n Mb/s is

$$T_{\text{CTS}}^n = T_{\text{ACK}}^n = 192 + \frac{8 \times 14}{n}$$

$$= 192 + \frac{112}{n} \qquad \text{DSSS/CCK} \qquad (6.20)$$

If the short PPDU is used, then the 192 terms become 96, and the PSDU data rate must be 2 Mb/s or higher.

For OFDM, note that for an m Mb/s PHY, the number of bytes in an OFDM symbol is $m/2$. Aside from the 24-byte MPDU header and the 4-byte FCS, the OFDM PPDU data field also contains a 16-bit service field and 6 tail bits, for a total of 30.75 bytes (ignoring for now any additional pad bits). The OFDM RTS frame contains a 20-byte MPDU header, a 16-bit service field, and 6 tail bits, for a total of 22.75 bytes. Likewise, the OFDM CTS and ACK contain a 14-byte MPDU header, along with the 16-bit service field and 6 tail bits, for a total of 16.75 bytes. Therefore, the equivalent TX times in μs (not including the 6 μs signal extension time for 802.11g ERP-OFDM) are

$$T_{\text{data}}^m = t\text{PLCPPreamble} + t\text{PLCPHeader} + \left\lceil \frac{30.75 + L}{m/2} \right\rceil \times t\text{Symbol}$$

$$\text{OFDM} \quad (6.21)$$

$$= 20 + \left\lceil \frac{30.75 + L}{m/2} \right\rceil \times 4$$

$$T_{\text{RTS}}^n = 20 + \left\lceil \frac{22.75}{n/2} \right\rceil \times 4 \qquad \text{OFDM} \qquad (6.22)$$

and

$$T_{\text{CTS}}^n = T_{\text{ACK}}^n = 20 + \left\lceil \frac{16.75}{n/2} \right\rceil \times 4 \qquad \text{OFDM} \qquad (6.23)$$

The necessary OFDM pad bits are accounted for in the ceiling operator in Equations 6.21 to 6.23.

The maximum MSDU size in a data frame without WEP is 2,304 bytes, and we assume that a 24-byte MAC header and a 4-byte FCS form the 2,332-byte PSDU under these conditions. (Many examples in the literature limit MSDU length to 1,500 bytes to conform to size restrictions above the MAC layer.) Table 6-2 shows the times required for sending some of the most common frames. For 802.11b, the long preamble is assumed. The RTS, CTS, and ACK frames each have a fixed PSDU size, and for data frames the TX times are for the longest allowable PSDU.

Table 6-2

Wi-Fi frame transmission times (long preamble for DSSS and CCK)

Frame	Modulation	Data rate (Mb/s)	PSDU size (bytes)	MSDU size L (bytes)	PPDU transmission time (μs)
Data	DSSS	1	2,332	2,304	18,848
Data	DSSS	2	2,332	2,304	9,520
Data	CCK	5.5	2,332	2,304	3,584
Data	CCK	11	2,332	2,304	1,888
Data	OFDM	6	2,332	2,304	3,152
Data	OFDM	12	2,332	2,304	1,588
Data	OFDM	24	2,332	2,304	804
Data	OFDM	54	2,332	2,304	368
RTS	DSSS	1	20	0	352
RTS	DSSS	2	20	0	272
RTS	OFDM	6	20	0	52
RTS	OFDM	24	20	0	28
CTS or ACK	DSSS	1	14	0	304
CTS or ACK	DSSS	2	14	0	248
CTS or ACK	OFDM	6	14	0	44
CTS or ACK	OFDM	24	14	0	28

Recall that the CCA time for DSSS/CCK is 15 μs (4 μs for OFDM). At first glance it may appear that the short RTS and CTS frames are more susceptible to collisions because the CSMA vulnerability period a is higher (approximately 0.1) than it is for the longer data frames. This is not the case, though, because the offered rate is normalized to frame length in our CSMA performance calculations. For a fixed average channel utilization given in frames per second, the offered rate drops as frames become shorter.

Data throughput for either form of modulation is the number of bits per frame (8L) divided by the average time to transmit the frame and its ACK. The header and FCS are excluded. Throughput is therefore given by

$$T(L,m,n) = \frac{8L}{a\text{DIFSTime} + \overline{T}_{bk} + T_{\text{data}}^m(L) + a\text{SIFSTime} + T_{\text{ACK}}^n} \text{ Mb/s} \quad (6.24)$$

where SIFS and DIFS times are given in Table 6-1, and all times are expressed in μs. The average backoff time $\overline{T}_{bk}$ depends on the number of users attempting to access the channel. We assume that only two users are connected in a point-to-point configuration to simplify the calculations, and that a contention window with backoff is always entered. In this case

$$\overline{T}_{bk} = \frac{CW \min}{2} \times a\text{SlotTime} \quad (6.25)$$

where slot times are given in Table 6-1 and the minimum contention window size *CWmin* is 31 slots. For each form of modulation, Equation 6.24 simplifies to

$$T(L,m,n) = \frac{8L}{370 + T_{\text{data}}^m(L) + T_{\text{ACK}}^n} \quad \text{DSSS/CCK} \quad (6.26)$$

and

$$T(L,m,n) = \frac{8L}{189.5 + T_{\text{data}}^m(L) + T_{\text{ACK}}^n} \quad \text{OFDM} \quad (6.27)$$

A plot of MSDU throughput versus frame length is given in Figure 6-21. DSSS/CCK uses 11 Mb/s for data and 2 Mb/s for ACK frames, and OFDM uses 54 Mb/s for data and 24 Mb/s for ACK. Even when the channel is error-free, with the longest possible data frames sent at the maximum rate, MSDU throughput is only about 2/3 of the raw data rate for either DSSS/CCK or OFDM. Under realistic operating conditions in a relatively isolated environment, with the occa-

Figure 6-21
MAC frame body through-put versus length for DSSS/CCK and OFDM Wi-Fi transmissions over a perfect channel. DSSS/CCK uses 11 Mb/s for data and 2 Mb/s for ACK, whereas OFDM uses 54 Mb/s for data and 24 Mb/s for ACK.

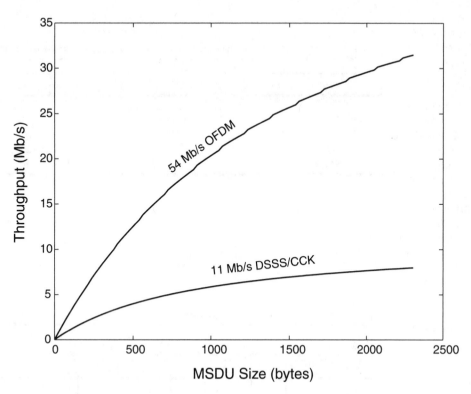

sional lost frame from a collision or from bit errors, throughput between the MAC and the layer above it is about half the raw data rate, which is a reasonable rule of thumb for wireless networks that employ some form of CSMA. Also, this throughput is divided among all the users within a BSS, or even across several BSS if they are operating in the same area and using the same carrier frequency.

Point Coordination Function (PCF)

For increased flexibility, the 802.11 MAC also provides for the optional use of a contention-free *point coordination function* (PCF), in which a *point coordinator* (PC) polls STA devices participating in PCF. The PC is always the AP, and it begins a *contention-free period* (CFP) by sending a beacon frame that contains a NAV designating the anticipated length of the CFP, as depicted in Figure 6-22. Wi-Fi nodes that aren't capable of participating in the CFP use this NAV as a virtual carrier sense. When the CFP is completed, a contention-based DCF begins, which may possibly encroach into the next CFP.

Figure 6-22

If a contention-free period is employed by a Wi-Fi AP, it immediately follows the beacon and precedes the DCF contention period. If necessary, a contention period may encroach upon the subsequent CFP.

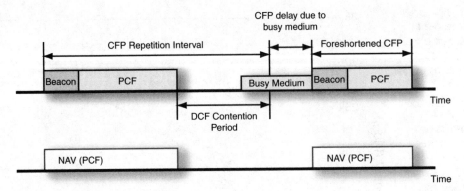

Figure 6-23

Frame exchange during the Wi-Fi CFP

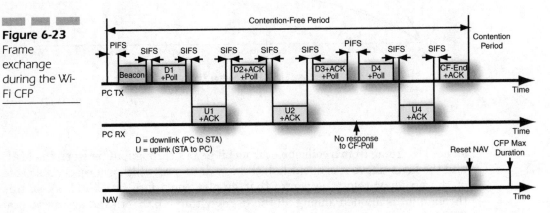

Timing details for one CFP are shown in Figure 6-23. The beacon signifying the start of the CFP begins PIFS after the last busy period. This allows any in-progress DCF exchanges to finish (because each DCF frame is separated by the shorter SIFS), but then the PC can seize the channel before DIFS has expired, giving it priority over the start of a new DCF busy period. The PC then polls participating STAs one by one for transmissions. Polling efficiency is improved by combining management and data frames to the maximum extent possible. For example, the PC can acknowledge the uplink frame from STA 1, send a downlink data frame to STA 2, and poll STA 2 for a response, all in a single frame.

If an STA fails to respond to its poll, the PC waits PIFS before sending its next poll to avoid preemption by a DCF STA that may not have acquired the PCF NAV. The PC can end the CFP early, in which case the NAVs are terminated with a CF-End frame, and the next contention-based DCF period begins. Aggregate throughput at the MAC layer using PCF increases about 10 percent over DCF, but the real strength of PCF is its capability to greatly reduce latency variance. Using PCF can cause coexistence problems for other nearby networks with overlapping bandwidth, however, because once the PC is able to seize the channel, polling replaces CSMA for access.

802.15.1 Bluetooth Operations

Bluetooth is unique among the major WLAN and WPAN systems in that it mitigates interference in part by mandating the use of FHSS. In return for the extensive overhead required for synchronization and hopping, the average bandwidth occupied by an FHSS transmission is limited only by the capability of its frequency synthesizer and, unlike DSSS and its variants, isn't tied to the data or chip rate. From a fabrication point of view, a FHSS transmitter can have an arbitrarily wide average bandwidth without necessarily requiring high-speed circuitry.

Although it's possible to implement CSMA in an FHSS system in which a device will defer transmission in a particular hop channel if it is otherwise occupied, Bluetooth avoids this complexity by using ALOHA-like channel access instead. Interference to other systems, whether to another Bluetooth piconet or a network operating on a fixed channel, is designed to be occasional rather than chronic. By using various other coexistence mechanisms in conjunction with FHSS, the Bluetooth piconet can improve its own operation while adhering to the good neighbor policy.

Where data integrity is of primary importance, *asynchronous communication* is used. This means that each incoming packet is checked for data bit errors by the destination node, and if errors are present then the destination asks the source to repeat the packet. Effective throughput is thus influenced greatly by the channel's *bit error rate* (BER). On the other hand, if low latency (small delay) is required, then *synchronous communication* is employed by the piconet. Real-time, two-way voice communication falls within this category. Digitized voice packets are assigned specific transmission time slots and repeat transmissions either aren't allowed or their number is strictly limited. The rate at which new packets are sent is therefore mostly unaffected by channel integrity; instead, voice packets sent over a channel with high BER will produce distortion when they are converted back to analog audio.

Packet Structure and Exchange

A Bluetooth data packet contains an *access code*, *header*, and *payload*, as shown in Figure 6-24. The access code (72 bits nominal) is used for the initial synchronization of the receiver circuitry, but it also contains other information such as the piconet identity, a portion of the sender's MAC address, or a portion of the recipient's MAC address, depending on the context of the application. The header (54 bits) includes the destination address, the type of payload to follow, and some error control information. This header is protected by both a (3,1) binary repetition FEC code and a *header error check* (HEC). The payload that follows is variable, with a maximum length determined by whether 1, 3, or 5 transmission slots are used.

In a manner consistent with most WPAN operations, Bluetooth baseband packets are sent directly from their source to their destination. The master is at one end of the link, and it can support up to seven active slaves. Slaves communicate only with the master, never with each other. Each packet is transmitted on a new hop frequency. Baseband packets are constructed to take advantage of the TDD process that provides for an orderly exchange of data between master and slave. The master and one of the slaves in the piconet take turns transmitting, so two or more transmitters are never on at the same time within a piconet. No listen-before-talk (CSMA) protocol exists in Bluetooth, but using a new hop frequency for each transmission reduces the disruption to the Bluetooth piconet from other users, and it reduces the disruption to other users from the Bluetooth piconet, regardless of whether the other users operate on a fixed channel or use FHSS.

Time is divided into slots that are nominally 625 μs in length and are numbered with consecutive integers. The master transmits to the slave in even-numbered time slots, and the slave transmits to the master in odd-numbered time slots. Each transmission takes place at a new hopping frequency, and a complete packet of data is sent in each time slot. This process is depicted in Figure 6-25. Each packet is allowed up to 366 μs for its transmission, equating to a maximum one-slot packet length of 366 bits. The additional 259 μs is used by the radio to change to the next frequency in the hop sequence and activate the appropriate transmitter or receiver.

Figure 6-24

General Bluetooth baseband packet structure

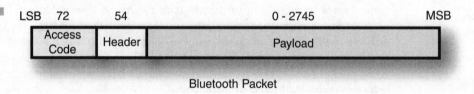

LSB 72	54	0 - 2745	MSB
Access Code	Header	Payload	

Bluetooth Packet

Figure 6-25
Bluetooth TDD communication process in a point-to-point configuration. The master transmits in an even time slot, starting at time (2k), where k is some integer index, and the slave receives in that slot. The slave transmits and the master receives in the odd-numbered slots [Mor02].

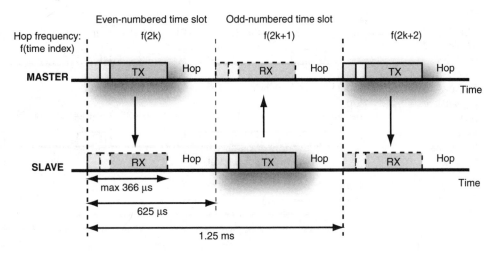

Figure 6-26
Multislave operation. Slaves communicate only with the master, and a slave can transmit only when addressed by the master in the previous time slot. Slaves not addressed can turn their receivers off after decoding the header [Mor02].

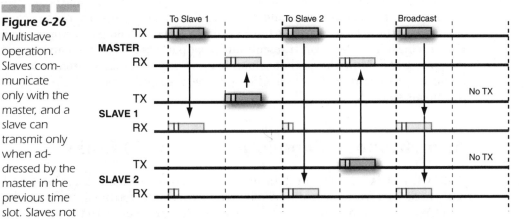

For multislave operations, depicted in Figure 6-26, the master transmits on even-numbered time slots as before, but for asynchronous communication a particular slave can transmit in the subsequent odd-numbered time slot only when it is specifically addressed by the master in the previous time slot. Active slaves are not allowed to transmit in the slave-to-master slot following a broadcast packet.

Figure 6-27
Single- and multislot baseband packets. All packets are sent on a single hop frequency [Mor02].

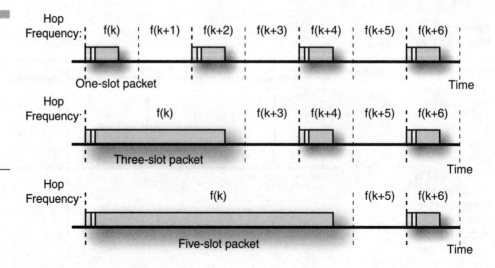

To enhance throughput, Bluetooth devices can transmit multislot packets of either three or five time slots in duration, as shown in Figure 6-27. Every baseband packet, regardless of its length, is transmitted on a single hop frequency. When transmission of a multislot packet is completed, the hop sequence resumes at the channel it would have used had no multislot transmission occurred, so these longer packets have minimal disruption on piconet operation. Either a master or slave can transmit multislot packets if the other can support their reception. Multislot packets shouldn't be used on an error-prone channel because a longer packet is more likely to be corrupted, requiring a lengthy retransmission.

Asynchronous Connectionless (ACL) Links

The ACL link is used where data integrity is more important than latency. A packet received with uncorrectable bit errors is normally retransmitted until it's error-free. The average number of retransmissions increases with the increasing channel BER, so latency is variable and can occasionally be quite long.

ACL packets have a three-character identifier consisting of two letters followed by a number. The first character is always a *D*, which means *data*. The second character can be either an *H* for *high speed* or *M* for *medium speed*, and the last character can be either a *1, 3,* or *5,* which identifies the number of time slots used by the packet. The packet is high speed if it has no FEC, and it is medium speed if the (15,10) shortened Hamming code is applied to the payload. A packet of type AUX1 is also available for transporting raw data (no FEC or CRC) within a single time slot.

As we mentioned at the beginning of this chapter, user data throughput comparisons must take into account only actual payload data contained in each ACL packet. That is, no access code, header, or error control (detection or correction) overhead should be included in the throughput calculations. If the master and slave are both using the same type of packet for their data exchange, the channel is *symmetric*. In many situations, such as a file transfer, one node will transmit data in multislot packets, but the other node will return only a single-slot packet containing the ACK/NAK information in the header, or perhaps a DH1 or DM1 packet if a small amount of data is to be returned as well. This *asymmetric channel* has a *forward direction* for the long packets and a *reverse direction* for the single-slot return packets. Maximum user data throughput values for the six most commonly used ACL packet types are given in Table 6-3 for a perfect communication channel. Asymmetric data is assumed to contain single-slot packets in the return channel that have the same FEC as the packets in the forward channel.

SCO and eSCO Links

If low latency is more important than data integrity, a *synchronous connection-oriented* (SCO) or *extended SCO* (eSCO) link is established between master and slave. Latency is the time between the creation of a new packet at the transmitting node and its successful reception at the destination node. The SCO link is a circuit-switched, point-to-point link between a master and a single slave. Fortunately, voice reproduction from a digitized bit stream can tolerate a fairly high percentage of bit errors, so under most conditions the lack of packet retransmissions shouldn't be a significant detriment to performance. Higher data rates and limited retransmission capabilities are supported by an eSCO link.

Table 6-3

Maximum user data throughput (kb/s) of ACL symmetric and asymmetric channels (BER = 0)

Packet type	Symmetric channel	Forward asymmetric channel	Reverse asymmetric channel
DM1	108.8	—	—
DH1	172.8	—	—
DM3	258.1	387.2	54.4
DH3	390.4	585.6	86.4
DM5	286.7	477.8	36.3
DH5	433.9	723.2	57.6

Table 6-4

Structure of
SCO and eSCO
HV packets
with maximum
user data
throughput
(BER = 0)

Type	Number of time slots	Payload size (bits)	FEC rate	CRC	Symmetric maximum user data throughput (kb/s)
HV1	1	80	1/3	No	64
HV2	1	160	2/3	No	64
HV3	1	240	None	No	64
EV3	1	8–240	None	Yes	96
EV4	1–3	8–960	2/3	Yes	256
EV5	1–3	8–1440	None	Yes	384

The structures of the types of SCO and eSCO packets are shown in Table 6-4. SCO packets all have identical payload sizes of 240 bits (30 bytes), but the amount of digitized voice contained in each one differs due to FEC overhead. The eSCO packets are suited for synchronous or isochronous data transfers, so their payloads are variable and a CRC is provided for error checking. A bad eSCO packet can be retransmitted within a subsequent retransmission window. A DV1 packet payload contains both synchronous and asynchronous data.

It's obvious from Table 6-4 that a single HV1 SCO link will fully load the Bluetooth piconet, with packet timing depicted in Figure 6-25. This type of link is often used when analyzing the effect of Bluetooth interference on another wireless network, because the HV1 link offers a worst-case channel utilization situation.

802.15.3 WiMedia Operations

As its name implies, 802.15.3 WiMedia was developed to transport media-based information over short distances. Most applications will involve isochronous streaming data, in which video or audio material is transferred from the source to destination. To prevent disruptions, the stream is buffered at the destination before playback commences. If the link experiences a short-term outage, the buffer continues the playback stream at the destination until communication is reestablished and the buffer filled again. For this method to work correctly, the link must support a data rate in excess of that required by the playback media, which can be tens of megabits per second in the case of *high-definition television*

(HDTV). WiMedia can support raw data rates up to 55 Mb/s in the 2.4 GHz PHY, and the maximum data rate over the UWB PHY will be about 480 Mb/s.

WiMedia PHY Frame Structure

The standard WiMedia 2.4 GHz PHY frame is depicted in Figure 6-28 for data rates of 22, 33, 44, and 55 Mb/s. The preamble is a known sequence of symbols that the receiver uses as an indication that a frame has arrived. The last symbol in the preamble is inverted to designate that the PHY header follows. This header contains length, frame data rate, and descrambling information.

The MAC header follows the PHY header, and this is where addressing information is located. Although neither header is protected by an FEC code, their combined integrity is checked via a single HCS. The frame payload follows, which can be up to 2,044 bytes long. If a frame payload exists, it is protected with a 4-byte FCS. If necessary, *stuff bits* (SB) are added to obtain the correct multiple of symbols for the selected modulation method, and *tail symbols* (TS) terminate the encoded trellis sequence to a known state.

For transmissions at 11 Mb/s, WiMedia uses a slightly different PHY frame, as depicted in Figure 6-29. The first part of the frame is identical to that used for faster data rates, but the part of the frame sent at 11 Mb/s includes a duplicate PHY and MAC header, along with their common HCS, ahead of the usual frame payload and FCS. No stuff bits are needed because 11 Mb/s *quadrature phase shift keying—trellis coded modulation* (QPSK-TCM) sends only 1 bit per symbol. The transmission times for a maximum payload frame at each data rate are

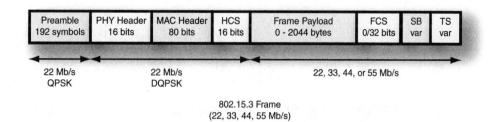

Figure 6-28

For raw data rates of 22, 33, 44, and 55 Mb/s in the 2.4 GHz band, the WiMedia PHY frame consists of a preamble, a PHY and MAC header protected by a single header check sequence (HCS), a frame payload protected by its own FCS, and stuff bits and tail symbols to accommodate the convolutional FEC process and modulation using TCM.

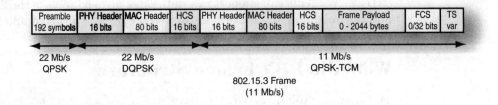

Figure 6-29
WiMedia frames sent at 11 Mb/s in the 2.4 GHz band include a duplicate copy of the PHY and MAC headers, along with their common HCS, as part of the frame body. The opening part of the frame is still sent at 22 Mb/s.

Table 6-5

WiMedia maximum payload TX times (excluding SB and TS fields)

Data rate (Mb/s)	TX time (µs)			
	Preamble	Headers + FCS	Max. payload + FCS	Total
11	17.5	5.1	1,500	1,522
22	17.5	5.1	745	767
33	17.5	5.1	496	519
44	17.5	5.1	372	395
55	17.5	5.1	298	320

given in Table 6-5, not including the relatively short SB and TS fields. The payload TX time at 11 Mb/s includes the repeated headers.

WiMedia MAC Functionality

The 802.15.3 WiMedia specification lists several functions that are to be supported by the MAC layer:

- Fast connection time
- Ad hoc networking
- Data transport with QoS
- Security

- Dynamic membership
- Efficient data transfer

A *device* (DEV) with *piconet controller* (PNC) capabilities starts a piconet by first scanning the available channels to find one that is unused. The channel must remain clear for about 65 *milliseconds* (ms) or longer before it can be classified as available. The PNC then begins transmitting beacons on the selected channel to notify another DEV that a WiMedia piconet exists. The time between beacons ranges from 1,000 to 65,512 μs. These beacons provide timing signals to other users and contain essential piconet information such as the PNC address and maximum TX power. If no clear channel is available, the PNC can start a dependent piconet, sharing the channel with another WiMedia piconet.

WiMedia Superframe   Timing in the WiMedia piconet is based on the superframe, which is composed of three parts:

- **Beacon,** which is used to set timing allocations and manage the piconet
- **Contention access period (CAP),** the optional time period for asynchronous data communication
- **Channel time allocation period (CTAP),** which is used for commands, isochronous data streams, and asynchronous data in a contention-free or slotted ALOHA environment

A timing diagram of a sample superframe is shown in Figure 6-30. The length of the CAP is determined by the PNC and is included in a beacon parameter field. This period is used for command and nonstream data as directed by the PNC. DEV access during CTAP is via TDMA, where each DEV has its own *channel time allocation* (CTA) window for sending a frame to another DEV. CTA periods can be shared, and case access is via slotted ALOHA. Piconet management communication with the PNC occurs during *management CTA* (MCTA) periods,

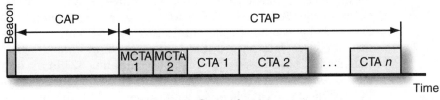

802.15.3 Superframe

Figure 6-30
The 802.15.3 WiMedia superframe begins with a beacon, followed by a contention access period (if available) and a channel time allocation period.

which can be placed anywhere within the CTAP. If the CAP is present, channel access is via CSMA/CA. When streaming data is being sent, superframes follow each other in sequence with a short guard time separating them.

Within the CAP, a DEV is allowed to transmit only when the channel is sensed idle for a random backoff period. Because WiMedia always operates with a CTAP, during which time several DEV intending to use the CAP may accrue frames to send, the CAP backoff algorithm must be used prior to every transmission within the CAP. The duration of a backoff slot's time is PHY dependent and is set at 17.3 μs for the 2.4 GHz band. Contention windows operate in a manner similar to the Wi-Fi process shown in Figure 6-18, but with size 7, 15, 31, and 63 backoff slots. Only one frame can be sent for each backoff period, and all frames and their associated acknowledgments must be completed prior to the beginning of the CTAP. Some frames can be acknowledged immediately, and these Imm-ACK frames don't require a backoff.

802.15.4 ZigBee Operations

The purpose of ZigBee is to provide relatively low data rate communications using nodes that are simple, low cost, and consume little power. The operational duty cycle is also expected to be low (typically 1 percent) for such applications as sensors and industrial control. The raw data rate in the 2.4 GHz *industrial, scientific, and medical* (ISM) band is 250 kb/s.

ZigBee PHY Packet Structure

Two classes of ZigBee devices are specified: the *full function device* (FFD) and the *reduced function device* (RFD). The FFD can take the role of PAN coordinator (principal PAN controller), coordinator (provides synchronization beacons), or a device (basic PAN member). The RFD contains a minimum protocol that enables it to function as a device communicating only with an FFD.

The ZigBee data packet structure is shown in Figure 6-31. The synchronization header is composed of a preamble containing 32 binary zeros, followed by an 8-bit *start-of-frame delimiter* (SFD). The PHY preamble contains only a 7-bit PHY payload length field and a single reserved bit, with no error checking to save overhead. The payload contains a maximum of 127 bytes of data, including a 2-byte FCS. This maximum-length packet requires 4.26 ms to transmit. Note that if the length field in the PHY preamble is corrupted then the payload will also be corrupted, causing the packet to be rejected.

Figure 6-31
The ZigBee
data packet is
sent at 250
kb/s in the 2.4
GHz ISM band.

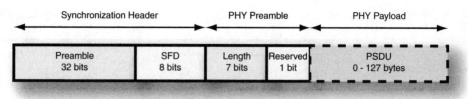

802.15.4 Data Packet

ZigBee MAC Functionality

The 802.15.4 ZigBee specification lists the following tasks provided by the MAC layer:

- Generating beacons if the device is the piconet coordinator
- Synchronizing to the beacons
- Supporting PAN association and disassociation
- Supporting device security
- Employing CSMA/CA channel access
- Maintaining the guaranteed time slot mechanism
- Providing a reliable link between two MAC entities

An FFD starts a piconet by first performing an *active scan* of the available channels to find one that is unused. The active scan sends a beacon request command on each channel. The FFD then records the information contained in any returned ZigBee beacons. If no beacon is forthcoming after a period of time has expired, the channel is considered to be available for a new ZigBee piconet. Of course, if non-ZigBee nodes are present on the channel, interference from the ZigBee piconet is possible. This situation will be examined further in subsequent chapters.

ZigBee CSMA/CA In light of its focus on simplicity, channel access within ZigBee is primarily via unslotted CSMA/CA, with an optional beacon-oriented superframe available when more structure is desired. The timing associated with the CSMA/CA algorithm is depicted in Figure 6-32. ZigBee measures interframe spaces in terms of symbol periods, and these values have widely varying limits. *Long frames* have an MPDU length greater than 18 bytes, and these are followed by a *long interframe space* (LIFS), either after the frame or after its associated ACK if it is acknowledged. *Short frames* are followed in a similar manner by a *short interframe space* (SIFS).

Figure 6-32
ZigBee
CSMA/CA
channel access
timing

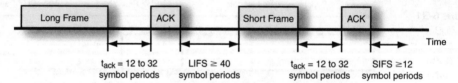

802.15.4 CSMA/CA with ACK

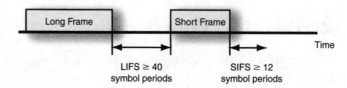

802.15.4 CSMA/CA without ACK

The CSMA/CA backoff procedure is similar to the others we've already examined. A single backoff time period is 20 symbols long, and the maximum number of backoff time periods grows in the usual exponential way after a packet transmission fails. The ZigBee specification provides some flexibility in selecting backoff parameters. For example, the first backoff is randomly selected from the default set {0, 1, . . . , 7} time periods; in other words, the backoff exponent is 3. However, the backoff exponent can be set to as low as 0, meaning that transmission will begin nearly immediately if the CCA senses the channel to be clear.

ZigBee Optional Superframe The PAN coordinator can optionally implement communication by using a superframe, which is composed of three parts:

- **Beacon,** which is used to set timing allocations and manage the piconet
- **Contention access period (CAP),** which is used for asynchronous data communication if present
- **Contention-free period (CFP),** which consists of *guaranteed time slots* (GTS), each with a unique device address and data direction

A timing diagram of a sample superframe is shown in Figure 6-33. The superframe is divided into a default value of 16 equally sized slots. Each slot duration is 60 symbols. The first slot is partly occupied by the beacon, and the remaining slots are used for the CAP and optional CFP. Channel access is via slotted CSMA/CA during the CAP, which has a minimum duration of 440 symbols. The PAN coordinator can allocate up to seven GTS periods within the CFP. Each GTS

Figure 6-33
The 802.15.4
ZigBee
optional
superframe
begins with a
beacon,
followed by a
CAP and a CFP
(if needed).

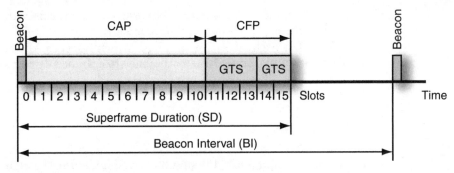

802.15.4 Superframe

is assigned a unique device address and data direction for low latency. A GTS can occupy more than one slot period.

Summary

The method by which a transmitting node accesses the medium is of critical importance for determining that node's potential to coexist within its own network, with other networks that follow the same protocols, and with all other users that share the communication channel. This channel can exist in the idle, communicating, or collision state. Network designers like to see the channel in the communicating state as much as possible to maximize aggregate throughput, but an individual user prefers to find the channel idle when a packet is available for transmission in order to minimize latency.

MAC is roughly divided into contention-free (scheduled) and contention-based (random) access. Contention-free MAC, either through FDMA or TDMA, prevents the channel from entering the collision state among the participants, but can be inefficient under asynchronous communication modes. Queuing has the best performance of scheduled access but can be difficult to implement in a distributed network. Contention-based access allows the possibility for collisions, but in return an arbitrary node has a greater chance of accessing the medium reasonably quickly if average loading isn't too high. As offered rates approach the level at which throughput is maximized (capacity), latency rapidly increases and the channel becomes unstable. The capacity of the ALOHA channel is somewhat low, so a listen-before-talk (CSMA) algorithm is superior when the CCA time is low compared to the length of a packet.

Wi-Fi, WiMedia, and ZigBee all make provisions to employ CSMA/CA with random backoff to allow relatively high channel loading without entering the region of instability. The random backoff algorithm is similar to *p*-persistent

CSMA with dynamic p. These three networks also allow contention-free channel access, either through TDMA (WiMedia or ZigBee) or polling (Wi-Fi). Bluetooth uses TDMA/TDD exclusively for its MAC, which appears ALOHA-like to other users. FHSS is employed by Bluetooth nodes to enhance its coexistence.

References

[Cho01] Choi, S., et al., "802.11a and 802.11b Maximum Throughput for Simulation Model Conformance," document IEEE 802.1101/055, January 2001.

[Hay84] Hayes, J., *Modeling and Analysis of Computer Communications Networks*. New York: Plenum Press, 1984.

[IEEE802.11] IEEE 802.11-1999, "Wireless LAN Medium Access Control (MAC) and Physical Layer (PHY) Specifications," August 20, 1999.

[IEEE802.11a] IEEE 802.11b-1999, "High-Speed Physical Layer in the 5 GHz Band," September 16, 1999.

[IEEE802.11b] IEEE 802.11b-1999, "Higher-Speed Physical Layer Extension in the 2.4 GHz Band," September 16, 1999.

[IEEE802.11g] IEEE 802.11g-2003, "Amendment 4: Further Higher Data Rate Extension in the 2.4 GHz Band," June 27, 2003.

[IEEE802.11h] IEEE 802.11h-2003, "Amendment 5: Spectrum and Transmit Power Management Extensions in the 5 GHz Band in Europe," October 14, 2003.

[IEEE802.15.1] IEEE 802.15.1-2002, "Wireless Medium Access Control (MAC) and Physical Layer (PHY) Specification for Wireless Personal Area Networks (WPAN)," June 14, 2002.

[IEEE802.15.2] IEEE 802.15.2-2003, "Coexistence of Wireless Personal Area Networks with Other Wireless Devices Operating in Unlicensed Frequency Bands," August 28, 2003.

[IEEE802.15.3] IEEE 802.15.3/D17, "Wireless Medium Access Control (MAC) and Physical Layer (PHY) Specifications for High Rate Wireless Personal Area Networks (WPAN)," draft standard, February 2003.

[IEEE802.15.4] IEEE 802.15.4-2003, "Wireless Medium Access Control (MAC) and Physical Layer (PHY) Specifications for Low-Rate Wireless Personal Area Networks (LR-PANS)," October 1, 2003.

[Kle75] Kleinrock, L., and Tobagi, F., "Packet Switching in Radio Channels: Part 1—Carrier Sense Multiple-Access Modes and Their Throughput-Delay Characteristics," *IEEE Transactions on Communications*, December 1975.

[Kle76a] Kleinrock, L., *Queueing Systems Volume I: Computer Applications*. New York: John Wiley and Sons, 1976.

[Kle76b] Kleinrock, L., *Queueing Systems Volume II: Computer Applications*. New York: John Wiley and Sons, 1976.

[Mor02] Morrow, R., *Bluetooth Operation and Use*, New York: McGraw-Hill, 2002.

[Tan81] Tannenbaum, A., *Computer Networks*. New Jersey: Prentice-Hall, Inc., 1981.

[Tob74] Tobagi, F., "Random Access Techniques for Data Transmission Over Packet Switched Radio Networks," PhD Dissertation, University of California at Los Angeles, 1974.

Passive Coexistence

The most rudimentary way for a wireless network to coexist with others located in the same area is simply to operate independently without considering whether or not access to the medium is shared with other transmissions. We call this *passive coexistence*. As we discovered in the last chapter, networks often demonstrate reasonable throughput under these conditions, especially if the *clear channel assessment* (CCA) mechanism can sense the presence of users in other networks to reduce the probability of collisions.

In this chapter we study the capability of a *wireless local area network* (WLAN) or *wireless personal area network* (WPAN) to operate over a communication channel that is also occupied by at least one other network, but without necessarily making use of various collaborative and/or noncollaborative methods to enhance coexistence. (We reserve that for the next chapter.) In other words, performance figures that we develop in this chapter are for the most part somewhat pessimistic, and the actual performance should be better if at least one of the networks involved employs coexistence enhancement techniques beyond those found within the basic *media access control* (MAC).

This chapter emphasizes coexistence between devices operating in the 2.4 GHz *industrial, scientific, and medical* (ISM) band since that is where most of the interference is. We also consider coexistence between *ultra-wideband* (UWB) and 802.11a since FCC rules require UWB emissions to reside primarily between 3.1 and 10.6 GHz. Within the 2.4 GHz band are two WLANs (802.11b and 802.11g) and three WPANs (802.15.1, 802.15.3, and 802.15.4). If all possible combinations of just two of these are analyzed against each other for coexistence performance, that still presents 25 different perspectives, an impractically large number. Therefore, we concentrate mainly on those networks likely to exist together, such as 802.11b Wi-Fi and 802.15.1 Bluetooth, rather than attempting to cover all the bases.

General Analytic Model

A general analytic model of two networks coexisting in the same area can be developed by diagramming their overlap opportunities and then quantizing the parameters mathematically. When 802.11b Wi-Fi and 802.15.1 Bluetooth operate in the same vicinity, for example, their transmissions will sometimes overlap in time, frequency, or both time and frequency. Figure 7-1 shows typical channel activity when both Wi-Fi and Bluetooth are active [Chi03].

The Bluetooth time slot is T_{BI} long, containing an actual transmission that lasts T_{BP}, and the Wi-Fi transmission is T_W in duration. If x is the time period from the beginning of the first overlapping Bluetooth slot until the Wi-Fi transmission begins, the number of overlapping Bluetooth packets is given by

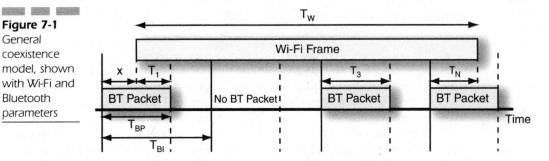

Figure 7-1
General coexistence model, shown with Wi-Fi and Bluetooth parameters

$$
N(x) = \begin{cases} \left\lceil \dfrac{T_w}{T_{BI}} \right\rceil & x \le \left(T_{BI} \left\lceil \dfrac{T_w}{T_{BI}} \right\rceil \right) - T_w \\[4mm] \left\lceil \dfrac{T_w}{T_{BI}} \right\rceil + 1 & \text{elsewhere} \end{cases}
\tag{7.1}
$$

The part of a particular Bluetooth slot that overlaps in time with the Wi-Fi transmission is given by

$$
T_i = \begin{cases} \max\left(T_{BP} - x, 0\right) & i = 1 \\ T_{BP} & i = 2, \ldots, N(x) - 1 \\ \min\left(x + T_W - (N(x) - 1)T_{BI}, T_{BP}\right) & i = N(x) \end{cases}
\tag{7.2}
$$

The first and last terms in Equation 7.2 account for the partial overlap in the first and last Bluetooth transmission slots, respectively. The other Bluetooth slots fully overlap the Wi-Fi transmission. Finally, we define δ_i as an indicator of whether or not a Bluetooth transmission exists in a particular time slot; that is,

$$
\delta = \begin{cases} 0 & \text{if the } i\text{th bluetooth slot is idle} \\ 1 & \text{otherwise} \end{cases}
\tag{7.3}
$$

For a statistical analysis, the δ_i can be replaced with the respective slot occupancy probabilities.

The probability that a Bluetooth slot overlaps with a Wi-Fi transmission in frequency is $h_f = 22/79 = 0.278$ when Bluetooth uses the 79-channel hopping sequence. Therefore, the number of symbols involved in a time-frequency overlap is [Chi03]

$$\eta(x) = h_f\left(T_1^{(s)}\delta_1 + \sum_{i=2}^{N(x)-1} T_i^{(s)}\delta_i + T_{N(x)}^{(s)}\delta_{N(x)} \right) \quad (7.4)$$

where $T_i^{(s)} = T_i/T_s$ and T_s is the symbol duration. Equation 7.4 is valid for both Bluetooth and Wi-Fi by substituting the appropriate symbol duration. Finally, the number of affected symbols and their associated *bit error rate* (BER) can used to determine the *packet error rate* (PER) or *frame error rate* (FER).

To reduce the collision probability, the quantities $N(x)$, h_f, or δ_i can be made small. The first is reduced by using faster data rates or shorter frames in the Wi-Fi network, the second is reduced by modifying the Bluetooth hop sequence to reduce overlapping hop frequencies, and the third is reduced by lowering the Bluetooth duty cycle. Notice that this model can also be used for Bluetooth-on-Bluetooth interference analysis by setting T_W equal to the desired Bluetooth transmission time. With other parameter changes, the model can apply to any coexistence situation.

Periods of Stationarity

The *Institute of Electrical and Electronics Engineers* (IEEE) 802.15.2 recommended practice uses a similar *physical* (PHY) model, as shown in Figure 7-2 for 802.15.1 and 802.11b [IEEE802.15.2]. For each given time offset, various *periods of stationarity* exist, during which parameters among interfering signals, such as data rates, modulation methods, and *carrier-to-interference ratio* (C/I) values, are fixed. Desired results, such as the average BER, PER, or FER, can be found by determining these values conditioned on each period of stationarity and then averaging these periods over all possible time offsets. In this manner, accuracy is improved at the expense of computation time and model complexity.

Figure 7-2

Periods of stationarity between 802.15.1 Bluetooth and 802.11b Wi-Fi for a particular time offset

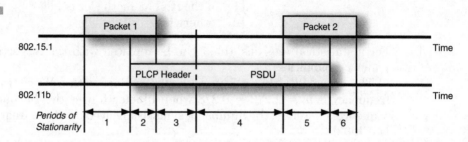

Simplifying Assumptions

To analyze the effect of interference, the most accurate and the most complex BER models derive BER values based on the precise interaction of the two signaling methods, such as *complimentary code keying* (CCK) and *Gaussian-filtered frequency shift keying* (GFSK), for example. An often-used simplification assumes that the interference can be approximated as *additive white Gaussian noise (*AWGN), so standard equations can be used that relate the BER to the *signal-to-noise ratio* (SNR) in AWGN. A further simplification presumes that the channel is useless whenever it is in the collision state, eliminating the use of BER values altogether.

Given the assumption that all packets involved in a collision are destroyed, finding the number of affected symbols is no longer required. Also, we no longer need to determine the *signal-to-interference ratio* (SIR), so the network topology variable is eliminated. Next, we can treat the partially overlapping situations at the beginning and end of the Wi-Fi frame transmission as fully overlapping. This assumption becomes more accurate for longer Wi-Fi frame durations. The latter two assumptions make the fact that a Bluetooth transmission only occupies part of a time slot irrelevant, because any time-frequency overlap results in a collision. By assuming that every time slot contains a Bluetooth transmission, the duty cycle variable is eliminated. We will often use these assumptions to develop simple worst-case approximations before presenting a more accurate and complex coexistence analysis.

Interference to the 802.15.1 Bluetooth Piconet

An interfering signal may disrupt the desired receiver if it is strong enough to exceed the receiver's C/I specification. For example, the Bluetooth specification states that the receiver must meet a co-channel C/I value of 11 dB for qualification, so if the desired signal's power is less than 11 dB stronger than a same-channel interfering signal's power then the BER at the desired receiver may exceed 10^{-3}. According to the Bluetooth specification, these interfering signals used for receiver qualification are Bluetooth transmissions, but we will assume that the desired receiver is equally vulnerable to any interfering signal producing sufficient *received signal strength indication* (RSSI) at the specified carrier offset.

Our approach here is to develop models that are simple and reasonably accurate, with an emphasis on situations likely to occur in the real world. Most of our simplifying assumptions are either pessimistic or worst case, without being too unrealistic. If an analysis shows that the network performs adequately under these assumptions, then actual performance will probably be acceptable as well. Accuracy can be enhanced slightly by following the detailed methods presented in some of the literature [How03], [Cor03]. Unfortunately, network performance analysis can rapidly reach diminishing returns, where a large increase in model complexity yields only a relatively modest improvement in accuracy. Furthermore, parameters that are unknown or poorly estimated may invalidate many aspects of a complex model anyway, so we may as well keep the model simple.

Asynchronous Connectionless (ACL) Throughput in AWGN

ACL packets transmitted on a channel in which AWGN, or AWGN-like interference, is present will experience occasional errors, each of which is independent of the others. If an error correction code is used that can correct t errors within a block of n bits, a packet is successful if every block has t or fewer errors in it. The probability that a single block will be successful, which we call Q_n, is given by

$$Q_n = \sum_{i=0}^{t} \binom{n}{i} p^i (1 - p)^{n-i} \tag{7.5}$$

Equation 7.5 is the probability that there are t or fewer errors within a block of n bits of data given a BER of p.

A packet of length L bits has L/n total blocks of data (where L/n is assumed to be an integer), so the probability that the entire payload is received successfully, Q_E, is equal to the probability that none of the blocks has more than t errors. This is given by

$$Q_E = (Q_n)^{L/n} \tag{7.6}$$

Notice that as L/n becomes larger, which it will for multislot packets, Q_E becomes smaller and performance may be diminished. We have $t = 1$ and $n = 15$ for *data medium speed* (DM) packets, and $t = 0$ and $n = 1$ for *data high speed* (DH) packets. The maximum number of payload bits L for various packet configurations is given in Table 7-1, along with values for t, n, and L/n. The reason L is sometimes different for DM and DH packets with the same slot duration is because the total number of bits (including the *forward error correction* [FEC]-encoded payload header and *frame check sequence* [FCS]) in a DM payload must be divisible by 15 to provide an integral number of (15,10) FEC codewords. The number of bits in a DH packet must be divisible by 8 for an integral number of bytes.

Table 7-1

Maximum payload length and error correction parameters for ACL packets with cyclic redundancy check (CRC)

Type	L	t	n	L/n
DM1	240	1	15	16
DH1	240	0	1	240
DM3	1,500	1	15	100
DH3	1,496	0	1	1,496
DM5	2,745	1	15	183
DH5	2,744	0	1	2,744

For DM packets, Equation 7.5 simplifies to

$$Q_n = (1 - p)^{15} + 15p(1 - p)^{14} \tag{7.7}$$

and then this result and the values for L/n from Table 7-1 are inserted into Equation 7.6 to obtain the Q_E. Throughput for a given BER is obtained by multiplying Q_E by the appropriate DM throughput values in Table 6-3 in Chapter 6. For DH packets, we insert the $L/n = L$ values from Table 7-1 into Equation 7.6 and multiply the resulting Q_E by the DH throughput values in Table 6-3. The results are shown in Figure 7-3 [Mor02]. The different packet types exhibit various degrees of robustness as channel conditions deteriorate.

Standard and Enhanced Receiver C/I Performance

The receiver is the key component that determines how well a node can operate in an interference-limited environment. The C/I performance for two different Bluetooth receivers is given in Table 7-2. The *standard receiver* C/I values represent the minimum required by the Bluetooth specification. We define an *enhanced receiver* as having an improved co-channel and adjacent-channel (1 MHz) C/I [Mei00]. Our analysis will take both types of receiver into account in an attempt to quantify the performance improvement offered by the enhanced receiver. Empirical measurements on an Ericsson Bluetooth receiver showed that its co-channel and adjacent-channel (1 MHz) C/I performance was close to that of the enhanced receiver [How03].

Figure 7-3

Throughput for various maximum-length ACL packet types across a channel with AWGN causing independent bit errors [Mor02]

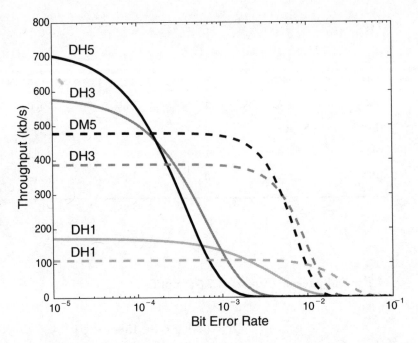

Table 7-2

Carrier-to-interference performance for the standard and enhanced Bluetooth receivers

Requirement	Standard receiver	Enhanced receiver
Co-channel C/I	11 dB	8 dB
Adjacent-channel (1 MHz) C/I	0 dB	−10 dB
Adjacent-channel (2 MHz) C/I	−30 dB	−30 dB
Adjacent-channel (≥3 MHz) C/I	−40 dB	−40 dB

Interfering Transmitter Disruption Distance   The next step in characterizing the interference problem is to calculate how close an interfering transmitter needs to be before it has the potential to affect the desired receiver. We assume that both desired and interfering transmitters have the same output power and omnidirectional antenna pattern. Suppose the interfering transmitter is physically close enough to just begin affecting the desired receiver's BER. If the desired transmitter is d_D meters away from the desired receiver, we can find the distance to the interfering transmitter, which we call d_I, by expressing the respective path losses given by Equation 2.5 in Chapter 2, "Indoor RF Propagation and Diversity Techniques," in terms of C/I. The result is

$$C/I = 40 + 10n \log(d_I) - [40 + 10n \log(d_D)] \qquad (7.8)$$

and by rearranging terms we get

$$\log\left(\frac{d_I}{d_D}\right) = \frac{C/I}{10n} \qquad (7.9)$$

Table 7-3 gives the ratio of d_I to d_D for standard and enhanced receiver C/I values in Table 7-2 with a path loss exponent $n = 3.0$ to represent typical room clutter. As an example, suppose the desired transmitter is 10 m from the desired receiver conforming to standard C/I performance. An interfering Bluetooth transmitter just begins affecting the desired receiver from 23 m away when they both hop into the same channel and the interfering node transmits while the desired node is receiving an incoming packet. The desired signal is affected when the interfering transmitter is 10 m away and it transmits either on the same channel or in either of the two adjacent channels. From 0.1 m away, the interferer can affect the desired hop channel along with two channels on either side of the desired hop channel, and from 0.05 m away the interferer has the potential to completely block the desired receiver. The latter situation is of critical importance when Bluetooth is collocated with another wireless service such as Wi-Fi or a cell phone.

In noncollocated situations, only the co-channel and perhaps adjacent-channel (1 MHz) C/I values come into play under most scenarios. Figure 7-4 depicts these

Table 7-3

Maximum disruption distances

Type	C/I Value	Number of hop channels affected	d_I/d_D
Co-channel	11 dB	1	2.3
Co-channel	8 dB	1	1.8
Adjacent (1 MHz)	0 dB	3	1.0
Adjacent (1 MHz)	−10 dB	3	0.46
Adjacent (2 MHz)	−30 dB	5	0.10
Adjacent (≥3 MHz)	−40 dB	79	0.05

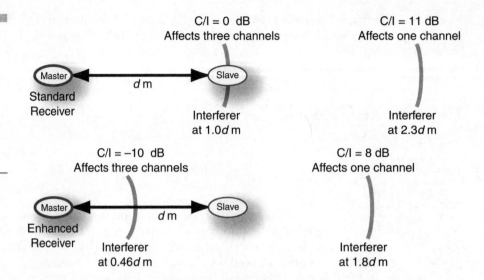

Figure 7-4
Maximum disruption distance of interfering nodes from a standard and enhanced Bluetooth receiver

two C/I situations for both the standard and enhanced receiver at one end of the link. Compared to the standard receiver, an enhanced receiver enables interfering nodes to be about 25 percent closer before co-channel C/I begins to affect performance. For adjacent-channel (1 MHz) C/I, the interfering node must be about 50 percent closer. Because the area of vulnerability to interference is proportional to the square of the radius of vulnerability, the enhanced receiver is significantly less susceptible to interference within a typical operating environment.

Coexistence Among Multiple Bluetooth Piconets

Bluetooth-on-Bluetooth PER and throughput analysis are often easier to do via an analytical method as opposed to an empirical approach. A theoretical analysis or computer simulation is facilitated because the coexisting piconets all operate in a similar manner, and most studies show that several piconets can coexist without causing a significant degradation in individual throughput. This feature, while good news for the Bluetooth user, makes the empirical approach to coexistence analysis more difficult because of the need to deploy several piconets within an area and move them around in such a way that meaningful throughput results are obtained.

Bluetooth Packet Error Rate

By making a few simplifying approximations, we can obtain some preliminary PER values under different C/I scenarios. First, we assume all piconets are using single-slot packets, and that they are fully loaded; that is, a packet is sent in every time slot. Because two independent piconets aren't time synchronized, we can make a pessimistic assumption that, during the time a desired packet is being received, each interfering piconet transmits two packets on two different hop channels, each of which has the potential to partially overlap the desired packet. (This assumption is pessimistic because the maximum duty cycle of a single-slot packet is only $366/625 \approx 0.6$ within a hop channel to provide time to hop to a new frequency.) If C/I parameters are violated, BER rapidly increases with decreasing C/I until the entire overlap period results in randomized data within the desired packet. This characteristic leads to another simplifying assumption: If C/I is violated, then partial packet overlap causes sufficient bit errors to destroy all involved packets, and they must be retransmitted. Even with FEC, an ACL packet is lost when two or more bit errors occur within a single 15-bit block, an event that is still likely even during a short interfering packet overlap time.

Equation 4.8 in Chapter 4, "Advanced Modulation and Coding," gives the probability of *frequency hopping spread spectrum* (FHSS) packet error P_E as a function of the number of total users K and the number of available hop channels M. This equation is based on the premise that two or more simultaneous transmissions within one hop channel destroy all affected packets. As discussed earlier, that equation only accounts for co-channel C/I and assumes hop time synchronism among all users. The lack of time synchronism in actual independent Bluetooth piconets can be worst case approximated by doubling the number of interfering users that are actually present. For adjacent-channel C/I at 1 or 2 MHz offset, one interfering piconet affects three or five channels for each of its two hops, respectively, which equates to a "hit" probability of 3/79 or 5/79. All these situations can be expressed mathematically as

$$P_E = 1 - \left(1 - \frac{n}{M} \right)^{2(K-1)} \tag{7.10}$$

where n is 1, 3, or 5 for co-channel, adjacent-channel (1 MHz), or adjacent-channel (2 MHz) C/I violations, respectively, and M is 79 available hop channels. Figure 7-5 shows a plot of the PER versus the number of interfering piconets for each C/I value. Equation 7.10 can be used in its various C/I configurations to find an overall PER when interfering piconets are at different distances from the desired receiver. This is done by finding the packet success probability $(1 - \text{PER})$

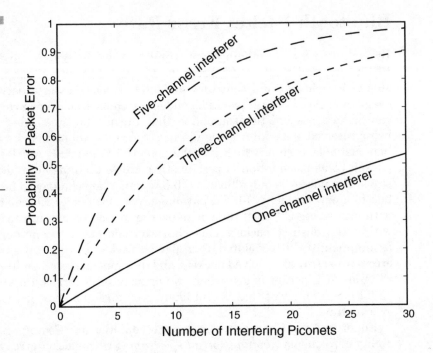

Figure 7-5
Approximate PER versus the number of interfering piconets for different C/I situations. All piconets use single-slot packets.

for each value of n and its associated K, multiplying the probabilities together, and then subtracting the result from 1 for the overall PER.

PER calculation accuracy can be improved by including the effect of all interfering piconets in the C/I calculation for a particular hop channel [Cor03]. This can be done at the desired receiver by adding the powers of all co-channel and adjacent-channel interfering signals and comparing the results to the power in the desired signal. The PER is the probability that the appropriate ratio of any of these two values is below some threshold. When the number of piconets is relatively small compared to the number of available hop channels, the probability is high that only one interfering signal will affect either co-channel or adjacent-channel (1 MHz) C/I. Furthermore, if multiple interfering signals do influence the C/I, one of these will often have significantly higher power at the desired receiver, allowing the others to be ignored. Holding to either of these last two assumptions simplifies the analysis significantly.

Bluetooth Packet Throughput

Obtaining average packet throughput figures for a desired piconet operating in the vicinity of several interfering piconets requires, among other things, plotting the distribution of possible locations for both the desired and interfering piconet members within the area, determining which C/I parameters are violated, finding the corresponding throughput values, and then averaging the results. Furthermore, each ACL data packet must also be acknowledged for the transfer to be successful, and this requires error-free packets in two consecutive time slots on two different hop frequencies. Although the return *acknowledgment* (ACK) packet consists of only an access code and header (if no return data is sent), this packet can still be destroyed if its hop channel contains interference. The result is that both the data packet and its associated ACK must be resent.

A computer simulation study performed by Motorola placed the desired piconet in the center of a circle with a 10 m radius [Mei00]. The desired signal's power at each receiver was set at -70 *decibels relative to 1 milliwatt* (dBm). Bluetooth interfering piconets were distributed randomly within the circle, as shown in Figure 7-6. All Bluetooth units use 0 dBm *transmit* (TX) power, and the path loss exponent is $n = 3.5$, modeling an environment of moderate to severe clutter. Receivers in the desired piconet are assumed to have a sensitivity limit of -80 dBm, and the master and slave are exposed to the same amount of interference. The latter assumption is somewhat unrealistic in practice, because the master and slave are usually separated by a few meters, but the results are easier to develop and interpret when asymmetries in the desired piconet are eliminated. To simulate a worst-case situation, all interfering piconets are considered to be fully loaded, transmitting in every available time slot. Both co-channel and adjacent-channel (1 MHz) interference are taken into account. An independent Rayleigh faded path loss is assumed for each interferer on each hop channel.

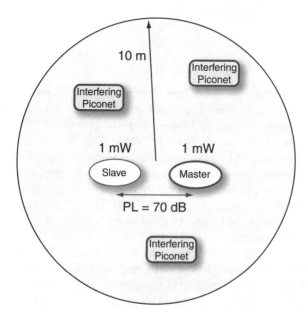

Figure 7-6
The Motorola study places a desired piconet at the center of a circle with a 10 m radius and places interfering piconets randomly within the circle.

Figure 7-7
Throughput as a function of the number of interfering piconets within the 10 m circle is plotted for a standard receiver (a) and an enhanced receiver (b) [Mei00].

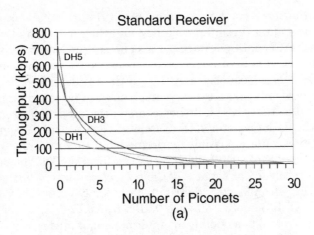

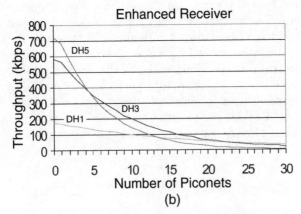

The results in Figure 7-7 show how throughput is affected as a function of the number of interfering piconets that exist within the 10 m circle. Plots are shown in Figure 7-7 for one-, three-, and five-slot packets without FEC, using standard and enhanced receivers as defined in Table 7-2. Notice that the enhanced receiver, while having a C/I advantage of only a few dB, significantly outperforms the standard receiver when interfering piconets are present. Also, switching from DH5 to DH3 packets improves performance once interference becomes sufficient to reduce DH5 throughput below about 400 kb/s. This is due to the higher retransmission penalty that must be paid when a DH5 packet or its subsequent ACK is destroyed by a collision.

A comparison of DH1 and DM1 packet throughput values is shown in Figure 7-8a, along with the throughputs of DH3 and DM3 packets in Figure 7-8b, again plotted as a function of the number of interfering piconets within the 10 m circle. Earlier in Figure 7-3, we discovered that the DM3 packet has the highest

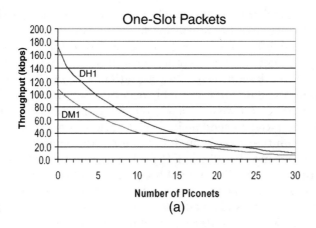

Figure 7-8
Throughput as
a function of
the number of
interfering
piconets within
the 10 m circle
is plotted for
single-slot
packets (a) and
three-slot
packets (b).
(Note
throughput
scale change.)
Packets
without FEC
have higher
throughput
when inter-
ference is the
primary cause
of packet loss
[Mei00].

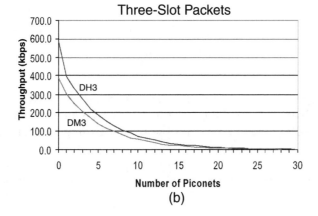

throughput when the BER is between about 10^{-4} and 10^{-2} when bit errors are caused by random noise. However, when packets are affected by collisions with other Bluetooth piconets instead, Figure 7-8 shows that the available FEC is overwhelmed and the packet is lost anyway. Therefore, it makes little sense to use FEC when most packets are destroyed from collisions instead of from random noise.

A relatively easy way to improve packet throughput (and simplify the analysis, for that matter) is to ensure that the distance between communicating members in the desired piconet is less than the distance to any interfering node for the standard receiver, or less than twice the distance to any interfering node for the enhanced receiver. This is usually convenient to do when the desired piconet consists of a point-to-point link, where the two participating users can move close to each other. Under these conditions only co-channel C/I becomes an issue because the interfering nodes are too far away to violate adjacent-channel C/I. If desired piconet members are separated by d meters, then according to Table 7-3, co-channel C/I is violated by interfering nodes within 2.3d meters for the standard receiver, and 1.8d meters for the enhanced receiver. The proximity of the desired piconet members to each other also allows us to model the channel as Rician with a large Rician factor (see Equation 2.21 in Chapter 2) so we can ignore fading and still obtain reasonably accurate results.

For multislot desired transmissions, collision probability is highest when the $K - 1$ interfering piconets send a single-slot packet in every time slot. Since we assume once again that a collision destroys a packet regardless of FEC capability, the desired piconet should use DH packets for better performance. With all piconets operating on the 79-channel hopping sequence, a desired packet occupying m time slots will experience at most $m + 1$ opportunities for collision from each interfering piconet, and an additional two collision opportunities during the associated ACK. Equation 7.10 can be adjusted to account for these factors by writing

$$P_E^{(b)} = 1 - \left(1 - \frac{1}{79}\right)^{(m+3)(K-1)} \qquad (7.11)$$

Packet throughput is determined by multiplying $(1 - \text{PER})$ by the error-free throughput given in Table 6-3 using the forward asymmetric channel values for multislot packets. Figure 7-9 shows the packet throughput for DH1, DH3, and

Figure 7-9
Desired packet throughput as a function of the number of interfering piconets for DH packets when the desired piconet members are close enough that only co-channel C/I applies

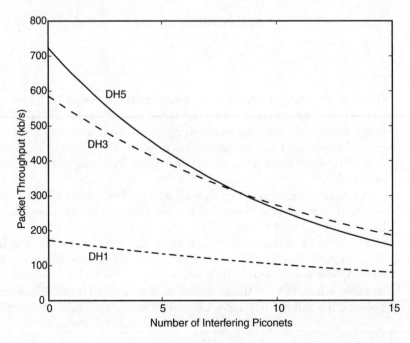

DH5 packets as a function of the number of interfering piconets that affect only co-channel C/I. Compared to the results in Figure 7-7, coexistence is significantly improved by moving desired piconet members close enough to each other that adjacent-channel C/I violations are avoided. Also, when more than seven piconets are causing interference, changing from DH5 to DH3 will improve performance slightly.

Bluetooth Network Throughput

As we learned in Chapter 6, throughput can be viewed in terms of an individual "desired user" (piconet) or in terms of total network (multipiconet) performance. Individual piconet throughput is maximized when no signal impediments (noise or interference) are present and the values in Table 6-3 are ideally realized. As more piconets begin operating in the vicinity, aggregate network throughput increases at the expense of reduced throughput experienced by each individual piconet, at least up to the point of network capacity.

To determine total network throughput when all piconets use the 79-channel hopping sequence, we begin by assuming that members of each piconet are distributed within an area such that only co-channel C/I affects performance. We also assume that all piconets use the same packet size, either DH1, DH3, or DH5, each with an associated ACK, to maintain network symmetry. For fully loaded piconets, collision opportunities by an interferer depend on the DH packet size. For DH1 packets, at most four collision opportunities exist: two for the data packet and two more for the associated ACK, as we've already discussed. When multislot packets are used, up to three collision opportunities per interferer exist for the data packet and up to an additional two for the associated ACK. For example, collision opportunities on a desired DH5 packet can come from the last part of an interfering DH5 packet, the associated ACK, or the first part of the following DH5 packet, each of which are sent on a different hop channel. Aggregate network throughput can be determined by multiplying individual throughput by the total number of piconets in use. This is given by

$$S = AK\left(1 - \frac{1}{79}\right)^{j(K-1)} \tag{7.12}$$

where S is the network throughput, K is the total number of independent piconets in the area, and A is the DH error-free throughput given in Table 6-3. The parameter j, representing the number of worst-case collision opportunities, is 4 for DH1 packets and 5 for DH3 or DH5 packets.

Figure 7-10 plots S as a function of K when all network members are using either DH1, DH3, or DH5 packets. The abscissa scale has been expanded to show the characteristic rise and fall of network throughput as total traffic increases,

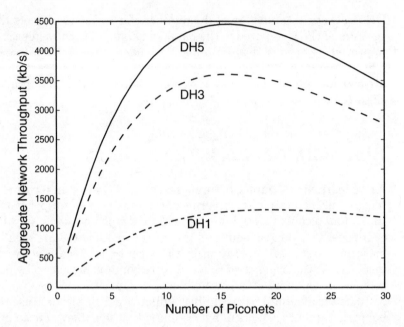

Figure 7-10
Network
throughput as
a function of
the number of
operating
piconets when
all piconet
members are
distributed
such that only
co-channel C/I
can be violated

which we discovered in Chapter 6. Network throughput is highest when DH5 packets are used, regardless of the total number of piconets affecting each other. Network capacity is about 4.5 Mb/s, reached when 16 independent piconets are in use. The throughput for each piconet under these conditions is about 280 kb/s, which is about 40 percent of the maximum possible in a perfect channel.

Synchronous Connection-Oriented (SCO) Performance in Interference

Bluetooth *synchronous connection-oriented* (SCO) performance in an imperfect channel is more difficult to analyze than ACL performance because audio quality is subjective to the listener and is most affected by how many bits are randomized within a burst error event. As such, packet loss criteria may not be particularly applicable for SCO performance analysis because a packet is considered lost over a wide range of bit error conditions. Even during catastrophic collisions, general disagreement exists regarding the SCO packet loss rate and its affect on audio quality. Some claim a loss rate of 2 percent is unacceptable, whereas others claim adequate audio quality remains with packet loss rates as high as 30 percent [Mei01].

The probability that an SCO packet is successful is equal to the probability that neither of the (worst-case) two collision opportunities takes place for each interfering transmission. Only single-slot transmissions are considered because all SCO packets occupy one slot and no associated ACK occurs. As before, the PER is one minus the packet success probability. If the desired link topology is such that only co-channel C/I can be violated, the PER is given by Equation 7.10 with $n = 1$.

For an *extended synchronous connection-oriented* (eSCO) packet with a single allowable retransmission, a packet is successful if either the original packet or retransmission is collision-free, so the PER is the probability that both transmission attempts experience a collision. Because both of these attempts experience independent opportunities for interference, we have (for single-slot eSCO)

$$P_E^{(eSCO)} = [P_E^{(SCO)}]^2 \qquad (7.13)$$

Whether or not the ACK or *negative acknowledgment* (NAK) associated with the first eSCO packet experiences a collision doesn't affect link performance. If the ACK is lost, an irrelevant retransmission occurs because the receiving node already has one good copy of the affected eSCO packet. If the NAK is lost, the sending node retransmits the eSCO packet anyway. The two quantities in Equation 7.13 are plotted as a function of the number of interfering users in Figure 7-11. The single retransmission in our eSCO scheme significantly improves reliability over simple SCO in which no retransmissions are allowed.

Figure 7-11
Packet error probability as a function of interfering piconets for SCO and eSCO with one retransmission. The desired piconet members are assumed to be close enough that only co-channel C/I can be violated.

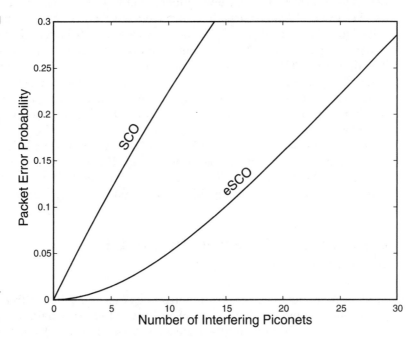

Coexistence Among Multiple 802.11 Wi-Fi Networks

Modeling coexistence among several Wi-Fi networks within range of each other is, in several ways, easier than doing the same for multiple Bluetooth piconets. Because Wi-Fi doesn't employ any interference avoidance mechanisms, the simplest coexistence model assumes that the CCA will prevent overlapping transmissions most of the time. In this way, several independent Wi-Fi networks can operate in the same vicinity and on the same channel without significantly interfering with each other through collisions. Instead, they effectively share a single network's throughput among them.

The situation is made more complex when hidden terminals exist. These are nodes that cannot hear the transmissions by some of the other Wi-Fi nodes, either within its own network or across other independent networks. For example, *station* (STA) 1 and STA 2, both associated with the same *access point* (AP), may not be able to hear each other. STA 1 may sense an idle channel when STA 2 is actively sending a packet to the AP. If STA 1 also begins transmitting, a collision at the AP may occur, and the destruction of both frames is the usual result. The hidden terminal situation is even more prevalent in ad hoc (*independent basic service set* [IBSS]) Wi-Fi networks, where all packet exchanges are peer to peer and significant path obstructions are common.

Because the number of channels available for Wi-Fi networks is limited, it's useful to determine how far away independent networks can be and still operate on the same channel without interfering with each other. This is especially important for 802.11b and 802.11g networks, which have only three assigned nonoverlapping channels. Two general approaches can be used to find this range. The first uses CCA as the determining factor. That is, a desired network is considered interference-free if all other co-channel networks are far enough away that none of the desired nodes' CCA mechanisms are activated by members of another BSS. The second approach is to find the FER for the modulation being considered (*direct sequence spread spectrum* [DSSS], CCK, or *orthogonal frequency division multiplexing* [OFDM]) and to consider the desired network interference-free if co-channel C/I is large enough that FER values are below some threshold. The former is easier to calculate, but the latter is more appropriate when interference is caused by hidden terminals.

Co-Channel Interference (CCI) Performance Using CCA

When a Wi-Fi network is operating in a noiseless and collision-free environment, the frame throughput values given in Figure 6-21 in Chapter 6 are achieved. In such an environment, the general rule is that the frame throughput in bits/second is about half the raw data rate across the channel. This rule is slightly pessimistic to allow for the occasional frame loss from noise or collisions within the BSS.

If several co-channel BSSs are close enough to activate each other's CCA mechanism, the maximum throughput is divided among the competing WLANs. Furthermore, additional users increase the probability of collision. To find the minimum distance to an interfering node beyond which it won't activate a desired node's CCA mechanism, first refer to Table 5-9 in Chapter 5, "Radio Performance," which gives the *energy detect* (ED) threshold values for 802.11b devices operating with different data rates and TX powers. Using the simplified *path loss* (PL) model and an interfering TX power of +20 dBm, these distances are given in Table 7-4. It's evident from these results that independent 802.11b networks operating on the same channel must be separated by several rooms before they will each have a chance to operate at full speed.

The 802.11g node must be able to detect the presence of low-rate, high-rate, or OFDM signals at an ED threshold of −76 dBm, equating to a minimum interfering node distance of 110 m. An 802.11a node's CCA mechanism will activate to an OFDM frame preamble with an ED threshold of −82 dBm at 5.x GHz. If an interfering node's TX power is +20 dBm, then its minimum separation distance is 102 m, and if the TX power is +15 dBm then the minimum separation distance is 72 m.

Hidden Terminals and Signal Capture

A typical hidden terminal situation within an IBSS is depicted in Figure 7-12. STA 2 is within range of both STA 1 and STA 3, but the latter two cannot hear

Table 7-4

Minimum distance between 802.11b nodes for no CCA activation (+20 dBm interfering TX power)

Desired node's TX power	Data rate	ED threshold	Required path loss	Minimum distance to interfering node
+17 dBm or lower	1 or 2 Mb/s	−70 dBm	90 dB	72 m
+17 dBm or lower	5.5 or 11 Mb/s	−70 dBm	90 dB	72 m
Above +17 dBm to +20 dBm	1 or 2 Mb/s	−76 dBm	96 dB	110 m
Above +17 dBm to +20 dBm	5.5 or 11 Mb/s	−73 dBm	93 dB	89 m
Above +20 dBm	1 or 2 Mb/s	−80 dBm	100 dB	145 m
Above +20 dBm	5.5 or 11 Mb/s	−76 dBm	96 dB	110 m

Figure 7-12
IBSS hidden
terminal
scenario.
Terminal 2 can
hear both
terminals 1 and
3, but the latter
two terminals
cannot hear
each other.

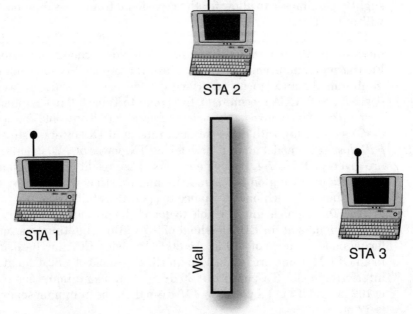

each other. This model can also apply to an infrastructure BSS in which STA 2 is the AP and STA 1 and STA 3 are part of its BSS.

Suppose STA 2 is already synchronized on a frame from STA 1. We consider two possible collision scenarios at the common receiver [War00]:

- A *request-to-send* (RTS) frame from STA 3 collides with a data frame from STA 1.
- An RTS frame from STA 3 collides with an RTS frame from STA 1.

At this point we depart slightly from our worst-case assumption that a collision destroys all involved frames and instead allow for the possibility of *capture*. Capture in this context refers to the capability of a receiver that is actively decoding a frame to accept some level of *co-channel interference* (CCI) from a later-arriving frame without losing synchronization on the active frame.

If STA 1 and STA 3 are both attempting to send frames that overlap in time with STA 2, the link that provides the stronger RSSI at STA 2 will eventually transfer its frame successfully, perhaps after several backoff periods [War00]. If a strong RTS arrives while STA 2 is receiving a weaker data frame, for example, the data frame will probably be corrupted and a retransmission will be needed. Unfortunately, the result is that the stronger link can effectively preempt the weaker link, disrupting network operation.

To quantify the effect of capture on network performance, we assume that the receiver at STA 2 has acquired a frame y and is actively decoding it when a disrupting frame x arrives. The relative signal strength of x to y is given by

$$\delta = \frac{E_{sx}}{E_{sy}} \qquad (7.14)$$

which is a ratio of the respective symbol energies in each signal. For identical symbol rates, this is essentially the reciprocal of C/I. The BER values versus δ, as determined by both AWGN power and spreading sequence correlation values, are plotted in Figure 7-13 when the collision is between two frames encoded with a Barker sequence [Pur77]. The results are almost identical when a later-arriving Barker-coded RTS collides with a CCK data frame being received at either 5.5 or 11 Mb/s while optimistically assuming that a symbol error almost always results in a single bit error [War00]. (Keep in mind that, for a given RSSI, the 11-chip Barker sequence has a higher symbol energy than an 8-chip CCK sequence.) Generally, when $\delta < 0$ dB, the BER is determined by AWGN, but for values greater than about 4 dB the BER exceeds 10^{-2} from the interfering transmission,

Figure 7-13
BER values when a Barker-coded RTS arrives while a receiver is actively decoding a low-rate data frame [War00]

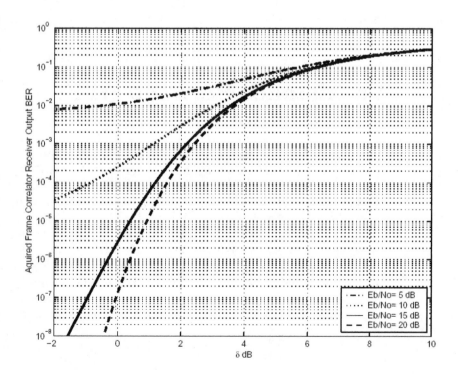

resulting in a high FER. We should point out that the results in Figure 7-13 are optimistic because the BER formulas that were used apply to nondifferential *phase shift keying* (PSK) detection, which exhibits greater interference immunity when the receiver locks onto the phase of the desired signal.

If a receiver can capture an incoming signal in such a way that its interference immunity is improved, network performance usually improves as well. This is especially true if, as in the previous example, all incoming frames are "desired" by the target receiver. In this case, network throughput is enhanced by the proportion of frames that are captured. Of course, if no hidden terminals existed, performance would be higher still.

On the other hand, if some of the arriving frames are not desired—coming from a different BSS, for example—then incorporating the effect of capture into the network model is more difficult. The distribution of desired to undesired frames and their respective arrival times must be determined, since performance is improved only if a desired frame arrives first and is captured. If an interfering frame arrives first, the desired receiver will consider the channel busy for the duration of the extracted *network allocation vector* (NAV), and it may not attempt to acquire any new transmissions until the NAV expires. The good news is that in most cases the desired frames are sent by nodes closer to the target receiver, and hence their RSSI is stronger than interfering frames from another BSS. If different BSSs are confined to different rooms in a building, the ratio of bit energies given by Equation 7.14 is such that capture could enhance performance in each BSS.

Coexistence Between Wi-Fi and Bluetooth

As mentioned earlier, the increasing consensus within the wireless industry is that a viable market exists for both Bluetooth and Wi-Fi devices, and their respective strengths are such that they will be required to coexist in many applications. It is therefore important to discover the effect that each will have on the other when they operate together. Researchers have embraced this problem with enthusiasm, not only because of its importance, but also because Bluetooth uses FHSS and Wi-Fi uses a modified DSSS, making the analysis more challenging.

In this section we lay some groundwork for analyzing both Wi-Fi–on–Bluetooth and Bluetooth–on–Wi-Fi interference. Some simplifying assumptions are made that lead to fast, reasonably accurate performance calculations. We also present some research results that use more complex models in an effort to improve accuracy.

Wi-Fi Interfering with Bluetooth

When a Bluetooth piconet hops to a channel in which a Wi-Fi network is operating, some of the Wi-Fi signal energy will enter the passband of the Bluetooth receiver. For an 802.11b Wi-Fi bandwidth of 22 MHz, roughly $\frac{1}{22}$ of the energy at the antenna will pass through the Bluetooth receiver's filters, so Bluetooth has a built-in 13 dB average rejection of this signal. For 802.11g Wi-Fi, the rejection is about 12 dB. We conclude, then, that for a C/I of 11 dB to exist at a desired Bluetooth receiver, the total Wi-Fi interfering signal power at the receiver's antenna can exceed the desired Bluetooth signal by about 2 dB before it begins to disrupt the Bluetooth piconet during hops within the Wi-Fi passband. Unfortunately for Bluetooth, many Wi-Fi networks operate at 30 to 100 *milliwatts* (mW, or +15 to +20 dBm) of TX power, so they can be fairly far away and still potentially interfere with Bluetooth operation.

Worst-Case Analysis: 802.11b on Bluetooth For a simple worst-case analysis, we can assume that a single 802.11b Wi-Fi network is always transmitting and is close enough to violate the desired Bluetooth receiver's co-channel C/I. We'll also assume that if the piconet hops anywhere within the 22 MHz Wi-Fi signal bandwidth, the affected packet is lost. Under these conditions, the probability that two successive hops, representing the data and its associated ACK, are outside the Wi-Fi signal bandwidth is given by

$$\Pr(2 \text{ hops OK}) = \left(\frac{79 - 22}{79} \right)^2 = 0.52 \qquad (7.15)$$

Thus, the throughput degradation is 48 percent compared to perfect operation. For an 802.11g bandwidth of about 17 MHz, Bluetooth throughput degradation is about 36 percent. Under these worst-case assumptions, Bluetooth ACL performance is severely, but not catastrophically, affected by the presence of interference on one Wi-Fi channel. The situation gets significantly worse, however, if interfering 802.11b WLANs exist on Wi-Fi channels 1, 6, and 11. Under these conditions, only 13 Bluetooth hop channels ($79 - 66$) are in the clear, equating to a worst-case throughput reduction of 97 percent.

Empirical Approach: 802.11b on Bluetooth We will now examine an empirical study, performed by Texas Instruments, in which the effect of Wi-Fi on a Bluetooth piconet was analyzed and ACL performance figures obtained. The study used the following equipment in a "real-world" test [Sho01]:

- Wi-Fi equipment
    - **AP** Cisco Aironet Model 340, TX power 30 mW

- **STA** Cisco Aironet Series 340 card in laptop computer, TX power 30 mW
- Bluetooth equipment
  - **Master** Digianswer card in laptop computer, TX power 100 mW
  - **Slave** Digianswer card in laptop computer, TX power 100 mW

To provide a maximum level of interference to the Bluetooth piconet, the Wi-Fi devices were placed about 0.3 m apart, so a Bluetooth piconet member will receive interference from WLAN communications in both directions. The Bluetooth slave's position was fixed directly between the Wi-Fi AP and STA in a collocated topology, or 10 m away from the Wi-Fi users in a separated topology, as shown in Figure 7-14. In each case, the distance between master and slave was varied and throughput measurements taken. The results are given in Figure 7-15 for both topologies, along with a baseline throughput for the piconet without any Wi-Fi interference. All paths were *line of sight* (LOS).

When one of the Bluetooth nodes is collocated with an operating Wi-Fi network, throughput is low even for small separations between master and slave. On the other hand, when the Bluetooth piconet is separated by 10 m from any Wi-Fi node, throughput is affected only slightly. Other studies have shown that separations of as little as 2 m between Bluetooth and Wi-Fi still allow the Bluetooth piconet to perform acceptably well [Lan00].

Figure 7-14
The Bluetooth slave can be either collocated with the Wi-Fi network or separated from the network by 10 meters (TI Study).

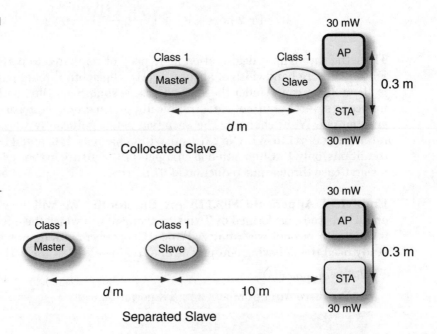

Figure 7-15
Bluetooth throughput as a function of master-slave distance for both collocated and separated topologies (TI Study) [Sho01]

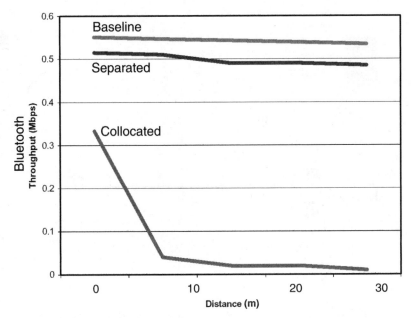

Audio-quality degradation in the presence of an 802.11b network has been examined to some degree. One study demonstrated that as long as a class 3 SCO link is shorter than 2 m, no audible degradation will occur from a WLAN operating in the vicinity [Haa99]. Another showed that an SCO link will perform well if it is located at least 2 m away from the nearest Wi-Fi AP or STA [Lan01].

Analytical Approach: 802.11b–on–Bluetooth As we mentioned in earlier chapters, using the analytical approach for finding coexistence performance involving dissimilar networks can be extremely complex, and numerous assumptions must be made regarding how two very different modulated signals affect each other. One such study employed the topology given in Figure 7-16, in which a Wi-Fi network interferes with a single point-to-point Bluetooth piconet [Con03]. The following issues are addressed:

- The path between devices is modeled as either Rician (LOS) or Rayleigh (*obstructed line of sight* [OLOS]).
- The path loss exponent is 2.0 for distances less than 8 m and 3.3 otherwise, corresponding to the IEEE breakpoint model.
- The channel exhibits slow, flat fading during the transmission of one frame or packet.
- Symbol errors are caused only by interference, which is modeled as AWGN.
- TX and *receive* (RX) antenna gains are 0 dB on all devices.

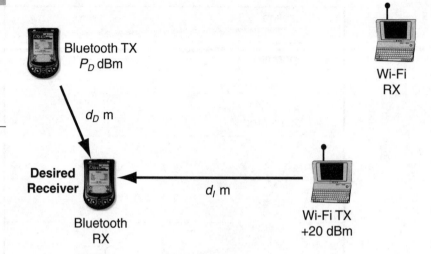

Figure 7-16
Network
topology for
an analytical
802.11b-on-
Bluetooth
coexistence
model

Bluetooth TX
P_D dBm

Wi-Fi
RX

d_D m

**Desired
Receiver**

d_I m

Bluetooth
RX

Wi-Fi TX
+20 dBm

Rather than assuming that $\frac{1}{22}$ of the Wi-Fi signal's energy always enters the Bluetooth receiver passband during any channel overlap, accuracy is improved by finding the proportion of interference power as a function of the relative carrier offset Δf between the two signals. A typical Wi-Fi TX spectrum is given in Figure 4-3 in Chapter 4, and the Wi-Fi TX mask is shown in Figure 5-3 in Chapter 5. Interference from the Wi-Fi network is greatest when the offset is zero and gradually decreases for Bluetooth hop channels having a greater offset. These values are given in Table 7-5 for a theoretical Wi-Fi spectrum using DSSS at 1 Mb/s and CCK at 11 Mb/s, and they are normalized to a carrier offset of 1 MHz. The Bluetooth receiver is assumed to use a 10-pole Butterworth filter with a -3 dB bandwidth of 0.85 MHz. It's evident that the interference to a Bluetooth receiver is about the same for a given carrier offset for either DSSS or CCK. For carrier offsets greater than 10 MHz, interference is considered to be negligible.

Figure 7-17 shows the mean Bluetooth DM1 PER as a function of carrier offset and the average C/I of a 1 Mb/s DSSS Wi-Fi interfering signal. Generally, a 10 dB reduction in interference power equates to an order-of-magnitude improvement in PER. For any offset, once the C/I exceeds about 13 dB, the Bluetooth PER is reduced to 0.1 or less. Due to the robust structure of the access code and header, almost all Bluetooth packet faults are from a corrupted payload.

Figure 7-18 plots the Bluetooth DM1 PER as a function of the separation ratio between the interfering Wi-Fi source and the desired Bluetooth source using various TX powers over link distances d_D. (PER values as the separation ratio approaches 0 should be ignored because the model doesn't account for receiver blocking.) The 1 Mb/s Wi-Fi signal is assumed always to be present at a TX power level of $+15$ dBm. The channel is modeled as Rician with factor $K = 5$, which

Table 7-5

802.11b Wi-Fi interference level on a Bluetooth receiver [Con03]

Carrier offset Δf (MHz)	Wi-Fi interference (dB)	
	DSSS (1 Mb/s)	CCK (11 Mb/s)
0	0.6	0.1
1	0.0	0.0
2	−0.4	−0.4
3	−1.1	−1.0
4	−2.0	−1.9
5	−3.8	−3.1
6	−5.7	−4.6
7	−6.9	−6.7
8	−9.6	−9.5
9	−13.5	−13.3
10	−19.9	−19.7

Figure 7-17
Bluetooth DM1 PER as a function of carrier offset and average C/I under 1 Mb/s 802.11b interference [Con03] © IEEE

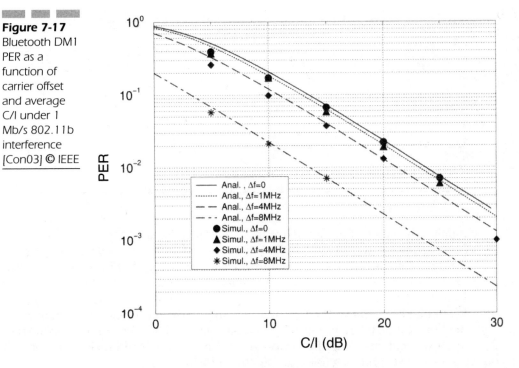

Figure 7-18
PER versus
separation ratio
between the
interfering Wi-
Fi source and
the desired
Bluetooth
source at the
desired
Bluetooth
receiver
[Con03]. See
Table 7-6 for
parameter list.
© IEEE

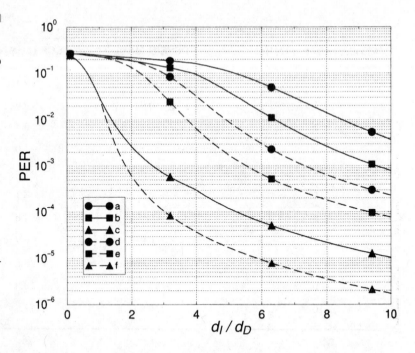

Table 7-6

Parameters for
plots in Figure
7-18 [Con03]

Plot	Desired link distance d_D (m)	Bluetooth TX power P_D (dBm)
a	2	0
b	2	+4
c	2	+20
d	8	0
e	8	+4
f	8	+20

reduces the effect of fading. Parameters for the plots are given in Table 7-6. As expected, when the Bluetooth TX power increases, the PER decreases significantly, but at the expense of greater interference to the Wi-Fi link. Perhaps less expected is the Bluetooth PER improvement when the desired link is extended from 2 to 8 m. This is due to the higher proportional path loss experienced by the Wi-Fi link at distances greater than 8 m due to the path loss exponent increasing from 2.0 to 3.3. In realistic situations, moving the Bluetooth devices farther apart without changing the location of the interfering Wi-Fi node often reduces the ratio d_I/d_D and increases the Bluetooth PER.

802.11g Interfering with Bluetooth   Due to the multicarrier nature of the 802.11g transmitted signal, its *power spectral density* (PSD) is much more uniform than the typical sinc(x) form exhibited in Figure 4-3 by the main lobe of 802.11b. As a result, the 802.11g TX power must be about 2 dB higher than 802.11b to achieve the same interference level at the Bluetooth receiver, and this level of interference is almost constant for carrier offsets up to about 8 MHz [Sel03]. Modeling 802.11g interference as AWGN is about 1 dB pessimistic at offsets of 2 MHz or less. In other words, AWGN at an RSSI that is 1 dB lower at the Bluetooth receiver will produce the same BER as 802.11g. For greater offsets, the AWGN model is accurate to within a fraction of a dB [Sel03].

Bluetooth Interfering with Wi-Fi

Although it's already clear that 802.15.1 Bluetooth and 802.11b Wi-Fi cannot be collocated without taking some action to alleviate interference to the Bluetooth piconet, it would also be instructive to determine whether a nearby Bluetooth piconet can significantly disrupt Wi-Fi throughput. And if it does, will the disruption be greater or less than Wi-Fi's effect on Bluetooth? To answer this question we turn once again to the two empirical studies and the analytical method presented in the preceding section, only this time we focus on the desired Wi-Fi receiver.

Empirical Approach: Bluetooth–on–802.11b   The TI study used the same equipment listed earlier, but for Bluetooth on Wi-Fi the network topologies were changed to those shown in Figure 7-19 [Sho01]. In these scenarios the Bluetooth master and slave are separated by 0.3 m, so the Wi-Fi network is affected about equally by all Bluetooth traffic. In the collocated topology, the Wi-Fi STA was placed between the Bluetooth nodes and the distance to the AP was varied. For the separated topology, the STA was moved to a position 10 m from the piconet, and the distance between it and the AP varied. Figure 7-20 plots the throughput figures for both of these situations, along with a baseline throughput that the Wi-Fi network can achieve without any Bluetooth interference. Once again, all paths were LOS.

When the Wi-Fi STA is collocated with a Bluetooth piconet, we notice that Wi-Fi throughput is significantly lower than the baseline, even for small AP-STA separations, but as the separation increases the throughput drops off gradually. Even with a 30 m separation, throughput is still 1 Mb/s. This is probably due to the processing gain inherent in the 2 Mb/s and 1 Mb/s data rates supported by Wi-Fi when confronted by narrowband interference. Although not mentioned in the study, the class 1 Bluetooth transmitters should have reduced their powers to 4 dBm or lower when communicating over a 0.3 m path to be compliant with the Bluetooth specification. When the Wi-Fi STA is separated from the Bluetooth piconet by 10 m, its throughput is at least half that achievable without Bluetooth interference, even for AP-STA separations approaching 80 m. However, the

Figure 7-19
The Wi-Fi STA can be either collocated with the Bluetooth piconet or separated from the piconet by 10 meters (TI Study).

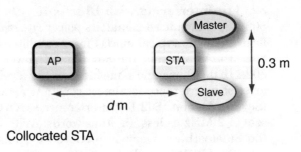

Figure 7-20
Wi-Fi throughput as a function of AP-STA distance for both collocated and separated topologies (TI Study) [Sho01]

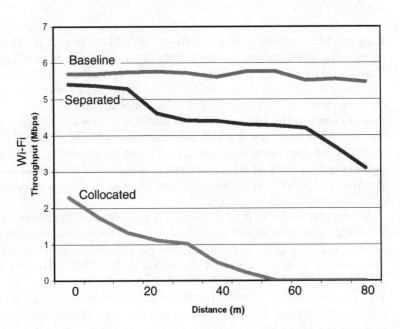

fractional throughput reduction exceeds that of a Bluetooth piconet being disrupted by Wi-Fi over similar separation distances.

Analytical Approach: Bluetooth–on–802.11b   For an analytical approach to Bluetooth–on–Wi-Fi interference, the scenario in Figure 7-16 is adjusted to account for the reversed roles of the two networks. This is shown in Figure 7-21, with the same path characteristics and antenna gains as before [Con03]. The following additional assumptions are made:

- The Wi-Fi payload is 1,024 bytes.
- A collision occurs over an entire Bluetooth single-slot packet (worst case).
- Each Bluetooth hop frequency is independently faded.

Under the second assumption, during a single-slot Bluetooth packet transmission time of 366 μs, a total of 366 low-rate or 504 high-rate Wi-Fi symbols are involved in any collision event that might occur. Although the 11-chip Barker sequence has a slightly greater processing gain than the 8-chip CCK sequence, the shorter frame duration at the higher data rate makes the two FER values nearly equivalent.

The proportion of interference power from the Bluetooth signal that enters the desired Wi-Fi receiver's passband also depends on the relative carrier offset between the two signals. These values are given in Table 7-7 by modeling the Bluetooth signal as a sinc(x) function sent through a Butterworth filter [How01]. Interestingly, the symmetrical nature of Bluetooth GFSK, coupled with the Barker code's processing gain and characteristic null at the carrier frequency, reduces its interference to low-rate Wi-Fi by about 10 dB when $\Delta f = 0$. For carrier offsets greater than 10 MHz, interference is negligible.

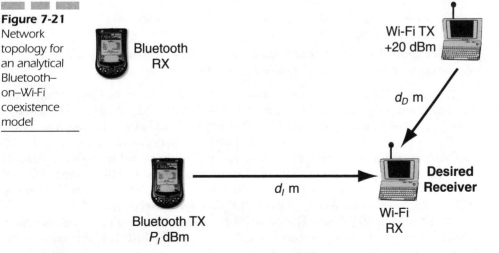

Figure 7-21
Network topology for an analytical Bluetooth–on–Wi-Fi coexistence model

Bluetooth
RX

Wi-Fi TX
+20 dBm

d_D m

d_I m

Desired
Receiver

Bluetooth TX
P_I dBm

Wi-Fi
RX

Table 7-7

Normalized
Bluetooth inter-
ference level
on a Wi-Fi
receiver
[Con03]

Carrier offset Δf (MHz)	Bluetooth interference (dB)	
	DSSS (1 Mb/s)	CCK (11 Mb/s)
0	−10.3	0.1
1	0.0	0.0
2	−0.4	−0.3
3	−1.0	−1.0
4	−1.2	−1.9
5	−13.4	−3.2
6	−4.6	−5.0
7	−6.7	−6.9
8	−9.5	−10.0
9	−13.5	−17.5
10	−19.9	−25.6

Rather than always assuming that the Bluetooth channel has maximum loading, three different realistic situations were considered:

■ **Bluetooth voice traffic** consists of an HV3 packet exchange, which occurs
during every third pair of single slots.

■ **Bluetooth file transfer traffic** consists of a fully loaded piconet using
single-slot packets.

■ **Bluetooth binomial traffic** consists of single-slot packets having a
binomial probability of transmission with average loading $1/6$.

The single HV3 voice channel is commonly used for a wireless link between the
headset and cell phone, and has a deterministic channel loading rate of $1/3$. When
the link is used for transferring files, the duty cycle will expand to approximately
100 percent, representing a worst-case interference situation. For other uses such
as Web browsing or piconet management, the binomial traffic model represents
the bursty nature of such exchanges. Figure 7-22 shows the effect of the different
Bluetooth traffic models on the Wi-Fi FER at 1 and 11 Mb/s transmission rates.
The parameters behind each of the eight plots are listed in Table 7-8.

When a single Bluetooth HV3 voice channel is active, Wi-Fi performance at 11
Mb/s is acceptable at any reasonable C/I value, but the FER is higher when operating at 1 Mb/s. This is primarily due to the shorter transmission time (851 μs) for

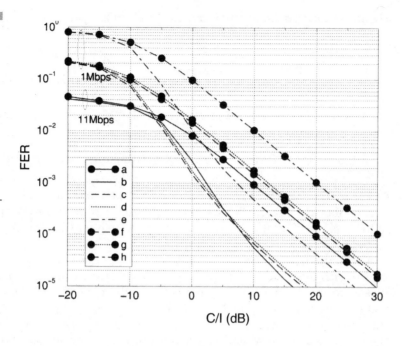

Figure 7-22
Wi-Fi FER as a function of average C/I under various renditions of Bluetooth interference [Con03]. See Table 7-8 for parameter list. © IEEE

Table 7-8

Parameters for plots in Figure 7-22 [Con03]

Plot	802.11 data rate (Mb/s)	Rice factor K	Bluetooth traffic
a	11	0	Voice
b	11	5	Voice
c	1	5	Voice
d	1	5	Binomial
e	1	5	File transfer
f	1	0	Voice
g	1	0	Binomial
h	1	0	File transfer

each 1,024-byte Wi-Fi payload sent at 11 Mb/s. Neglecting hop times, Bluetooth HV3 transmissions occupy 1,250 μs during every 3,750 μs interval. Therefore, the worst-case probability of partial time overlap on an 11 Mb/s Wi-Fi frame is

$$\text{PR(time overlap)} = \frac{851 + 1250}{3750} = 0.56 \qquad (7.16)$$

The worst-case probability of frequency overlap is

$$\Pr(\text{frequency overlap}) = 1 - \left(\frac{79 - 22}{79}\right)^2 = 0.48 \qquad (7.17)$$

and thus the worst-case probability of collision is

$$\Pr(\text{collision}) = \Pr(\text{time overlap}) \times \Pr(\text{frequency overlap}) = 0.27 \qquad (7.18)$$

At 1 Mb/s, the same Wi-Fi frame requires 8,384 μs to send, so a time overlap with at least four Bluetooth HV3 packets is guaranteed, with a resulting worst-case collision probability of 0.73, almost three times higher than for the same payload sent at 11 Mb/s.

As expected, interference from Bluetooth is greatest when the piconet is fully loaded with file transfer traffic. The worst Wi-Fi performance results when the Rice factor is 0 so that the channel is Rayleigh faded (OLOS). Fortunately, even under these conditions, the Wi-Fi 1 Mb/s FER is only 0.1 when the C/I is 0 dB, equivalent to a 90 percent raw throughput level. This demonstrates the ability of Wi-Fi to reject narrowband interference through the modest processing gain delivered by the 11-chip Barker spreading sequence. Finally, when the Bluetooth piconet loading is randomized through the binomial distribution, the Wi-Fi PER at 1 Mb/s drops below 0.1 for C/I values greater than about -10 dB.

Finally, Figure 7-23 plots the Wi-Fi PER against the separation ratio between the interfering Bluetooth source and the desired $+15$ dBm Wi-Fi transmitter at 11 Mb/s with various Bluetooth TX powers and different separation distance ratios [Con03]. The Bluetooth piconet is fully loaded with file transfer traffic, and the channel is modeled as LOS Rician with factor $K = 5$. Parameters for the plots are given in Table 7-9. It's no surprise that for higher Bluetooth TX powers, the Wi-Fi PER increases accordingly. These results show that Bluetooth significantly interferes with Wi-Fi only when the Bluetooth TX power approaches $+20$ dBm. At this power level, the Wi-Fi FER drops below 0.1 when d_I/d_D is greater than 1. In other words, the Wi-Fi network works well when the distance between communicating Wi-Fi nodes is less than the distance from any Wi-Fi node in the BSS to any Bluetooth device operating at high power. Figure 7-23 also demonstrates why Bluetooth class 1 transmitters are required to implement power control.

Bluetooth Interfering with 802.11g The performance of OFDM is enhanced by the FEC coding used on the data modulated onto each subcarrier. Prior to FEC, the information bit stream is interleaved so that the bit randomization of a single subcarrier due to a deep fade or narrowband interference results in only occasional isolated errors after deinterleaving. The Viterbi convolutional decoding process works well with these scattered bit errors, increasing the probabil-

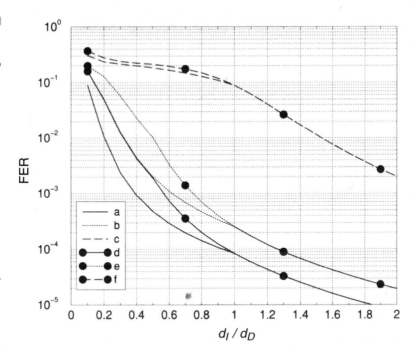

Figure 7-23
FER versus the separation ratio between the interfering Bluetooth source and the desired Wi-Fi source at the desired Wi-Fi receiver operating at 11 Mb/s [Con03]. See Table 7-9 for parameter list. © IEEE

Table 7-9

Parameters for plots in Figure 7-23 [Con03]

Plot	Desired link distance d_D (m)	Bluetooth TX power P_I (dBm)
a	8	0
b	8	+4
c	8	+20
d	16	0
e	16	+4
f	16	+20

ity that it produces error-free results. Interference can be significantly more disruptive to 802.11g if it falls on one of the four pilot subcarriers, because these are used as phase references for decoding the other subcarriers and as channel-fading indicators for setting the proper *quadrature amplitude modulation* (QAM) demodulation thresholds.

Table 7-10 shows the relative power required at the 802.11g receiver by a Bluetooth transmission to produce a BER of 10^{-3} at 48 Mb/s [Sel03]. The interference

Table 7-10

Bluetooth interference to 802.11g OFDM at 48 Mb/s [Sel03]

Carrier offset (MHz)	Relative Bluetooth TX power (dB) for 802.11g BER = 0.001 @ 48 Mb/s
0	−11
1	−11
2	−22
3	−12
4	−11
5	−11
6	−12
7	−13
8	−8
9	−5
10	−3

effect is fairly consistent for offsets to about 8 MHz, except for offsets at which the pilot subcarriers are affected. When the Bluetooth hop channel is 2 MHz away from the 802.11g channel center frequency, the pilot subcarrier at ±2.19 MHz is affected and the system is about 11 dB more sensitive to interference. This susceptibility is less pronounced at offsets of 6 and 7 MHz, both of which affect the 802.11g pilot subcarrier at 6.56 MHz. This subcarrier is located between two Bluetooth hop channels, reducing its susceptibility to disruption.

IEEE 802.15 TG2 Coexistence Analysis

The IEEE 802.15 TG2 addresses only interference between 802.11b Wi-Fi and 802.15.1 Bluetooth, but the methods the group used to form the basis of their work can apply to other forms of interference. For this reason, and because the comprehensive TG2 study is endorsed by the IEEE, we devote an entire section to the models and the associated results.

The PHY models developed by TG2 provide a basis for a mathematical performance analysis and the results from computer simulation. The IEEE breakpoint propagation model is used for these results, which assumes free-space propagation for the first 8 m between nodes and a PL exponent of 3.3 thereafter. The analysis and simulations use BER figures that exist during different periods of stationarity between two competing transmissions, and they apply the requisite BER to the packet or frame structure to obtain a PER and FER for Bluetooth and Wi-Fi, respectively.

Models for Finding BER in Interference

Finding the SIR, which is a term used by 802.15 TG2 that is synonymous with C/I, can be done using the multistep process depicted in Figure 7-24. First, the associated TX power from each desired and interfering transmitter is multiplied by the reciprocal of its associated PL (dB subtraction). The resulting desired signal power forms the numerator of the SIR. Next, the interfering signal levels are each adjusted by the fraction of power that enters the desired receiver's passband, and the powers are added together. This value forms the denominator of the SIR. The BER is then calculated according to the modulation used in the desired link. This calculation is simplified by assuming that the interference can be modeled as AWGN; thus, the SIR becomes SNR in the BER calculation formulas. Simulations have shown that modeling 802.11b interference as AWGN in the Bluetooth receiver is pessimistic by about 1 dB at most [Sel03]. Because the model is assumed to be interference limited, thermal noise is not included.

Spectrum Factor (SF)   The SF determines the fraction of interfering signal power that actually enters the desired receiver's passband, thus disrupting its capability to correctly decode the desired signal. Three parameters are used to determine this factor: the TX mask of the interfering transmitter, the RX mask

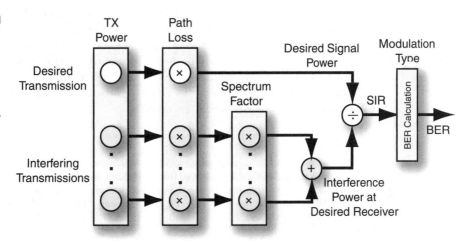

Figure 7-24
BER calculation process in an interference-limited environment

Table 7-11

TX and RX masks for 802.15.1 Bluetooth and 802.11b Wi-Fi

| | TX mask | | RX mask | |
Standard	Carrier offset (MHz)	Attenuation (dB)	Carrier offset (MHz)	Attenuation (dB)
802.15.1	0	0	0	0
	1	20	1	11
	2	40	2	41
	3	60	>2	51
802.11b	0–10	0	0–10	0
	11–21	30	11	12
	>21	50	12–20	36
			>20	56

of the desired receiver, and the carrier offset between the two. The TX mask is normalized so the area under the curve is 1. The RX mask, however, is not normalized. The SF is found by multiplying the two masks together at the desired offset, integrating, and converting the result to dB. The spectrum masks used by IEEE 802.15 TG2 are listed in Table 7-11. The TX masks are based on values in the respective 802.11 and 802.15.1 standards. The RX masks reflect C/I values published in the respective standards that are required for radio certification, along with good engineering practice.

From these masks the resulting SF values are shown in Table 7-12 for a variety of carrier offsets.[*] Out to a carrier offset of 10 MHz, all the Bluetooth signal power enters the Wi-Fi receiver, but at offsets greater than 11 MHz the Wi-Fi receiver rejects Bluetooth transmissions effectively. (Additional processing gain at the Wi-Fi receiver during DSSS or CCK demodulation isn't accounted for in the SF.) Due to the relatively wideband nature of the Wi-Fi signal, the Bluetooth receiver has 12.6 dB of attenuation built in through its 1 MHz passband filter for offsets up to about 10 MHz. For Wi-Fi carrier offsets beyond 10 MHz, very little Wi-Fi power enters the Bluetooth receiver.

[*]The SF values listed in Table 7-11 differ from the interference levels in Tables 7-5 and 7-7 because the latter were based on typical TX PSD rather than on a specification mask, and they are normalized to an offset of 1 MHz rather than on total TX mask area. Generally, the 802.15 TG2 SF values are easier to calculate, and they produce higher interference powers because actual TX PSD levels are usually well below limits set by the mask.

Table 7-12

SF values for
802.15.1
Bluetooth and
802.11b Wi-Fi

Carrier offset (MHz)	SF (dB)	
	802.15.1 on 802.11b	802.11b on 802.15.1
0–9	0.0	−12.6
10	0.0	−12.9
11	−11.4	−24.2
12	−30.1	−41.8
13	−35.9	−42.0
14–20	−36.0	−42.0
21	−52.9	−42.3
22	−55.6	−49.1
23–35	−55.7	−50.7
36–40	−55.8	−50.7
41–42	−55.8	−51.0
43–48	−55.9	−51.0

Wi-Fi and Bluetooth BER in Mutual Interference

The IEEE 802.15 TG2 performed extensive simulations of the BER as a function
of SIR in an interference-limited environment between Bluetooth and Wi-Fi at
various carrier offsets. These results were used in another simulation of PER and
FER values as a function of distance in a simple network topology. The simulation approach has the advantage that interaction between both desired and
interfering transmissions can be directly modeled without necessarily resorting
to an AWGN approximation for the interference. A disadvantage includes making sure that run times are sufficiently long to achieve good confidence intervals.

BER Simulation Results   The BER of low-rate 802.11b in the presence of
Bluetooth interference from the simulation is plotted in Figure 7-25. Due to the
null at the center frequency of the Barker code PSD, interference from Bluetooth
at a 1 MHz carrier offset is more significant than when the offsets are 0. This is
corroborated by Table 7-7, which shows that low-rate 802.11b has a 10 dB reduction in Bluetooth interference when the piconet hop frequency is identical to the
Wi-Fi carrier frequency, and when compared to a carrier offset of 1 MHz. (The
TG2 simulation doesn't show a BER drop at an offset of 5 MHz at a Wi-Fi data

Figure 7-25
802.11b Wi-Fi
BER in the
presence of
802.15.1
Bluetooth
interference for
low-rate data
[IEEE 802.15.2]
© IEEE

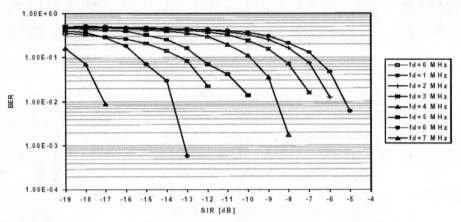

Bluetooth on 1 Mb/s 802.11b Wi-Fi

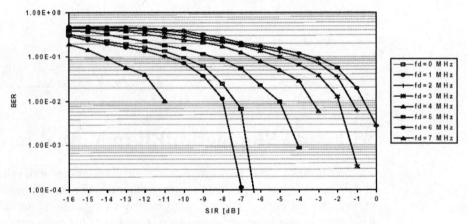

Bluetooth on 2 Mb/s 802.11b Wi-Fi

rate of 1 Mb/s that is also implied by Table 7-7.) The transition from 1 Mb/s *differential binary phase shift keying* (DBPSK) to 2 Mb/s *differential quadrature phase shift keying* (DQPSK) carries about a 5 dB SIR penalty.

When 802.11b is operating in its high-rate mode using 5.5 Mb/s or 11 Mb/s CCK, the BER values in the presence of Bluetooth interference are shown in Figure 7-26. Under these conditions, Wi-Fi no longer exhibits interference rejection at a carrier offset of 0. This is supported by Table 7-7 for a Wi-Fi data rate of 11 Mb/s. Generally, faster Wi-Fi data rates require higher SIR values for similar BER performance, as expected.

802.11b Wi-Fi
BER in the
presence of
802.15.1
Bluetooth
interference for
high-rate data
[IEEE 802.15.2]
© IEEE

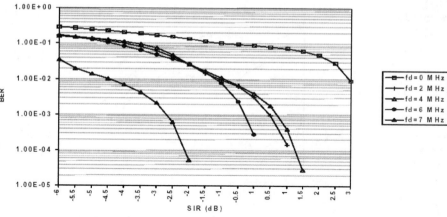

Bluetooth on 5.5 Mb/s 802.11b Wi-Fi

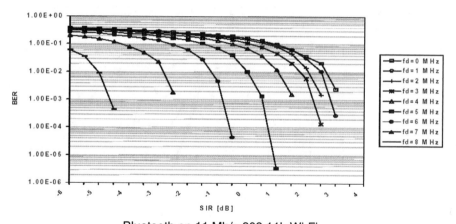

Bluetooth on 11 Mb/s 802.11b Wi-Fi

From another perspective, the results in Figures 7-25 and 7-26 can be used to adjust the hop collision probability between a Bluetooth and a Wi-Fi transmission as a function of SIR. First, we assume that all Wi-Fi frames are destroyed when the BER is above some threshold; otherwise, the frame is error-free. For example, if we use a threshold BER of 10^{-3}, then for 1 Mb/s Wi-Fi frames, an SIR of higher than -12 dB is sufficient to effectively eliminate Bluetooth interference for carrier offsets greater than 6 MHz. Now the collision probability for a single Bluetooth transmission is reduced from about 22/79 to 13/79. For 11 Mb/s Wi-Fi frames, the same result occurs when the SIR is above about -7 dB.

Simulations of Bluetooth affected by 802.11b Wi-Fi interference were performed by 802.15 TG2 for a Bluetooth receiver using a *limiter-discriminator with*

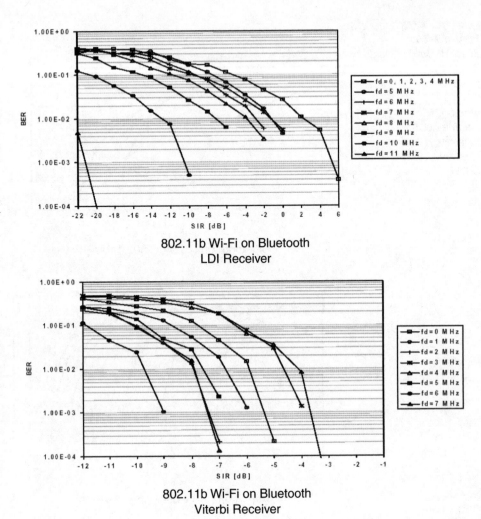

802.11b Wi-Fi on Bluetooth
LDI Receiver

802.11b Wi-Fi on Bluetooth
Viterbi Receiver

integrate and dump filtering (LDI) for a low-cost implementation, and with a
Viterbi receiver for higher interference immunity. BER performance as a function of SIR for both of these receivers is shown in Figure 7-27.

The Viterbi receiver is superior to LDI methods because it takes advantage of
the phase trellis created by the GFSK transmitter [Sol02]. To do this, the receiver
requires memory and a fixed modulation index known to the receiver. The Bluetooth modulation index is allowed by the 802.15.1 specification to vary from 0.28
to 0.35, which can degrade performance. In this simulation, the modulation index
is assumed to be fixed at 0.32, in which case the Viterbi receiver performs significantly better than the LDI receiver.

Wi-Fi FER and Bluetooth PER in Mutual Interference

For FER and PER simulation modeling, TG2 developed a simple topology consisting of one point-to-point 802.15.1 Bluetooth piconet and one point-to-point 802.11b Wi-Fi BSS. Figure 7-28 depicts the relationship between each of the four nodes involved in the simulation, which is a common topology when independent, noncollocated Bluetooth and Wi-Fi networks are operating in the same area.

The Bluetooth nodes are separated by 1 m and operate with a TX power of 0 dBm. The Wi-Fi AP is located 15 m "north" of the Bluetooth slave, and this distance is considered to be far enough that its transmissions will have no detrimental affect on the Bluetooth receivers. Both Wi-Fi nodes operate with a TX power of +14 dBm. The Wi-Fi STA is located between the Bluetooth piconet and the AP, and its distance d from the piconet can be varied. This situation is common in actual applications, because the relatively portable STA can improve both its own SIR and the SIR of the Bluetooth piconet by moving away from the interference source and closer to the AP.

The Bluetooth piconet can send voice, consisting of HV1 packets that fully load the channel with single-slot packets, or it can send data, consisting of DM5 packets with a 12.5 ms packet interarrival time. The Wi-Fi network sends data at

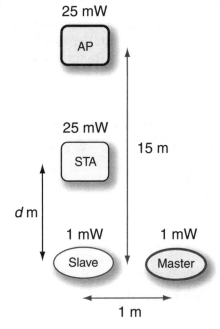

Figure 7-28
Coexistence topology used for Wi-Fi packet loss measurements (IEEE 802.15 TG2 study)

1 Mb/s with a 24.8 ms frame interarrival time, or at 11 Mb/s with a 2.6 ms frame interarrival time.

Interference analysis is focused on the Wi-Fi STA and the Bluetooth slave. The Wi-Fi STA can be sending or receiving data frames at 1 or 11 Mb/s. When sending data frames, the STA experiences interference on the associated ACKs arriving from the AP. Interference comes from both master and slave in the Bluetooth piconet. While receiving either voice or data packets, the Bluetooth slave can experience interference from either data frame or ACK transmissions from the Wi-Fi STA.

The 802.15.2 recommended practice provides several plots of FER and PER as a function of distance $d > 0.5$ m for various combinations of Wi-Fi and Bluetooth transmissions. Instead of reproducing these plots, we summarize the results by listing the separation distances at which the FER or PER drops below 20 and 10 percent under the various interference conditions studied. The assumption here is that noticeable performance degradation occurs at FER/PER values above 20 percent, but degradation is minimal when the error rate drops below 10 percent. Smaller separation distances d for a given FER or PER equate to greater interference immunity.

Table 7-13 lists the distance d to the nearest half-meter for FER values of 20 and 10 percent at the Wi-Fi STA under Bluetooth voice and data interference. This interference is smallest during ACK reception, which isn't surprising because the ACK frame is short, providing fewer opportunities for a Bluetooth collision, and it uses the more robust Barker spreading code. Performance is worst when the STA is receiving a data frame at 11 Mb/s due to the reduced interference immunity provided by CCK.

Table 7-14 shows the distance d to the nearest half-meter for PER values of 20 and 10 percent at the Bluetooth slave under interference from the Wi-Fi STA. Bluetooth PER remains below 20 percent in all situations at the minimum

Table 7-13

Bluetooth-on-Wi-Fi interference results from IEEE 802.15 TG2 simulations

Desired signal	Interfering signal	Minimum separation distance (m) for FER:	
		<20 percent	<10 percent
STA RX ACK @ 1 Mb/s	Bluetooth TX voice	<0.5	1.0
STA RX ACK @ 1 Mb/s	Bluetooth TX data	<0.5	1.5
STA RX data @ 1 Mb/s	Bluetooth TX voice	2.0	2.5
STA RX data @ 1 Mb/s	Bluetooth TX data	1.5	2.0
STA RX data @ 11 Mb/s	Bluetooth TX voice	3.5	4.0
STA RX data @ 11 Mb/s	Bluetooth TX data	<0.5	3.0

Table 7-14

Wi-Fi-on-Bluetooth interference results from IEEE 802.15 TG2 simulations

		Minimum separation distance (m) for PER:	
Desired signal	**Interfering signal**	**<20 percent**	**<10 percent**
Bluetooth slave RX voice	Wi-Fi STA TX data @ 1 Mb/s	<0.5	1.5
Bluetooth slave RX voice	Wi-Fi STA TX data @ 11 Mb/s	<0.5	1.5
Bluetooth slave RX voice	Wi-Fi STA TX ACK @ 1 Mb/s	<0.5	<0.5
Bluetooth slave RX data	Wi-Fi STA TX data @ 1 Mb/s	<0.5	2.5
Bluetooth slave RX data	Wi-Fi STA TX data @ 11 Mb/s	<0.5	5.0
Bluetooth slave RX data	Wi-Fi STA TX ACK @ 1 Mb/s	<0.5	<0.5

simulated d of 0.5 m, and interference from the Wi-Fi STA is small whenever it is transmitting short ACK frames. The worst interference comes from the STA sending data frames while the Bluetooth slave is receiving its own DM5 frame, which has a longer five-slot duration and less robust FEC code than the single-slot HV1 voice packet. Interestingly, 11 Mb/s CCK is more disruptive to the Bluetooth piconet than 1 Mb/s DSSS at the same distance and TX power.

Enhancing Passive Coexistence Between Wi-Fi and Bluetooth

Comparing empirical and analytical results in the previous section can be a bit risky due to the different sets of assumptions and methods used, but we can at least offer the following observations brought forth by the various studies:

- When Bluetooth and 802.11 Wi-Fi devices are collocated, Bluetooth packet loss and Wi-Fi frame loss are both high, sometimes exceeding 50 percent, but performance improves rapidly as the two networks are moved away from each other.
- If Bluetooth and Wi-Fi use comparable TX powers, Bluetooth PER is acceptably low if the distance to the nearest interfering Wi-Fi node is greater than the separation between communicating piconet members.
- Bluetooth–on–Wi-Fi interference is relatively low for Bluetooth TX power classes 1 and 2, provided the Bluetooth node isn't so close that the Wi-Fi receiver is blocked.

- When a Bluetooth class 1 transmitter is operating at full power, Wi-Fi FER is acceptably low if the distance to the nearest class 1 Bluetooth node is greater than the separation between Wi-Fi nodes.

- Using FEC in a Bluetooth ACL packet reduces the PER over an interference-limited channel, but often not enough to offset the increased coding overhead.

The Two-Meter Rule

Finally, we offer a way that users can improve the passive coexistence performance of noncollocated Bluetooth and Wi-Fi operating in the same area. This method, which we call the two-meter rule, doesn't guarantee interference-free operation, but it works well under most real-world situations [Gal01], [Mob01]. The rule consists of three parts:

- If a Wi-Fi AP and STA are separated by less than 2 m, then the presence of Bluetooth will have a negligible effect.

- If a Bluetooth master and slave are separated by less than 2 m, then the presence of Wi-Fi will have a negligible effect.

- If all Wi-Fi nodes are separated from all Bluetooth nodes by at least 2 m, then both networks will operate reasonably well.

The third part of the two-meter rule is less stringent than an earlier recommendation that desired link lengths should be shorter than interfering link lengths, but the latter can often be implemented by moving communicating nodes closer together and away from sources of interference.

Coexisting with 802.15.3 WiMedia

Because 802.15.3 WiMedia shares the same 2.4 GHz ISM band with several other wireless systems, the potential exists for interference between networks. Most WiMedia applications consist of high-speed data transfers over distances of less than 10 m, and sometimes as short as 1 m. The close proximity and relatively low power of participating WiMedia nodes will help reduce mutual disruption with other users. On the other hand, the low bit energy and lack of significant processing gain means that, when interference is present, the WiMedia FER can be quite high.

The 802.15.3 WiMedia superframe is divided into the contention-based *contention access period* (CAP), in which channel access is via *carrier sense multiple access with collision avoidance* (CSMA/CA), and into a contention-free *channel time allocation period* (CTAP) where either *time division multiple access* (TDMA) or a combination TDMA/ALOHA channel access is employed. The coexistence studies performed by the 802.15 TG3 included the following general performance assumptions:

- WiMedia RX sensitivity is −75 dBm and bandwidth is 15 MHz for 22 Mb/s DQPSK.
- The average desired signal RSSI for all receivers is 10 dB above the sensitivity limit.
- The WiMedia TX power is +8 dBm with PSD conforming to the mask given in the specification.
- The propagation PL exponent is 2.0 for a TX-RX separation distance of less than 8 m; otherwise, the PL exponent is 3.3, conforming to the IEEE breakpoint PL model.
- The interfering signal energy is modeled as AWGN so the BER plot in Figure 4-22 applies by substituting SIR for SNR.
- WiMedia uses the 802.11b coexistence channel plan.

WiMedia and Wi-Fi

Co-channel interference between WiMedia and Wi-Fi can disrupt both networks, especially if they aren't capable of cross-network CCA. The following 802.11b assumptions were made by the 802.15 TG3 for calculating its coexistence capabilities with WiMedia:

- Wi-Fi RX sensitivity is −76 dBm and the bandwidth is 22 MHz for 11 Mb/s CCK.
- Wi-Fi TX power is +14 dBm with the PSD conforming to the mask given in the specification.

The WiMedia and Wi-Fi channel structure is such that they can operate on co-channels, adjacent channels, or alternate channels. These are defined in Table 7-15. The minimum separation distances from an interfering node when a desired node experiences a PER of 10 percent under different conditions are shown in Table 7-16. The assumption here is that neither network's CCA mechanism can sense the other's presence.

When co-channel interference exists, the WiMedia and Wi-Fi nodes must be separated by relatively large distances to achieve reasonable throughput in both networks. Wi-Fi interferes with 22 Mb/s WiMedia more severely than vice versa due to the lack of processing gain in the WiMedia signal. Also apparent is the reduced Wi-Fi processing gain as data rates increase, manifested by a greater separation distance needed from a co-channel WiMedia node to maintain an FER

Table 7-15

Adjacent and alternate channel operation for WiMedia and Wi-Fi interference

WiMedia channel	Adjacent Wi-Fi channel	Alternate Wi-Fi channel
1	6	11
3	1 or 11	None
5	6	1

Wi-Fi channel	Adjacent WiMedia channel	Alternate WiMedia channel
1	3	5
6	1 or 5	None
11	3	1

Table 7-16

WiMedia and Wi-Fi interference results from IEEE 802.15 TG3 simulations

Desired signal	Interfering signal	Type	Minimum separation distance (m) for FER <10 percent
WiMedia @ 22 Mb/s	Wi-Fi	Co-channel	80
WiMedia @ 22 Mb/s	Wi-Fi	Adjacent channel	6
WiMedia @ 22 Mb/s	Wi-Fi	Alternate channel	0.6
Wi-Fi @ 1 Mb/s	WiMedia	Co-channel	40
Wi-Fi @ 1 Mb/s	WiMedia	Adjacent channel	0.4
Wi-Fi @ 1 Mb/s	WiMedia	Alternate channel	0.2
Wi-Fi @ 11 Mb/s	WiMedia	Co-channel	60
Wi-Fi @ 11 Mb/s	WiMedia	Adjacent channel	0.7
Wi-Fi @ 11 Mb/s	WiMedia	Alternate channel	0.4

of 10 percent. This demonstrates the necessity of using nonoverlapping channels when simultaneous transmissions can't be avoided.

The interference situation improves dramatically for adjacent and alternate channel operation. In most of these cases, when the two networks are separated by more than 1 m, essentially no interference occurs. The exception is when the Wi-Fi and WiMedia networks are operating on adjacent channels, and a separation of more than 6 m is necessary for the WiMedia FER to drop below 10 percent.

WiMedia and Bluetooth

The 802.15.3 WiMedia specification includes a study of mutual interference in the presence of a Bluetooth piconet when no attempt is made to improve the situation through collaborative or noncollaborative methods [IEEE802.15.3]. In this analysis, the following assumptions about the Bluetooth signal are made:

- Bluetooth RX sensitivity is -70 dBm and the bandwidth is 1 MHz for 1 Mb/s GFSK.
- Bluetooth TX power is 0 dBm with the PSD conforming to the mask given in the specification.

Table 7-17 shows WiMedia and Bluetooth separation distances for an FER (WiMedia) or PER (Bluetooth) of 10 percent when Bluetooth is using the 79-channel hopping sequence. Assuming neither receiver is blocked, the worst-case FER/PER for either piconet is about 0.19 (15/79), which is the probability that a Bluetooth hop channel enters the WiMedia passband. As before, Bluetooth ACL packet throughput can drop below the level implied by the average PER because a successful exchange requires error-free performance in two successive hop channels. Substantial separation is required between the two piconets before the FER/PER drops from 0.19 to below 0.10. The Bluetooth piconet performs well when all piconet members are more than 17 m from the nearest WiMedia node, and the WiMedia piconet performs well when all participants are separated from Bluetooth nodes by more than 30 m.

WiMedia is affected more by Bluetooth than the other way around because the entire Bluetooth TX signal's energy enters the WiMedia receiver's passband during a hop "hit," and WiMedia at 22 Mb/s carries with it essentially no processing gain. On the other hand, only about $\frac{1}{15}$ of the WiMedia TX signal enters the Bluetooth receiver's passband when the TX and RX roles are reversed. This equates to a 12 dB processing gain at the Bluetooth receiver during a hit, which

Table 7-17

WiMedia and Bluetooth interference results from IEEE 802.15 TG3 simulations

Desired signal	Interfering signal	Minimum separation distance (m) for FER/PER <10 percent
WiMedia @ 22 Mb/s	Bluetooth	30
Bluetooth	WiMedia	17

more than compensates for the higher WiMedia TX power used in the analysis. Even so, separation distances must still be quite high to achieve a reasonable PER, leading us to conclude that additional coexistence mechanisms must be employed when the two piconets are operating near each other.

Coexisting with 802.15.4 ZigBee

The modulation method used by ZigBee is one of the ways passive coexistence with other services is enhanced. As we discussed in Chapter 4, the 2.4 GHz Zig-Bee PHY uses a quasi-orthogonal modulation scheme in which 4 data bits are encoded into 1 of 16 spreading sequences, each 32 chips long. This power-efficient modulation scheme is expected to provide a PER of about 1 percent for SNR values of 5 to 6 dB [IEEE802.15.4].

The ZigBee channel set includes channels 11 to 26 for use in the 2.4 GHz band, and these have the potential to interfere with, and experience interference by, the other 2.4 GHz wireless services. Fortunately, the relatively high processing gain of approximately 15 dB inherent in its 32-chip DSSS signal structure reduces the ZigBee signal's susceptibility to interference. Nominal ZigBee TX power of 0 dBm enhances its status as a good neighbor, along with its low duty cycle in many applications. On the other hand, if a large number of ZigBee devices are configured as a mesh network, then store-and-forward transmissions could raise the duty cycle, and hence the interference level, significantly.

The 802.15.4 ZigBee superframe is divided into the contention-based CAP, in which channel access is via slotted CSMA/CA, and a contention-free CFP where TDMA channel access is employed. If the optional superframe is not used, then channel access is via unslotted CSMA/CA. The coexistence studies performed by the 802.15 TG4 included the following general performance assumptions:

- ZigBee RX sensitivity is -85 dBm and the bandwidth is 2 MHz for 250 Kb/s DSSS.
- The average desired signal RSSI for all receivers is 10 dB above the sensitivity limit.
- The ZigBee TX power is 0 dBm with the PSD conforming to the mask given in the specification.
- The average ZigBee frame length is 22 bytes.
- The propagation PL exponent is 2.0 for the TX-RX separation distance of less than 8 m; otherwise, the PL exponent is 3.3, corresponding to the IEEE breakpoint PL model.
- The interfering signal energy is modeled as AWGN so the BER plot in Figure 4-22 applies by substituting SIR for SNR.

ZigBee and Wi-Fi

Several 802.15.4 ZigBee and 802.11b Wi-Fi coexistence performance calculations have been included in the 802.15.4 specification [IEEE802.15.4], in which the following assumptions for the Wi-Fi signal are made:

- Wi-Fi RX sensitivity is −76 dBm and the bandwidth is 22 MHz for 11 Mb/s CCK.
- Wi-Fi TX power is +14 dBm with the PSD conforming to the mask given in the specification.
- The average Wi-Fi frame length is 1,024 bytes.

Because the ZigBee TX bandwidth is narrow enough to support 16 channels in the 2.4 GHz band, modeling interference with a Wi-Fi WLAN is best done using carrier offsets. Table 7-18 shows the minimum separation distances from an interfering node when a desired node experiences a PER (ZigBee) or FER (Wi-Fi) of 10 percent under different conditions. Once again, the assumption is that neither network's CCA mechanism can sense the other's presence.

For a frequency offset of 3 MHz, the ZigBee node must be separated from an interfering Wi-Fi node by more than 60 m before performance begins to become acceptable. The separation distance is reduced to about 7 m for an offset of 22 MHz and is only 3 m when the offset is 47 MHz. If a Wi-Fi network is operating on Wi-Fi channel 6, it occupies a bandwidth of 22 MHz centered at 2,437 MHz. If the ZigBee piconet is on ZigBee channel 1 (2,405 MHz), the carrier offset is 32 MHz and all Wi-Fi nodes must be more than 5 m from the ZigBee piconet

Table 7-18

ZigBee and Wi-Fi interference results from IEEE 802.15 TG4 simulations

Desired signal	Interfering signal	Carrier offset	Minimum separation distance (m) for FER <10 percent
ZigBee	Wi-Fi	3 MHz	60
ZigBee	Wi-Fi	22 MHz	7
ZigBee	Wi-Fi	47 MHz	3
Wi-Fi @ 1 Mb/s	ZigBee	3 MHz	15
Wi-Fi @ 1 Mb/s	ZigBee	22 MHz	3
Wi-Fi @ 1 Mb/s	ZigBee	47 MHz	0.3
Wi-Fi @ 11 Mb/s	ZigBee	3 MHz	22
Wi-Fi @ 11 Mb/s	ZigBee	22 MHz	5
Wi-Fi @ 11 Mb/s	ZigBee	47 MHz	0.8

members before the ZigBee PER is acceptable. If the ZigBee piconet is operating on ZigBee channel 26 (2,480 MHz), the piconet will operate reasonably well if all active Wi-Fi nodes on Wi-Fi channel 6 are more than about 3 m away due to the greater carrier offset of 43 MHz. Even when a Wi-Fi network is operating on channel 11 (2,462 MHz), the highest designated U.S. carrier frequency, a ZigBee piconet using the highest ZigBee channel (26) has an offset of 18 MHz.

Table 7-18 also shows 802.11b Wi-Fi FER under ZigBee interference. A carrier offset of 3 MHz requires a separation of more than about 20 m for acceptable Wi-Fi performance. For an offset of 22 MHz, the networks require a separation of more than 5 m, and for offsets of 47 MHz the required separation is about 1 m. We conclude that, when the two networks operate on overlapping channels, Wi-Fi affects ZigBee more than the other way around primarily due to the higher TX power in the Wi-Fi node. Even so, the higher processing gain from the ZigBee 32-chip spreading sequence, coupled with its 2 MHz signal bandwidth, partially compensates for this TX power disadvantage. From the Wi-Fi perspective, the anticipated low duty cycle of the standard ZigBee piconet will cause only occasional WLAN disruption, even for relatively small separation distances.

ZigBee and Bluetooth

Coexistence between ZigBee operating in the 2.4 GHz band and Bluetooth should be adequate under nonblocking situations, mainly because the Bluetooth hit probability is only about 3/79 per hop. A co-channel hit occurs when the Bluetooth hop channel is identical to the ZigBee carrier, and the two adjacent-channel hits have a 1 MHz offset and partially overlapping signals. Thus, the worst-case nonblocking PER from this interference is only about 4 percent. The PER will be higher if the two piconets are close enough that receiver C/I figures at greater offsets are violated.

Table 7-19 shows the separation distance required between the ZigBee and Bluetooth piconets for a co-channel hit to cause a PER of <10 percent. As before, the following assumptions about the Bluetooth signal are made:

■ Bluetooth RX sensitivity is −70 dBm and the bandwidth is 1 MHz for 1 Mb/s GFSK.

Table 7-19

ZigBee and Bluetooth interference results from IEEE 802.15 TG4 simulations

Desired signal	Interfering signal	Minimum separation distance (m) for FER/PER <10 percent during a co-channel hit
ZigBee	Bluetooth	22
Bluetooth	ZigBee	13

Table 7-20

ZigBee and
WiMedia inter-
ference results
from IEEE
802.15 TG4
simulations

Desired signal	Interfering signal	Carrier offset	Minimum separation distance (m) for FER <10 percent
ZigBee	WiMedia	2 MHz	42
ZigBee	WiMedia	17 MHz	6
ZigBee	WiMedia	27 MHz	2
WiMedia	ZigBee	2 MHz	30
WiMedia	ZigBee	17 MHz	4
WiMedia	ZigBee	27 MHz	0.7

- Bluetooth TX power is 0 dBm with the PSD conforming to the mask given in the specification.

ZigBee and WiMedia

The WiMedia signal used in the simulation has the following characteristics:

- WiMedia RX sensitivity is −75 dBm and the bandwidth is 15 MHz for 22 Mb/s DQPSK.
- WiMedia TX power is +8 dBm with the PSD conforming to the mask given in the specification.

The passive coexistence results between ZigBee and WiMedia are given in Table 7-20. The carrier offsets given in this table are smaller than the offsets used when comparing ZigBee and Wi-Fi. For example, if WiMedia is operating on channel 1 (2,412 MHz), a ZigBee piconet can be on channel 18 (2,440 MHz) through channel 26 (2,480 MHz) to achieve a carrier offset greater than the maximum in the simulation.

Coexistence Between 802.11a and UWB

The FCC in the United States allows the operation of UWB transmitters between 3.1 and 10.6 GHz at an average *equivalent isotropic radiated power* (EIRP) of −41.3 dBm/MHz, as discussed in Chapter 1, "Introduction." This bandwidth completely contains the 5.x GHz *Unlicensed National Information Infrastructure* (U-NII) bands where 802.11a resides, so it would be interesting to analyze the coexistence between one or more UWB systems and 802.11a. Due to the unpredictable nature of interference entering the UWB receiver, we will examine only UWB-on-OFDM performance.

As a quick preliminary analysis, the maximum average UWB TX power within the 16.6 MHz bandwidth of an 802.11a OFDM transmission is -29.1 dBm, or 1.2 *microwatts* (μW) ($-41.3 + 10 \log (16.6)$). In free space, the signal power drops to -100 dBm at a distance of about 16 m from the UWB transmitter. This is 18 dB below the required minimum sensitivity level of an 802.11a receiver operating at 6 Mb/s, and 35 dB below the minimum sensitivity level at 54 Mb/s. We conclude that reasonable passive coexistence exists between one UWB transmission and 802.11a WLAN operating on any channel within the U-NII bands. If multiple UWB nodes exist in the same area as the 802.11a WLAN, performance will suffer accordingly.

Analytical Modeling of UWB-on-OFDM Interference

The PSD of a short pulse of RF energy appears similar to the PSD of AWGN, regardless of the pulse shape used to form the UWB transmission. Therefore, it could be argued that modeling UWB interference on each OFDM carrier as AWGN will give accurate BER results. For uncoded BPSK, QPSK, 16-QAM, and 64-QAM, the BER is given by [Sol03]

$$P_b^{(\text{BPSK})} = P_b^{(\text{QPSK})} = Q\left(\sqrt{F \cdot \frac{2E_b}{N_0 + N_{\text{UWB}}}} \right) \tag{7.19}$$

where N_{UWB} is the PSD of UWB in W/Hz, and $F = 0.74$ is the fraction of the OFDM symbol containing information.

For M-ary QAM, where M is 16 or 64, the symbol error rate P_M is

$$P_M = 1 - (1 - P_{\sqrt{M}})^2 \tag{7.20}$$

where

$$P_{\sqrt{M}} = 2\left(1 - \frac{1}{\sqrt{M}} \right) Q\left(\sqrt{F \cdot \frac{3}{M-1} \cdot \frac{kE_b}{N_0 + N_{\text{UWB}}}} \right) \tag{7.21}$$

For Gray-coded QAM symbols and reasonably high SNR/bit values,

$$P_b = \frac{1}{k}P_M \tag{7.22}$$

where $k = \log_2 M$.

Figure 7-29 shows a plot of the BER versus SNR/bit for these uncoded symbols in the presence of AWGN and UWB interference. For a BER of 10^{-3}, the presence of UWB causes a penalty of a few dB, and the penalty is more severe for higher k. As expected, when the UWB interference at a particular level is significantly greater than the AWGN, the latter has little influence on the BER. Figure 4-15 in Chapter 4 compares the BER of uncoded BPSK/QPSK, 16-QAM, and 64-QAM modulation with their coded renditions. Similar performance advantages should be realized over a UWB-interference-limited channel.

Summary

When wireless networks passively coexist in an area of mutual interference, they take no extraordinary measures to improve performance but instead simply operate as they would in an interference-free environment. By far, the greatest numbers of coexisting networked devices exist within the 2.4 GHz band. These include 802.11b and 802.11g Wi-Fi, 802.15.1 Bluetooth, 802.15.3 WiMedia, and 802.15.4 ZigBee.

It could be argued that 802.15.1 Bluetooth has the best passive coexistence mechanism by using FHSS, which spreads its operation over 79 MHz of the 2.4 GHz band. ZigBee is next, with its 32-chip sequence providing a processing gain of 15 dB, followed by 802.11b with a PG of 10.4 dB at 1 Mb/s, which decreases as the data rate increases. Finally, 802.15.3 has no *processing gain* (PG) at 22 Mb/s, and mild PG via *trellis coded modulation* (TCM) at other data rates.

For those WLAN and WPAN systems operating on a fixed channel (all except Bluetooth), most require significant separation distances between the desired and interfering nodes when both operate on the same channel. For WiMedia operating on the Wi-Fi coexistence channel set, the adjacent and alternate channels are both nonoverlapping, so coexistence between the two networks under these conditions is acceptable even for small separation distances. To generalize, when no signal overlap exists, coexistence is acceptable for noncollocated nodes belonging to independent networks.

For Bluetooth and Wi-Fi operation using passive coexistence, performance degradation will occur when the two networks are near each other due to the random hopping of Bluetooth FHSS. When a hit occurs, both networks experience reduced throughput from the resulting collision. Empirical research has shown that the two-meter rule provides reasonably good throughput. This rule states that communicating nodes should be separated by less than 2 m or each desired node should be more than 2 m away from any interfering node. Alternatively, frame or packet throughput is usually acceptable if communicating nodes are placed closer to each other than they are to any interfering source. This rule is generally applicable to fixed-channel devices, provided that no channel overlap exists between independent networks.

Figure 7-29
BER versus
SNR/bit for
uncoded
OFDM as used
in 802.11a
[Sol03]

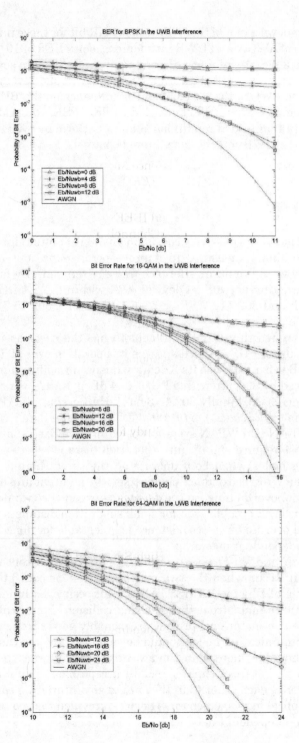

Finally, noncollocated UWB transmissions that conform to U.S. FCC regulations should have only a mild detrimental effect on 802.11a operations due to the low PSD of the UWB transmitted signal.

References

[Chi03] Chiasserini, C., et al., "Coexistence Mechanisms for Interference Mitigation in the 2.4-GHz ISM Band," *IEEE Transactions on Wireless Communications*, September 2003.

[Con03] Conti, A., et al., "Bluetooth and IEEE 802.11b Coexistence: Analytical Performance Evaluation in Fading Channels," *IEEE Transactions on Communications*, February 2003.

[Cor03] Cordeiro, C., et al., "Interference Modeling and Performance of Bluetooth MAC Protocol," *IEEE Transactions on Wireless Communications*, November 2003.

[Eli01] Eliezer, O., "802.15.3 Coexistence with 802.15.1 (Bluetooth)," IEEE 802.15 Document IEEE 802.1501/083 r1, January 2001.

[Gal01] Galli, S., et al., "Bluetooth Technology: Link Performance and Networking Issues," Telcordia white paper, 2001.

[Haa99] Haarsten, J., and Zürbes, S., "Bluetooth Voice and Data Performance in 802.11 DS WLAN Environment," Ericsson White Paper, May 1999.

[How01] Howitt, I., et al., "Empirical Study for IEEE 802.11 and Bluetooth Interoperability," *Proceedings Vehicular Technology Conference, Vol. 2*, Athens, Greece, May 2001.

[How03] Howitt, I., "Mutual Interference Between Independent Bluetooth Piconets," *IEEE Transactions on Vehicular Technology*, May 2003.

[IEEE802.11] IEEE 802.11-1999, "Wireless LAN Medium Access Control (MAC) and Physical Layer (PHY) Specifications," August 20, 1999.

[IEEE802.11a] IEEE 802.11b-1999, "High-Speed Physical Layer in the 5 GHz Band," September 16, 1999.

[IEEE802.11b] IEEE 802.11b-1999, "Higher-Speed Physical Layer Extension in the 2.4 GHz Band," September 16, 1999.

[IEEE802.11g] IEEE 802.11g-2003, "Amendment 4: Further Higher Data Rate Extension in the 2.4 GHz Band," June 27, 2003.

[IEEE802.15.1] IEEE 802.15.1-2002, "Wireless Medium Access Control (MAC) and Physical Layer (PHY) Specification for Wireless Personal Area Networks (WPAN)," June 14, 2002.

[IEEE802.15.2] IEEE 802.15.2-2003, "Coexistence of Wireless Personal Area Networks with Other Wireless Devices Operating in Unlicensed Frequency Bands," August 28, 2003.

[IEEE802.15.3] IEEE 802.15.3/D17, "Wireless Medium Access Control (MAC) and Physical Layer (PHY) Specifications for High Rate Wireless Personal Area Networks (WPAN)," draft standard, February 2003.

[IEEE802.15.4] IEEE 802.15.4-2003, "Wireless Medium Access Control (MAC) and Physical Layer (PHY) Specifications for Low-Rate Wireless Personal Area Networks (LR-PANS)," October 1, 2003.

[Lan00] Lansford, J., "Wi-Fi (802.11b) and Bluetooth Simultaneous Operation: Characterizing the Problem," Mobilian Corporation White Paper, 2000.

[Lan01] Lansford, J., "Adaptive Hopping," presented at the Bluetooth Developers Conference, San Francisco, CA, December 2001.

[Mei00] Meihofer, E., "The Performance of Bluetooth in a Densely Packed Environment," Bluetooth Developers Conference, December 4-7, 2000.

[Mei01] Meihofer, E., "Enhancing ISM Band Performance Using Adaptive Frequency Hopping," Motorola White Paper, December 2001.

[Mil03] Miller, L., "Validation of 802.11a/UWB Coexistence Simulation," National Institute of Standards and Technology white paper, October 2003.

[Mob01] "Wi-Fi (802.11b) and Bluetooth: An Examination of Coexistence Approaches," Mobilian White Paper, 2001.

[Mor02] Morrow, R., *Bluetooth Operation and Use*, New York: McGraw-Hill, 2002.

[Pur77] Pursley, M., "Performance Evaluation for Phase Coded Spread Spectrum Multiple Access Communication—Part 1: System Analysis," *IEEE Transactions on Communications*, August 1977.

[Sel03] Selby, S., "Co-Existence Warrants a Second Glance," *Wireless Systems Design*, October 2003.

[Sho01] Shoemake, M., "Wi-Fi (IEEE 802.11b) and Bluetooth: Coexistence Issues and Solutions for the 2.4 GHz ISM Band," Texas Instruments White Paper, Version 1.1, February 2001.

[Sol02] Soltanian, A., "Performance of the Bluetooth System in Fading Dispersive Channels and Interference," National Institute of Standards and Technology white paper, 2002.

[Sol03] Soltanian, A., "Coexistence of Ultra-Wideband System and IEEE 802.11a WLAN," National Institute of Standards and Technology white paper, revised by L. Miller, October 2003.

[War00] Ware, C., et al., "On the Hidden Terminal Jamming Problem in IEEE 802.11 Mobile Ad Hoc Networks," Telecommunications Research Centre white paper, University of Wollongong, Australia.

Active Coexistence

The results in the last chapter showed that significant increases in *packet error rates* (PERs) can sometimes occur when two or more wireless networks operate in the same area and create mutual interference. When a communication channel is rendered useless from packet collisions, all the involved packets are destroyed. Even without this worst-case assumption, long packets experiencing even a relatively mild increase in the *bit error rate* (BER) from interference may still have a high PER, especially if no error correction capability exists and a single bit error destroys the packet. Therefore, it makes sense to try to reduce the level of interference when a desired frame or packet is being received. To do this, each *wireless local area network* (WLAN) and *wireless personal area network* (WPAN) specification includes several methods to improve coexistence, mostly through noncollaborative techniques that go beyond simply using *carrier sense multiple access* (CSMA) or, with Bluetooth, 79-channel *frequency hopping spread spectrum* (FHSS). For collocated nodes, defined as nodes operating within 0.5 m of each other, various collaborative methods have been proposed to enhance throughput even further.

We refer to these collaborative and noncollaborative coexistence methods as *active coexistence*. Although it may be argued that CSMA itself is a form of active coexistence, this listen-before-talk channel access method was selected for 802.11 devices primarily to reduce the probability of collision among members of a single *basic service set* (BSS) using contention-based *media access control* (MAC). Depending on the *clear channel assessment* (CCA) mode employed, non-802.11 transmissions may not even activate the 802.11 CCA mechanism.

In this chapter we will address methods that exist mainly to enhance coexistence with other networks. The IEEE 802.15.2 recommended practice addresses passive, active noncollaborative, and active collaborative coexistence between 802.11b Wi-Fi and 802.15.1 Bluetooth. The 802.15.3 WiMedia and 802.15.4 Zig-Bee specifications also include an analysis and discussion of methods for improving coexistence with the other WLAN and WPAN standards.

802.11 Wi-Fi Noncollaborative Coexistence Mechanisms

One of the most common coexistence scenarios is a Wi-Fi WLAN operating in an interference-limited environment over which it has little or no control. In the last chapter we assumed that the WLAN takes no additional measures to improve performance other than the usual *CSMA with collision avoidance* (CSMA/CA) MAC available for normal operation, and we concentrated on finding the resulting *frame error rate* (FER) under these conditions. Several remedial methods are given in the 802.11 specification that can be used for actively improving WLAN coexistence. Among these are dynamic frequency (channel) selection, transmit power control, rate scaling, and frame fragmentation. Additionally, the 802.15.2 recommended practice includes adaptive packet selection as a noncollaborative method for enhancing WLAN performance in interference. Finally, when a WLAN consists of mixed 802.11b and 802.11g devices, operations are modified slightly to increase the aggregate throughput. We now look at each of these enhancements to coexistence in greater detail.

Dynamic Channel Selection

Wireless networks operating on a fixed channel are susceptible to interference from another nearby fixed-channel network. One obvious way to escape such interference is to change channels such that frequency orthogonality is achieved. Different renditions of the 802.11 specification describe the means by which an *access point* (AP) can send its BSS to another channel periodically or if it senses that the present channel is unreliable. Channel agility is available for 802.11b/g WLAN in the 2.4 GHz band, and dynamic frequency selection can be used in the 5.x GHz band for 802.11a networks.

Channel Agility for 802.11b/g   As the level of interference in the 2.4 GHz *industrial, scientific, and medical* (ISM) band increases, *channel agility* may find its way into 802.11b and 802.11g Wi-Fi implementations. This optional method isn't suitable for avoiding Bluetooth transmissions because the Wi-Fi channel cannot be changed quickly enough. However, channel agility can prove useful against tone interference, or even against a competing independent network operating on a fixed channel. Channel agility avoids interference in a manner similar to FHSS in that the changes occur regularly and aren't based on an evaluation of existing interference.

When channel agility is enabled, the BSS periodically hops to a new frequency according to a predetermined channel set. In North America, a nonoverlapping set consisting of channels 1, 6, and 11, or an overlapping set composed of channels 1, 3, 5, 7, 9, and 11, can be selected. The corresponding sets in Europe are channels 1, 7, and 13, or channels 1, 3, 5, 7, 9, 11, and 13. The channel dwell time isn't specified, but the maximum time to change from one operating channel to another is 224 μs.

Dynamic Frequency Selection (DFS) for 802.11a   For legal operation in the European 5.x GHz band, and in the U.S. *Unlicensed National Information Infrastructure* (U-NII) band segments at 5,250–5,350 MHz and 5,470–5,725 MHz, *dynamic frequency selection* (DFS) is mandated. Generally, a DFS-capable node monitors the operating channel for a radar signal. If such a signal is present, the AP moves its BSS to a different channel. If no vacant channel is available, the WLAN must enter the sleep mode.

The DFS requirements for 802.11a are spelled out in the 802.11h standard, published in October 2003. This service provides the following functions [IEEE802.11h]:

- Association of *stations* (STAs) with an AP in a BSS based on the STA's supported channels
- Quieting the current channel so it can be tested for the presence of radar with reduced interference from other STAs
- Testing channels for radar before using a channel and while operating in a channel
- Discontinuing operations after detecting radar in the current channel to avoid interference with the radar
- Detecting the radar in the current channel and others based on regulatory requirements
- Requesting and reporting of measurements in the current channel and others
- Selecting and advertising a new channel to assist the migration of a BSS or *independent BSS* (IBSS) after the radar is detected

An AP can request a quiet time for channel assessment by transmitting Quiet elements in its beacon frame. Channel measurements can be exchanged through Measurement Request and Measurement Report frames or beacon elements. The AP sends the BSS to a new channel using a Channel Switch Announcement frame. This is sent after the channel is idle for a *point coordination function interframe space* (PIFS) instead of a *distributed coordination function interframe space* (DIFS), allowing the AP to preempt the channel between DCF contention windows.

Transmit Power Control (TPC)

The level of *transmit* (TX) power used in the communication link is one of the major factors affecting the ability of independent networks to coexist. As we discovered in earlier chapters, changing TX power by 10 dB affects the maximum range by a factor of 2 in moderate indoor clutter, which in turn affects the area in which interference is significant by a factor of 4. The good neighbor policy insists that the minimum TX power be used to provide reliable communication. Obviously, if all wireless systems operated in this manner, the *carrier-to-interference ratio* (C/I) would be minimized for everyone. On the other hand, TX power should not be increased as a response to higher levels of interference or a "power war" could result.

Transmit Power Level Control for 802.11b/g   The 802.11 specification requires the implementation of *transmit power level control* if the node has the capability to transmit with an *equivalent isotropic radiated power* (EIRP) greater than +20 *dB relative to 1 milliwatt* (dBm), or 100 *milliwatts* (mW). The power control mechanism must be able to reduce TX power to a level at or below +20 dBm, and a maximum of four power levels may be provided. The primary function of transmit power level control is to insure compliance with regulatory domain power limits.

TPC for 802.11a For legal operation in the European 5.x GHz band, and in the U.S. U-NII band segments 5,250–5,350 MHz and 5,470–5,725 MHz, TPC is sometimes required. The IEEE 802.11h specification provides the following characteristics for TPC [IEEE802.11h]:

- Association of STAs with an AP in a BSS based on the STA's power capability

- Specification of regulatory and local maximum TX power levels for the current channel

- Selection of a TX power for each transmission in a channel within constraints imposed by regulatory requirements

- Adaptation of TX power based on a range of information, including path loss and link margin estimates

Link margin and TX power information can be exchanged through TPC Request and TPC Report frames. A TPC Report element is included in the AP Beacon frame.

Rate Scaling and Fragmentation

Wi-Fi can support several data rates, and the selected rate can be based on such criteria as the PER, the *signal-to-interference-and-noise ratio* (SINR), or the *received signal strength indication* (RSSI). Using the lower bit rates in an interference environment increases the energy per data bit, which in turn lowers the BER. On the other hand, under sporadic interference conditions, using faster data rates implies shorter frame durations and a higher probability that the frame can be successfully transferred between interference bursts.

The 802.11 specification provides a means for fragmenting a large data packet into several smaller pieces, each of which is transmitted and acknowledged separately. When interference is present, the shorter fragments have an increased probability of success, and overall throughput may be enhanced.

Frame Length Adaptation When interference is constant and is modeled as *additive white Gaussian noise* (AWGN), higher data rates require a higher *signal-to-interference ratio* (SIR) for a given BER. As such, a frame of a given length that is sent at a higher rate is more susceptible to being received in error. Both the frame length and data rate are adjustable within the various 802.11 specifications, so it would be interesting to study how these parameters affect data throughput when the BER is greater than 0 for the duration of the frame. All 802.11b calculations in this study are based on the long preamble, because its use is mandatory.

From Chapter 6, "Medium Access Control," we have

$$T(L,m,n) = \frac{8L}{370 + T_{\text{data}}^m(L) + T_{\text{ACK}}^n} \qquad \text{DSSS/CCK} \qquad (8.1)$$

and

$$T(L,m,n) = \frac{8L}{189.5 + T_{\text{data}}^m(L) + T_{\text{ACK}}^n} \qquad \text{OFDM} \qquad (8.2)$$

where $T(L,m,n)$ is the data throughput in Mb/s, $T_{\text{data}}^m(L)$ is the time required for transmitting a data frame of length L at rate m, and T_{ACK}^n is the time required to transmit the *acknowledgment* (ACK) frame at data rate n. All times are given in μs. The *PLCP protocol data unit* (PPDU) data frame requires

$$T_{\text{data}}^m(L) = 192 + \frac{8 \times (28 + L)}{m} \qquad \text{DSSS/CCK} \qquad (8.3)$$

and

$$T_{\text{data}}^m(L) = 20 + \left\lceil \frac{30.75 + L}{m/2} \right\rceil \times 4 \qquad \text{OFDM} \qquad (8.4)$$

to transmit, and the ACK frame requires a transmission time of

$$T_{\text{ACK}}^n = 192 + \frac{112}{n} \qquad \text{DSSS/CCK} \qquad (8.5)$$

and

$$T_{\text{ACK}}^n = 20 + \left\lceil \frac{16.75}{n/2} \right\rceil \times 4 \qquad \text{OFDM} \qquad (8.6)$$

This assumes that the average backoff time is half the minimum contention window, which is the case for a BSS with multiple STA as long as collisions are rare. To simplify the analysis, all transmissions are assumed to require a backoff. This also adds a bit of pessimism to the throughput calculations.

We'll assume that the *direct sequence spread spectrum* (DSSS)/CCK ACK is sent at 1 Mb/s for a transmission time of 304 μs, and the *orthogonal frequency division multiplexing* (OFDM) ACK is sent at 6 Mb/s for a transmission time of 44 μs. Equations 8.1 and 8.2 now become

$$T(L,m) = \frac{8L}{674 + T^m_{\text{data}}(L)} \quad \text{DSSS/CCK} \tag{8.7}$$

and

$$T(L,m) = \frac{8L}{233.5 + T^m_{\text{data}}(L)} \quad \text{OFDM} \tag{8.8}$$

Equations 8.7 and 8.8 denote the throughput in a perfect channel. When the channel is imperfect, the *MAC protocol data unit* (MPDU) frame body is more susceptible to corruption than either the header or ACK for two reasons. First, the frame body is likely to be sent at a faster rate than the ACK or header, so the BER is higher. Second, the frame body usually contains a far greater number of bits. For a BER of $P_b(m)$ associated with a data rate m, the throughput becomes

$$T(L,m,P_b(m)) = \frac{8L}{674 + T^m_{\text{data}}(L)}[1 - P_b(m)]^{8(28+L)} \quad \text{DSSS/CCK} \tag{8.9}$$

and

$$T(L,m,P_b(m)) = \frac{8L}{233.5 + T^m_{\text{data}}(L)}[1 - P_b(m)]^{8(30.75+L)} \quad \text{OFDM} \tag{8.10}$$

Figure 8-1 depicts throughput at 11 Mb/s at a BER of 1×10^{-4}, 5.5 Mb/s at a BER of 5×10^{-5}, and 2 Mb/s at a BER of 1×10^{-5} as a function of *MAC service data unit* (MSDU) length.[*] Frames sent at 11 Mb/s have a higher throughput than those sent at 5.5 Mb/s for MSDU lengths up to about 800 bytes, beyond which throughput is higher for frames sent at 5.5 Mb/s. For an MSDU longer than about 1,800 bytes, frames sent at even 2 Mb/s have a greater throughput

---

[*]The relationship among BER values for each data rate is a function of the type and level of interference as it affects a particular receiver. The BER values used in this paragraph are chosen somewhat arbitrarily to illustrate certain throughput characteristics.

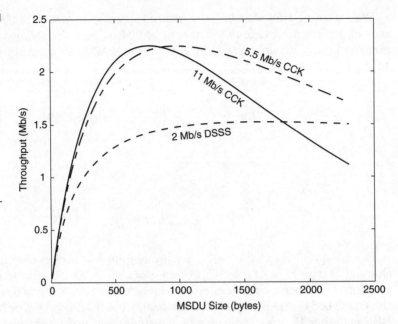

Figure 8-1
IEEE 802.11b
throughput at
11 Mb/s at a
BER of 1 ×
10⁻⁴, 5.5 Mb/s
at a BER of 5 ×
10⁻⁵, and 2
Mb/s at a BER
of 1 × 10⁻⁵ as
a function of
MSDU length

than those sent at 11 Mb/s. Finally, when the MSDU length approaches its maximum, throughput at 5.5 Mb/s only slightly exceeds that at 2 Mb/s. The same general characteristics are exhibited by OFDM (see Figure 8-2) for 54, 24, and 6 Mb/s at the same respective BER values. We conclude that the BER can affect the data rate and frame size under which maximum throughput occurs [Ole01]. With slight modifications to the throughput equations, similar results can be shown when frames are fragmented.

Adaptive Interference Suppression

The IEEE 802.15.2 recommended practice includes adaptive interference suppression as a means of reducing 802.15.1 Bluetooth interference to 802.11b Wi-Fi. This noncollaborative approach uses signal processing on the *physical* (PHY) layer via an adaptive filter that has no *a priori* knowledge of the timing or hop frequency used by the interfering Bluetooth piconet.

One rendition of the adaptive interference suppression system is shown in Figure 8-3 [IEEE802.15.2]. The received signal $x(n)$ containing the desired Wi-Fi and interfering Bluetooth transmissions is delayed and sent to the adaptive notch filter. The uncorrelated nature of the wideband 802.11b signal is exploited to predict the structure of the Bluetooth interference $y(n)$, which is subtracted from the composite to create the prediction error signal $e(n)$. The latter is an approximation to the desired Wi-Fi signal, and this is used to adjust, or adapt,

Figure 8-2
IEEE
802.11a/g
throughput at
54 Mb/s at a
BER of 1 ×
10^{-4}, 24 Mb/s
at a BER of 5 ×
10^{-5}, and 6
Mb/s at a BER
of 1 × 10^{-5} as
a function of
MSDU length

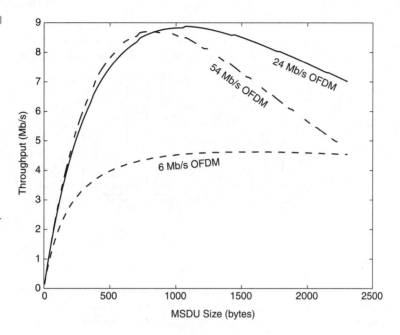

Figure 8-3
Adaptive
interference
suppression to
remove an
undesired
802.15.1
Bluetooth
signal from a
desired
802.11b Wi-Fi
signal

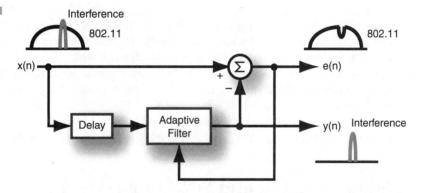

notch filter parameters for improved accuracy over time. Incidentally, this concept can be used in the front end of other receivers as well, and the filter can
operate with multiple interferers on different carrier frequencies and with varying degrees of overlap, limited only by the speed and precision of the *digital signal processor* (DSP) engine.

For reduced computation time and complexity, a *recursive least-squares lattice*
(RLSL) delay block and filter could be used, as shown in Figure 8-4. The order of
the lattice is M, which is 3 for this example, and the filter memory is designated
by the parameter λ. The *infinite impulse response* (IIR) filter has $\lambda = 1$, but the

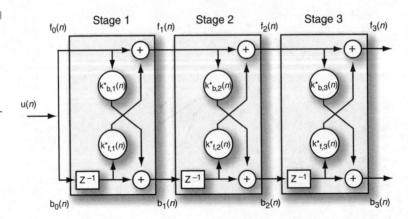

Figure 8-4
Lattice delay block and filter for adaptive interference suppression

adaptive filter in Figure 8-4 performs best when $\lambda = 0.97$ to give past data an exponentially deceasing influence on filter response [IEEE802.15.2]. Adaptation is accomplished by updating the forward and backward coefficients $k_{f,i}$ and $k_{b,i}$ [Hay96]. The forward prediction value $f_3(n)$ is equivalent to $y(n)$ in Figure 8-3.

Figure 8-5 shows the results from a simulation of 802.11b Wi-Fi adaptive interference suppression performance with a Bluetooth interferer and a lattice filter [IEEE802.15.2]. The Bluetooth packet is a short transmission that fully overlaps a longer Wi-Fi frame. When the Wi-Fi data rate is 1 Mb/s, adequate performance is achieved with filter order $M = 3$ even with an extremely low SIR of about -20 dB. Because the 11 Mchip/s Barker sequence already has a spectral notch at the carrier frequency, the adaptive filter performs better at 0 offset than for offsets from 1 to 5 MHz. When the WLAN is operating at 11 Mb/s, the processing gain is lower and a lattice filter with $M = 4$ or higher is needed for adequate interference rejection. Furthermore, the results aren't quite as impressive.

Mixed 802.11b and 802.11g Networks

Several additional coexistence issues become important when a BSS contains both 802.11b and 802.11g nodes. The 802.11b CCA mechanism may not be able to detect the presence of 802.11g operating with OFDM. Also, 802.11b nodes can significantly slow the effective frame throughput of the 802.11g devices by taking longer to transmit an equivalent frame. These factors can place an 802.11g node at a significant disadvantage when competing for channel access using DCF within a mixed BSS, sometimes lowering 802.11g throughput nearly to that of 802.11b. The disparity is apparent even when communication is quite orderly, such as when an AP is transferring data to multiple STAs within a single BSS [Zyr03].

Figure 8-5
Performance of
a Wi-Fi receiver
at 1 Mb/s and
at 11 Mb/s
with Bluetooth
interference
reduced by an
adaptive
interference
suppression
filter
[IEEE802.15.2]
© IEEE

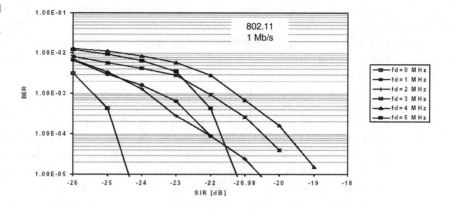

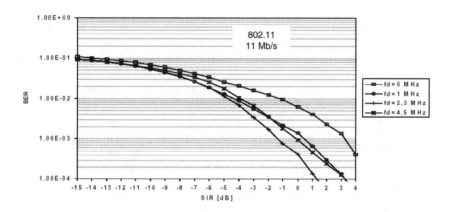

802.11g Protection Mechanisms An 802.11g node's CCA mechanism is required to sense the presence of a legacy 802.11b transmission, but the reverse isn't necessarily true. An 802.11g sending data using OFDM will not be protected from collisions if the legacy 802.11b nodes cannot sense the presence of such a signal. The 802.11g *protection mechanism* is defined in the specification as a procedure that updates the *network allocation vector* (NAV) of all receiving STAs prior to the transmission of a frame that may not be understood by the receivers. The protection mechanism is not required when using DSSS-OFDM or *extended-rate PHY packet binary convolutional coding* (ERP-PBCC) because both these frames begin with a DSSS header and are thus sensed by all members of the BSS. However, if the AP in a 802.11g BSS determines that a neighboring co-channel BSS contains legacy 802.11b nodes, it may implement protection to reduce foreign interference to its own BSS.

When ERP-OFDM frames are used by a pair of 802.11g nodes, they can obtain protection from legacy 802.11b transmissions by first exchanging *request to send/clear to send* (RTS/CTS) frames at one of the 802.11b rates to notify all nearby nodes that the channel is reserved through the associated NAV. Unfortunately, the relatively slow data rates necessary for these RTS/CTS transmissions can add significant overhead. For example, an RTS/CTS exchange at 2 Mb/s using long preambles, including a *short interframe space* (SIFS) after each frame, requires 540 μs to complete, much longer than the 368 μs needed to send the maximum 2,346 bytes of data in a 54 Mb/s OFDM frame.

To reduce overhead, the node originating the frame exchange can send a single CTS-to-self to reserve the channel for the subsequent OFDM exchange. Although only a single CTS is needed for channel reservation, a hidden terminal may cause a collision anyway, especially if the originating node is an STA and not likely to be heard by all members of the BSS. Also, certain conditions exist in which the NAV will be reset by an STA if the CCA senses a channel to be clear for a period of time, potentially leading to a collision during ERP-OFDM transmission, but this latter situation is rare [IEEE802.11g].

Mixed Rate Throughput: Downlink   A common WLAN usage scenario is for an AP to have in its queue several frames of data to be sent to different STA in the BSS. For example, if two clients are downloading large files, the AP will usually alternate frames between the two associated STAs, as shown in Figure 8-6. For 1,500-byte data frames sent at 11 Mb/s and the associated ACKs sent at 2 Mb/s, transmission times in μs are

$$t_1^{(b)} = t_2^{(b)} = a\text{DIFSTime} + \overline{T}_{bk} + T_{\text{data}}^m(L) + a\text{SIFSTime} + T_{\text{ACK}}^n$$

$$= 50 + (15.5 \times 20) + 1303 + 10 + 248$$

$$= 1921 \tag{8.11}$$

and the cycle time t_{cycle} is 3,842 μs (1,921 + 1,921). Throughput is the total data transferred during a cycle divided by the cycle time. In this example, throughput is 6.25 Mb/s. Each node is allocated half the available throughput, or 3.12 Mb/s.

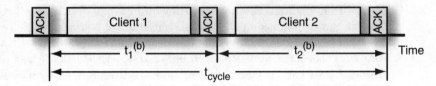

Figure 8-6
Downlink frames and uplink ACKs alternate between two 802.11b clients during long file transfers from an AP.

Figure 8-7
The 802.11g-
capable AP
downloads
frames to
Client 1 with
an 802.11b
node and to
Client 2 with
an 802.11g
node.

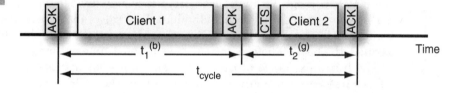

Figure 8-7
The 802.11g-capable AP downloads frames to Client 1 with an 802.11b node and to Client 2 with an 802.11g node.

Now consider the case where an 802.11g-capable AP is transferring 1,500-byte frames as before, but now one 802.11b STA (Client 1) is receiving frames at 11 Mb/s, and the other is an 802.11g STA (Client 2) receiving frames at 54 Mb/s, as shown in Figure 8-7. The time $t_1^{(b)}$ remains at 1,921 μs because the 802.11b transmission is unaffected. The time needed for the 802.11g node to receive its data frame at 54 Mb/s also includes the time occupied by a CTS frame sent at 2 Mb/s, and an ACK frame sent at 6 Mb/s in our example. This time in μs is given by

$$t_2^{(g)} = a\text{DIFSTime} + \overline{T}_{bk} + T_{\text{CTS}} + a\text{SIFSTime} + T_{\text{data}}^m(L)$$

$$+ \ T_{se} + a\text{SIFSTime} + T_{\text{ACK}}^n + T_{se}$$

$$= 50 + (7.5 \times 20) + 248 + 10 + 247 + 6 + 10 + 44 + 6 = 771 \qquad (8.12)$$

where T_{se} is the signal extension time. For reasons we'll discuss in the next subsection, the average backoff time $\overline{T}_{bk}$ for initiating an 802.11g ERP frame exchange is approximately half the average backoff time for a legacy 802.11b exchange.

The associated cycle time is now 2,692 μs (1,921 + 771) and throughput is 8.92 Mb/s, or about 43 percent higher than when both STAs operated with 802.11b nodes. It's important to note, though, that the throughput is still divided evenly between the two participating STA at 4.46 Mb/s each, even though the 802.11g STA is receiving data at 54 Mb/s. Put another way, when Client 2 increases its capability from 11 to 54 Mb/s (a factor of about 5), its throughput increases by a factor of only 0.43 when sharing a fully loaded downlink with an 802.11b node. This discrepancy is due to the high fraction of available time the AP devotes to transmitting at the slower rate. Incidentally, this unequal sharing of resources also applies to a homogeneous 802.11b BSS in which one SNR-challenged STA is communicating at 1 or 2 Mb/s and the others are operating at higher rates.

Mixed Rate Throughput: Uplink As we discovered in the last subsection, an AP handles downlink traffic to multiple STAs by effectively implementing a form of TDMA, in which each STA is sent a frame in turn. If all frames have the same MSDU length L, then throughput is evenly divided among all participating STAs, regardless of their individual data rate capabilities. Because uplink traffic is also possible during DCF, the AP must wait for the DIFS plus, we assume, the backoff before initiating each new frame exchange.

The frame exchange process is not as orderly for uplink traffic, because the STAs must compete with each other during the DCF contention window for access to the channel. Unfortunately, if every node has statistically equal access to the channel, slower nodes (with their longer transmission times) will again acquire a disproportionate share of access to the channel. To review, during DCF a node wanting access to a channel that is sensed busy must wait DIFS after it becomes idle, and then begin decrementing an internal backoff counter loaded with a random integer between 0 and the *contention window* (CW) value, where aCWmin $\leq$ CW $\leq$ aCWmax. The backoff counter is decremented at each slot boundary. If another node begins transmitting before the backoff counter reaches 0, the present value is retained until the other frame exchange has ended and an additional DIFS has elapsed. The node begins transmission when the backoff counter reaches 0.

A legacy 802.11b node initially loads its backoff counter with a uniformly distributed integer random variable between 0 and 31. If 802.11g nodes use an identical method for selecting their initial backoff counter value, the same throughput imbalance would occur on the uplink as on the downlink. To prevent this, 802.11g nodes select their initial backoff counter value between 0 and 15 when the BSS includes 802.11b nodes. This provides the 802.11g node with twice the number of transmit opportunities, on average, compared with a slower 802.11b node [Zyr03].

A cycle time is now $t_1^{(b)} + 2t_2^{(g)} = 3463$ μs for an aggregate throughput of 10.4 Mb/s. Of this aggregate, 6.93 Mb/s is claimed by the 802.11g node and 3.47 Mb/s goes to the 802.11b node. Recall that in our downlink scenario each node achieved 4.46 Mb/s throughput. For our example, the statistical channel access advantage given to Client 2 on uplink traffic equates to a 56 percent increase in throughput at the expense of a 22 percent decrease in throughput for Client 1. However, even with Client 2's channel access advantage, Client 1 still has higher throughput than if Client 2 had been an 802.11b node with equal channel access.

802.15.1 Bluetooth Noncollaborative Coexistence Mechanisms

Bluetooth is unique among the major short-range wireless networks in that it uses FHSS for interference avoidance. Although 802.11 and 802.15.4 both employ variants of DSSS, they are still quite susceptible to *co-channel interference* (CCI) due to their relatively low processing gains. It could be argued, then, that Bluetooth is the only specification that makes a serious attempt to mitigate interference from a variety of other sources as part of its basic operation. Furthermore, the FHSS hop sequence can be adjusted to avoid areas of known interference. Consequently, Bluetooth devices are often expected to bear the brunt of the coexistence burden by modifying their own behavior to reduce interference to other fixed-channel wireless networks.

The noncollaborative coexistence mechanisms available to Bluetooth piconets include transmit power control, adaptive packet selection, adaptive packet scheduling, and adaptive frequency hopping. All these coexistence mechanisms can improve operation when competing networks use a fixed carrier frequency, and adaptive packet selection and TPC enhance coexistence with other Bluetooth piconets as well.

Accurate channel classification is critical for the successful use of most noncollaborative coexistence methods, so we begin with the classification methods discussed in the IEEE 802.15.2 recommended practice [IEEE802.15.2], IEEE specification 802.15.1 [IEEE802.15.1], and Bluetooth specification 1.2 [BT1.2]. Although TX power control outlined in the Bluetooth specification is based on RSSI and not channel classification criteria, the latter provides additional information that may be useful for adjusting TX power to meet certain performance requirements.

Channel Classification

For packet scheduling to be successful, channels must be classified in some way to determine whether the channel should be utilized or skipped. Channel classification can be accomplished by every Bluetooth piconet member and is based on measurements that determine whether or not the channel contains interference. A particular channel is classified as "good" or "bad" based on RSSI, carrier sensing, PER, the *negative acknowledgment* (NAK) rate, or a combination of these [IEEE802.15.2]. For *adaptive frequency hopping* (AFH), Bluetooth specification 1.2 also includes the possibility of classifying a channel as "unknown."

Classification Methods   Channel classification can occur either actively or passively. *Active classification* can be done during the course of normal communications, or the devices can exchange dummy packets with the specific goal of building a classification list. *Passive classification* is accomplished by listening to channels while utilizing RSSI or carrier sense to determine if they are occupied. The channel classification list should be updated periodically to account for changes in the interference situation.

RSSI and PER are used together to determine whether an erroneous packet is due to poor SNR or SIR. If RSSI is low and a packet error occurs, its loss is probably due to propagation effects. If a packet is lost with high RSSI, interference is probably the reason. In either case, it's important to average results over time to remove the influence of transient interferers, such as other FHSS nodes, as sources of interference.

A Bluetooth packet can be received in error for a number of reasons. These include an inability to synchronize on the access code, a *header error check* (HEC) failure, a *cyclic redundancy check* (CRC) failure, or the lack of an expected ACK. A master or slave can use any of these to assist its determination of PER on each channel. The nodes build their respective classification lists using an implementation-dependent threshold between a good and bad channel. The Bluetooth specification requires that at least 20 hop channels be used, so under high interference conditions it may be necessary to retain some of the bad channels in the hop sequence. Thus, it may be more useful to rank-order the channels according to their interference levels rather than simply assign a bad or good assessment.

If the Bluetooth receiver is capable of recognizing non-Bluetooth signals, it can classify channels as bad if they are occupied by other fixed-channel networks such as Wi-Fi, WiMedia, or ZigBee. As the software-defined radio begins to proliferate, this cross-network capability will become more common. In an existing piconet, the master or slave can exit the piconet temporarily and use the time to passively scan other hop channels for the presence of interference.

Finally, Bluetooth specification 1.2 allows the host to direct its attached Bluetooth module to classify certain channels as bad. This feature is useful when the host also supports a collocated Wi-Fi link, for example, allowing it to classify the Bluetooth hop channels that the Wi-Fi node operates on as bad.

TX Power Control

Unlike 802.11, where TPC is directed primarily at regulatory compliance, Bluetooth TPC was established for TX powers greater than +4 dBm to enhance coexistence. A Bluetooth class 1 transmitter must operate with a feedback mechanism from the device at the other end of the link for power control implementation. TPC information between the master and slaves is exchanged through *link manager protocol* (LMP) *protocol data unit* (PDU) packets. The LMP_incr_power_req and LMP_decr_power_req are used for this purpose, commanding the local transmitter to increase or decrease power, respectively, to keep the other receiver's RSSI within the Golden Receive Power Range (Chapter 5, "Radio Performance").

The Bluetooth specification requires that RSSI alone be used for power control in accordance with the good neighbor policy. The downside to this, however, is that a Bluetooth node may react to a strong foreign interfering signal by generating an LMP_decr_power_req, further encumbering communication over the desired Bluetooth link. The good neighbor policy prohibits the Bluetooth node from using higher TX power to mitigate interference. One of the interference avoidance mechanisms, such as AFH, should be used instead.

Adaptive Packet Selection

Bluetooth uses different methods than Wi-Fi does for optimizing the packet size and rate to match channel interference conditions for two reasons. First, the amount of interference on different hop channels is often significantly different, so the best *synchronous connection-oriented* (SCO) or *asynchronous connection-less* (ACL) packet options become hop dependent. Second, Bluetooth's raw data rate is always 1 Mb/s, so it lacks the capability to send data faster on a good channel and slower on a poor channel.

SCO Packet Selection   During SCO link setup, the master can direct the use of either HV1, HV2, or HV3 packets, depending on the level of FEC desired. When operating near the SNR range limit, AWGN is the predominant cause of bit errors, and these tend to be random and uniformly distributed across the entire hop channel set. (If significant interference is present, performance near the SNR range limit is usually not possible.) Additional FEC protection will improve performance in AWGN, so HV1 packets are the most preferred, and HV3 packets are the least preferred [IEEE802.15.2]. RSSI and SNR can be evaluated to assist in selecting the best SCO packet type.

When communicating in an interference-prone environment, packet losses occur primarily from catastrophic collisions, and this significantly reduces the effectiveness of FEC as a recovery technique. As such, it is usually preferable to make use of HV3 packets for SCO communication to reduce overall channel traffic and associated interference to other users. The channel classification list can be used by the master to assist it in deciding which SCO link to set up.

ACL Packet Selection   As we already know, Bluetooth units have the capability to select one-, three-, or five-slot ACL packets and to decide whether or not to use the rate ⅔ FEC. Packet selection can affect throughput because the penalty for retransmitting a long packet is greater than that for resending a short packet. A Bluetooth device could attempt to mitigate interference by matching the ACL packet type against some performance criteria such as average delay, throughput, PER, or BER and adjust the packet type accordingly. Some of these criteria, such as average delay, are easier to monitor at the higher protocol layers than others such as BER. At the lower protocol layers, the channel classification list can be used to determine which type of ACL packet has the greatest chance of success. This capability is enhanced if the channel classification values include more information than a simple good or bad assessment. Figure 7-3 in Chapter 7, "Passive Coexistence," demonstrates that the ACL packet type that yields the best throughput is influenced by channel reliability.

Adaptive Packet Scheduling

If only a few hop channels contain significant interference, the piconet master can schedule packets for transmission on hop channels that are clear through a simple firmware modification. The master checks the next two hop frequencies used for a master-to-slave packet and an associated slave-to-master packet, and delays the transmission if either of those two channels is classified as bad. Both ACL and *extended synchronous connection-oriented* (eSCO) links have some flexibility in how their packet exchanges can be scheduled.

ACL Packet Scheduling A piconet master can schedule ACL packet transmissions in the presence of fixed-channel interference by classifying each available hop channel followed by implementing the appropriate delay policy [IEEE802.15.2]. Once the piconet master has determined the channel classification list, it can schedule each ACL packet and its associated ACK to occur on two consecutive good channels. For slave-to-master data transfers, the master can poll the slave for a transmission when the following two hop channels are classified as good. Furthermore, packet scheduling can be combined with packet selection to send, for example, multislot packets on good channels to compensate for not transmitting on bad channels.

It could be argued that packet scheduling will actually reduce a piconet's throughput, because some packets probably would have been successfully conveyed over bad channels. This may be true for certain single-slot packet exchange scenarios, but using packet scheduling to improve the reliability of multislot transmissions increases the piconet's throughput because the retransmission rate is lowered. An additional goal of ACL packet scheduling is directed toward implementing the good neighbor policy by reducing the level of interference to others.

Figure 8-8 shows the Bluetooth PER both with and without packet scheduling in the presence of 802.11b WLAN interference using the topology in Figure 7-28. Wi-Fi channel loading is 50 percent of the capacity and consists of the STA sending data frames to the AP and receiving ACK frames. Bluetooth loading is 30 percent of the capacity to enhance the packet scheduling flexibility. Under perfect adaptive scheduling, the Bluetooth PER is always 0, but without scheduling the PER is between 10 and 15 percent when separated from the 802.11b STA by 2 m. Under these same conditions, the Wi-Fi PER is less than 1 percent, even for separation distances as short as 2 m [IEEE802.15.2]. This is due to the relatively high AP TX power (+14 dBm), the low power (0 dBm) and channel loading (30 percent) of the Bluetooth piconet, and the short ACK frames being received by the STA.

SCO and eSCO Packet Scheduling Conveying real-time, two-way voice communication is one of the major uses for Bluetooth-enabled devices, and due to the synchronous nature of such communication, along with relatively high

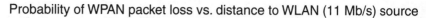

Figure 8-8
Bluetooth PER
with and
without ACL
packet
scheduling in
the presence of
an 802.11b
WLAN with
topology given
in Figure 7-28
[IEEE802.15.2]
© IEEE

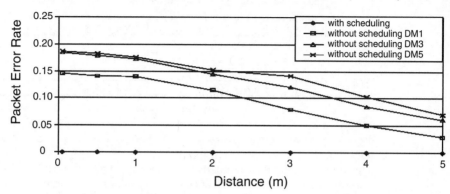

Probability of WPAN packet loss vs. distance to WLAN (11 Mb/s) source

channel loading, voice communication can be extremely susceptible to interference. Adequate *quality of service* (QoS) is necessary for customer acceptance of Bluetooth over wired audio.

During SCO operation using HV1 packets, the master and slave exchange a pair of packets in every available master-to-slave and slave-to-master time slot. Each packet contains 80 bits of audio data that expands to 240 bits after applying a rate ⅓ FEC. Obviously, HV1 packets can't accommodate any scheduling flexibility. When an HV3 SCO link is operating, a pair of packets is exchanged during two consecutive time slots, followed by four unused time slots. An HV3 packet contains 240 bits of audio data and no FEC. The normal HV3 SCO link doesn't include the capability to adjust transmission times based on any criteria such as channel assessment.

Bluetooth specification 1.2 includes the eSCO EV3 packet option for better scheduling flexibility while maintaining an ideal throughput of 64 kb/s in each direction. The master can select one of three master-to-slave slots to begin the EV3 exchange, and the slave-to-master transmission follows in the subsequent slot. Like HV3, the EV3 packet has no FEC and a 240-bit payload containing 3.75 ms of audio data encoded at 64 kb/s. Figure 8-9 depicts HV3 and EV3 packets exchanged on a channel set containing hypothetical classification data, showing how EV3 timing can be adjusted to increase the probability that the exchange will be successful.

Figure 8-10 shows Bluetooth throughput results in which a synchronous link using HV1, HV3, or EV3 packets begins 15 seconds into a simulation where a Wi-Fi node is operating within 1 m. EV3 eSCO throughput under these conditions is close to that of the HV1 SCO link with its powerful FEC, but at a much lower power consumption. Both HV1 and EV3 maintain their throughput close to the required 64 kb/s, so their audio quality will be noticeably better than the 40 kb/s

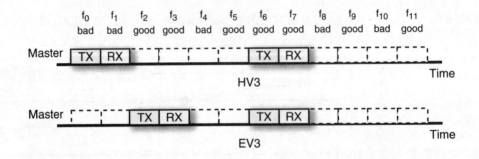

Figure 8-9

HV3 and EV3 packets exchanged on channels that have been classified as good or bad, showing the enhanced performance possible through EV3 packet scheduling

Figure 8-10

Bluetooth synchronous throughput using HV1, HV3, or EV3 packets with Wi-Fi inter-ference [IEEE802.15.2] © IEEE

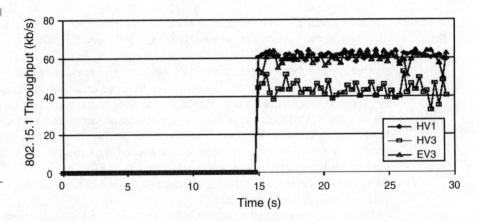

provided by HV3. The corresponding Wi-Fi throughput is given in Figure 8-11. When interference is from HV1 packets, the Wi-Fi channel becomes essentially useless, but reasonable throughput is attained under HV3 or EV3 interference due to reduced channel loading. Taken together, EV3 is the best choice for establishing a Bluetooth synchronous audio link when interference is high over a fixed part of the Bluetooth hop channel set.

Adaptive Frequency Hopping

Bluetooth specification 1.2 enables the use of AFH to improve performance in the presence of interference by avoiding portions of the 2.4 GHz ISM band that are causing a loss of throughput efficiency. If the interference is from another wire-

Figure 8-11
Wi-Fi
throughput
with Bluetooth
interference
using HV1,
HV3, or EV3
packets
[IEEE802.15.2]
© IEEE

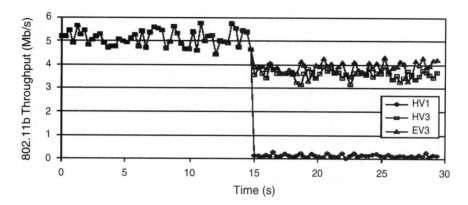

Figure 8-12
Bluetooth AFH
can be used to
avoid inter-
ference
to/from Wi-Fi
networks. (The
Bluetooth hop
channels are
not to scale.)

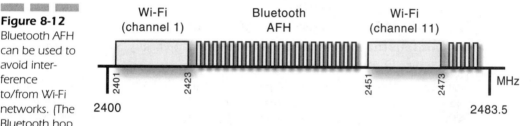

less system, by maintaining frequency orthogonality the performance of both networks is enhanced. Figure 8-12 shows an example of Bluetooth AFH used to avoid Wi-Fi networks operating on Wi-Fi channels 1 and 11.

The *basic channel hopping sequence* consists of the usual 79 hop channels, and the *adapted channel hopping sequence* can have as few as 20 channels. When the adapted channel hopping sequence is used, a slave-to-master slot operates on the same hop channel as the preceding master-to-slave slot. This rule applies to all packet exchanges, including multislot and synchronous links. This is different from operation on the basic channel-hopping sequence, in which a hop takes place between a master-to-slave slot and the following slave-to-master slot.

When the adapted hopping sequence is used, the hopping sequence on good channels is identical to that used in the basic hopping sequence. However, if the basic hop sequence lands on a bad channel, the adapted hop algorithm substitutes a good channel pseudorandomly. If an AFH-capable master is communicating with a mixture of AFH-capable and non-AFH-capable slaves, it can use an adapted channel-hopping sequence for the former and the basic channel hopping sequence for the latter, dynamically switching between the two channel sets depending on which slave is being addressed. Both sets of slaves will be synchronized on good channels in the basic hop sequence, so the master can still use these channels to send broadcast packets to all slaves in the piconet.

AFH Implementation The implementation of adaptive hopping must be accomplished with care, because any miscommunication will result in the loss of full hop synchronization between the master and associated slave(s). AFH information between the master and slaves is exchanged through LMP PDU packets. The piconet master must accomplish the following tasks when AFH is to be implemented:

- **Obtain channel classification information** Gather information from piconet members on whether or not interference is present on each channel.
- **Form the adapted channel hop sequence** Create the adapted channel hop sequence from channel classification information.
- **Distribute the adapted channel hop sequence** Send the new hop sequence to piconet members.
- **Initiate AFH** Transition to the AFH channel set.
- **Maintain the AFH channel sequence** Periodically reevaluate the channels to react to changing conditions.

Channel classification can be performed independently by all piconet members based on information local to each device. The piconet master maintains an AFH_channel_map containing channel classification information. A master with AFH capabilities can build the AFH_channel_map for the piconet based on any combination of active or passive channel classification from its own local measurements, channel classification information from its host, and/or channel classification information provided by the piconet slaves using LMP_channel_classification PDUs. The algorithm used by the master to combine information from these three sources into the AFH_channel_map is implementation dependent and not defined in the Bluetooth specification.

The master distributes the adapted channel hop sequence and initiates AFH by sending an LMP_set_AFH PDU to each slave in turn. This PDU contains three parameters: AFH_instant (the instant at which the hopset switch becomes effective), AFH_mode (stating whether AFH is being enabled or disabled at the AFH_instant), and AFH_channel_map. AFH channel maintenance can be accomplished by sending a new LMP_set_AFH PDU with a new AFH_instant and AFH_channel_map.

Bluetooth Packet Throughput During AFH The Bluetooth-on-Bluetooth interference situation may change significantly when the adapted channel hopping sequence is used. The number of channels M in the adapted hop sequence may be less than 79, which could increase the probability of collision among members of independent Bluetooth piconets, depending on the contents of each piconet's AFH_channel_map. A worst-case scenario occurs when the desired and interfering piconets create the same AFH_channel_map, which is certainly likely if existing non-Bluetooth interference affects all these piconets identically. For example, if Wi-Fi networks exist on 802.11 channels 1, 6, and 11, then as few as 13 clear hop channels could remain for Bluetooth devices that employ AFH and can sense the presence of the Wi-Fi signals. Because there must be at least 20 channels in the adapted channel-hopping sequence, some mutual interference between Bluetooth and Wi-Fi under these conditions is to be expected.

If data is being transferred from master to slave using the adapted hop sequence, the data packet and its associated ACK are both sent on the same hop channel. In this situation, $m + 2$ collision opportunities are possible at most from each interfering piconet on a desired data and ACK packet pair, given that the data packet occupies m time slots. Therefore, the packet error probability from catastrophic collisions is

$$P_E^{(a)} = 1 - \left(1 - \frac{1}{M} \right)^{(m+2)(K-1)} \tag{8.13}$$

when K piconets operate in the same vicinity.

Figure 8-13 shows throughput for DH5 packets when adapted hop sequences of 20, 40, and 79 channels are used, and all piconets have the same AFH_channel_map. A comparison of master-to-slave DH5 throughput for the 79-channel

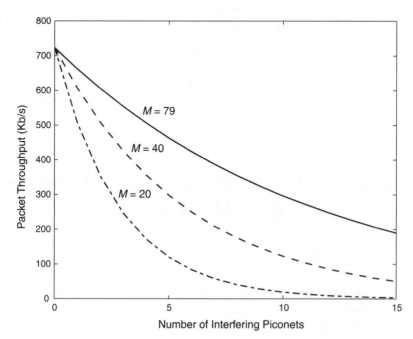

Figure 8-13
Desired packet throughput as a function of the number of interfering piconets for DH5 packets when desired piconet members are close enough that only co-channel C/I applies. Adapted channel-hopping sequences of 20, 40, and 79 channels are used where all piconets have the same AFH_channel_map.

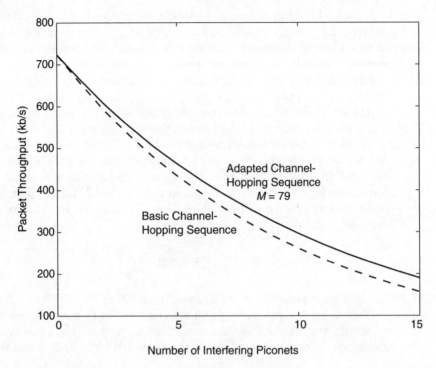

Figure 8-14
Desired master-to-slave packet throughput as a function of the number of interfering piconets for DH5 packets when desired piconet members are close enough that only co-channel C/I applies. The basic channel-hopping sequence is compared to the 79-channel adapted channel-hopping sequence.

basic and adapted channel-hopping sequences is given in Figure 8-14, which shows a slight throughput improvement for the 79-channel adapted hop sequence. This implies that perhaps the adapted hop sequence should always be enabled, even when no interference is present, if all piconet members are AFH-capable.

802.15.3 WiMedia Noncollaborative Coexistence Mechanisms

As a WPAN operating on a fixed channel, the major coexistence issue with WiMedia will most likely be with Wi-Fi in the 2.4 GHz ISM band. Other coexistence scenarios include operating in the presence of other WiMedia networks or Bluetooth piconets. We will examine all these situations from an analytical point of view.

General Methods for WiMedia Coexistence

The 802.15.3 standard provides several mechanisms to improve coexistence with other users in the 2.4 GHz ISM band. These include selecting the proper channel plan, passive scanning, link quality and RSSI, channel-quality information retrieval, dynamic channel selection, using lower TX power or power control, and implementing neighbor piconet capabilities. Many of these are described in terms of coexistence with Wi-Fi, but in most cases application can be extended to other wireless networks.

Channel Plan   As we mentioned earlier, two channel plans are available. The first is the high-density plan that includes four channels (1, 2, 4, 5) that can support four simultaneous 802.15.3 piconets at full speed without mutual interference. The second is the 802.11b coexistence plan that consists of three channels (1, 3, 5) that have identical carrier frequencies to 802.11b channels 1, 6, and 11. The latter plan should be used if the WiMedia *piconet controller* (PNC) detects the presence of 802.11b signals on any of the five available 802.15.3 channels during its passive scan. Because WiMedia channel 3 is not part of the high-density channel plan, if a PNC senses that another WiMedia piconet is using that channel it will assume that the 802.11b coexistence channel plan is in use and select channel 1 or 5 for its own operation.

Passive Scanning   All WiMedia *devices* (DEVs) with PNC capabilities are required to passively scan potential channels before starting a piconet. The DEV is not required to have the capability to detect non-WiMedia signals specifically, but a channel should be relatively quiet before beacon transmission commences. Coexistence is greatly improved if the WiMedia receiver can sense the presence of Wi-Fi or other network transmissions. WiMedia receivers are required to indicate a busy channel for any signals above -55 dBm RSSI. For a Wi-Fi transmitter operating at a power level of $+20$ dBm and a path loss exponent of 3.0, this equates to a radius of only 15 m. Therefore, an AP more than 15 m away may not trigger the CCA channel-busy indication in a WiMedia receiver.

Link Quality and RSSI   A DEV has the capability to report RSSI and, for data rates of 33 Mb/s and above, a *link quality indication* (LQI) using an SNR at the receiver's decision point. RSSI is defined here as the power relative to the -10 dBm maximum receiver input power level in 8 steps of 8 dB with ±4 dB step-size accuracy. At least 40 dB of RSSI range must be available. The SNR evaluation includes thermal noise, distortion, uncorrected interference, and other signal impairments. A 5-bit number is reported that covers SNR values between 6 and 21.5 dB. RSSI and LQI together help the node differentiate between either low signal power or interference causing a loss of frames. For example, poor FER coupled with low RSSI means that a request for increased TX power at the other end of the link should be considered. On the other hand, if RSSI is sufficient but LQI is low, interference is present and switching to another channel may be a better course of action.

Channel-Quality Information Retrieval   The PNC has three options for finding a new channel on which to operate. First, it may initiate a channel-scanning procedure by temporarily stopping beacons and checking other available channels. Second, it may ask another DEV in its piconet to perform a remote scan and report its results back to the PNC. Third, it can ask all its DEV members for channel status information in the form of FER values.

Dynamic Channel Selection   If a PNC determines that the FER is above some threshold, it can search for a new operating channel that has a lower level of interference. If a better channel is available, it will direct all DEVs in the piconet to change to the new channel through a command field in its beacon. The method for evaluating channel scan information and deciding whether or not to initiate a dynamic channel change is implementation dependent.

Lower Transmit Power   Most WiMedia transmitters are expected to operate with about +8 dBm EIRP to be legal in Europe, Japan, and the United States. For 802.11b/g nodes, TX power levels are typically between +15 and +20 dBm. Although 802.15.3 devices are at a TX power disadvantage compared to 802.11b, their operating ranges are anticipated to be much shorter, improving coexistence capabilities for both networks.

Transmit Power Control (TPC)   Three methods of TPC are provided in the 802.15.3 specification. First, the PNC can specify the maximum TX power allowed during the *contention access period* (CAP), beacons, and *management channel time allocations* (MCTAs) down to 0 dBm. The goal of TPC in the CAP is to prevent one DEV from having unfair access to the medium by virtue of its higher TX power, but coexistence with Wi-Fi networks is improved as well. Second, individual DEV in a CTA can request a TX power change from the DEV at the other end of the link. This is done by sending a message to the target DEV indicating the increase or decrease in TX power required. Two DEV that are close to each other can reduce their power for better coexistence without jeopardizing their own communication reliability. Finally, a DEV is able to change its TX power based on its own estimation of channel conditions. A special probe request command can be used to obtain information from another DEV to assist in determining the proper TX power to use.

Neighbor Piconet Capability   As implemented in the 802.15.3 specification, the neighbor piconet capability improves coexistence among multiple WiMedia piconets. However, this capability could also be built into Wi-Fi nodes so that the

AP could either direct WiMedia neighbor operation or perhaps even become a WiMedia neighbor itself. The WiMedia PHY has several similarities to the 802.11b PHY (some common channels, *differential quadrature phase shift keying* [DQPSK] modulation, and an 11 Mchip/s chip rate) such that the construction of dual-mode radios is feasible, at least to the point where they could detect each other's presence and support neighbor piconets. If an 802.11b node receives a WiMedia neighbor piconet request, it could use the PCF and NAV to set aside the appropriate amount of time for WiMedia while maintaining clear WLAN operation during the remaining time using either PCF or DCF.

WiMedia-on-WiMedia Interference Mitigation

Several coexistence mechanisms are provided in the 802.15.3 PHY and MAC layers that are directed at coordinating operations among several WiMedia piconets. After the piconet is established, the PNC should periodically monitor both the current channel and any overlapping WiMedia channels for the presence of other networks. If interference or another network is detected, the PNC can proceed in one of the following ways:

- Change channels to one that is unoccupied.
- Reduce the maximum TX power allowed within the piconet.
- Become a neighbor or child piconet to an existing WiMedia network for shared channel access.

Maximum power levels for CAP, beacons, and MCTAs are conveyed to other DEVs in a beacon field. All DEVs are required to maintain their TX power levels at or below the maximum, and the PNC will set its nominal TX power for beacons at or below the maximum. During a CTA, each participating DEV may request that the DEV at the other end of the link either increase or decrease its TX power level. The specification also allows a DEV to change its TX power level based on its own estimation of channel conditions during the CTA, but this algorithm must be implemented carefully to prevent a power war between DEVs competing for access.

A *child piconet* exists entirely within the CTA of another piconet, called the *parent piconet*, and is controlled by a device within the parent piconet. The parent piconet allocates time within the CTA both for itself and for the child, and the proportion of time for each can vary. The coverage area between the two piconets can also vary between totally congruent and mostly nonoverlapping. The child PNC is also a member of the parent piconet, so the two piconets can exchange information through the child PNC. This allows the range of the WiMedia network to be extended an arbitrary amount through a set of interlocked parent-child relationships.

A *neighbor piconet* is similar to a child piconet, but the neighbor PNC is not part of the parent piconet and the two piconets are thus totally independent. The neighbor piconet mechanism is used when no vacant channel is available and another independent WiMedia piconet must be established without creating interference problems. Both child and neighbor piconets are called *dependent piconets*.

Enhancing WiMedia Coexistence with 802.15.1 Bluetooth

A Bluetooth piconet uses narrowband FHSS, so its coexistence requirements with WiMedia are quite different from those needed for Wi-Fi. The primary methods within the 802.15.3 standard to improve coexistence with 802.15.1 are lower TX power and TPC, both of which were discussed in the preceding section. Also, the various collaborative coexistence mechanisms, described later in this chapter, may also be designed to operate with WiMedia.

At the PHY layer, adaptive interference suppression (noncollaborative) or deterministic interference suppression (collaborative) each use a variable notch filter to reduce the energy from a Bluetooth transmission within the passband of a WiMedia receiver. The latter method will be discussed later in this chapter.

Coexistence at the MAC layer can be enhanced by packet traffic arbitration and/or a wireless medium access method. The former provides frame-by-frame medium access arbitration through knowledge of frequency utilization, frame type, and frame priority. The latter uses timing criteria to ensure that one node doesn't transmit while the other is actively receiving a frame. Both of these will be covered in the section on collaborative coexistence.

802.15.4 ZigBee Noncollaborative Coexistence Mechanisms

As a WPAN aimed at low-power, low-rate, low duty cycle applications, 802.15.4 ZigBee devices may find viable markets in areas populated by other wireless networks sharing the same band. Unlike Wi-Fi, Bluetooth, and WiMedia, ZigBee can operate on channels in the 868 MHz European band and the 900 MHz ISM band in North America. Of course, operating in those bands provides frequency orthogonality to others in the 2.4 GHz band, but the ZigBee data rate is lower at those lower carrier frequencies and the bands aren't available worldwide. Therefore, it is likely that most ZigBee devices will use the 2.4 GHz band.

General Methods for ZigBee Coexistence

ZigBee has several built-in coexistence advantages, both for itself and for other networks sharing the same band. These include CCA, channel alignment, energy detection and link-quality indication, dynamic channel selection, low duty cycle, low transmit power, and neighbor piconet capabilities [IEEE802.15.4]. Most of these capabilities can be altered in such a way that coexistence is enhanced; hence, they are included in this chapter on active coexistence.

Clear Channel Assessment (CCA)   The CCA mechanism used by ZigBee can employ *energy detect* (ED) over a threshold, the detection of a signal with ZigBee characteristics, or a combination of the two. For improved coexistence, the ED option should be used as a good neighbor policy. The risk is that a noncommunication signal such as that emitted by a microwave oven will prevent ZigBee transmissions on that particular channel.

Channel Alignment   When 802.11b/g nodes are operating in the vicinity, ZigBee can usually select a channel that is well outside the Wi-Fi passband, except when all three nonoverlapping Wi-Fi channels are occupied. In North America, these channels are 1, 6, and 11. Figure 8-15 shows that four ZigBee channels (15, 20, 25, and 26) are available for reduced mutual interference, although channel 26 is the only one that is well clear of Wi-Fi. In Europe, the three nonoverlapping Wi-Fi channels are 1, 7, and 13, which leaves no room at the upper part of the 2.4 GHz band for ZigBee. Instead, two ZigBee channel pairs (15, 16, 21, and 22) can fit between the Wi-Fi occupants.

Energy Detection (ED) and Link Quality Indication (LQI)   The receiver ED measurement is an estimate of the RSSI within the operating channel and

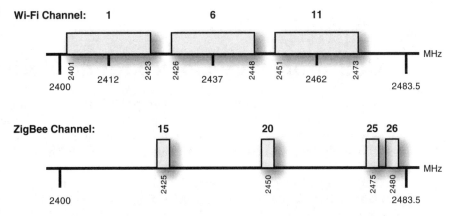

Figure 8-15
If Wi-Fi occupies the three nonoverlapping channels available in North America, ZigBee piconets can use channels that fall outside the passbands.

can be used as part of the channel selection algorithm at the network layer. The LQI measures the received energy level and/or SNR for each incoming packet. By combining these two measurements, the receiver can determine whether a corrupted packet was the result of propagation loss or interference.

Dynamic Channel Selection   Dynamic channel selection can be performed by a ZigBee device either at network initialization or in response to an outage. The device scans a set of channels given in the ChannelList parameter, which can be defined to exclude channels occupied by known Wi-Fi networks.

Low Duty Cycle   ZigBee is specifically tailored for applications that require low power, low data rates, and low duty cycles, such as remote measurements, alarms, and control. These applications typically have duty cycles well under 1 percent, reducing the probability that a ZigBee piconet will interfere with others.

Low Transmit Power   The low-cost nature of ZigBee will probably dictate a maximum TX power of +10 dBm, with 0 dBm being typical. The 802.15.4 specification requires that a device can operate with at least −3 dBm of TX power. Because the majority of Wi-Fi nodes use TX power levels between +15 and +20 dBm, and WiMedia TX power is typically +8 dBm or less, ZigBee is a good neighbor due to its power disadvantage. Fortunately, the processing gain and relatively narrow bandwidth of the ZigBee signal helps the receiver extract the desired signal in an interference-prone environment.

Neighbor Piconet Capability   Although interoperability with other systems is beyond the scope of the 802.15.4 specification, the ZigBee PAN coordinator can set aside *guaranteed time slot* (GTS) periods specifically for use by other wireless networks. These quiet periods provide time orthogonality for interference-free operation.

Collaborative Coexistence Solutions

Several computer manufacturers integrate both Bluetooth and Wi-Fi into the same host, and as WiMedia and ZigBee proliferate, they will no doubt be included as well. Considerable operational challenges exist in such a configuration due to the possibility that one receiver will be blocked while the other transmitter is active. To prevent this, an attempt can be made to operate the two systems in such a way that each encumbers the other as little as possible. Fortunately, collaborative control of collocated nodes can be accomplished by the host through a direct (wired) connection, so collaborative control is interference-free.

Wireless devices that aren't collocated on the same host can also use collaborative techniques to limit their interference to each other, but such collaboration will probably be controlled over the wireless link itself, decreasing reliability. Fortunately, at least in the case of Wi-Fi and Bluetooth, we've discovered earlier that even relatively short separations (2 m, for example) are sufficient to reduce interference to a tolerable level without adding the complexity of collaborative coexistence.

The goal of collaboration is to implement some type of reservation or interference cancellation scheme for transmission that is fair to both systems by enhancing the throughput of each without adding long delays. Collaboration can take place at the PHY or MAC layers. The 802.15.2 recommended practice lists three collaborative methods to improve performance between collocated Wi-Fi and Bluetooth nodes. These are deterministic interference suppression (PHY), alternating wireless medium access (MAC), and packet traffic arbitration (MAC). The first of these is used by Wi-Fi to reject Bluetooth interference, and the other two coordinate Wi-Fi and Bluetooth transmissions to reduce their cross-interference.

Remedial Methods for Collaborative Coexistence

Among the most basic (and least convenient) ways of implementing collocated wireless nodes is to switch one off while the other is operating. This can be done through manual switching, driver-level switching, or MAC-level switching.

Manual Switching The simplest collaboration method between Wi-Fi and Bluetooth, or another combination of wireless devices, is for the user to manually switch between them based on what is needed at any particular time. One common and cumbersome implementation of manual switching is through the use of separate *network interface cards* (NICs), each with a different protocol. The desired NIC is installed, the matching software launched, and communication can commence.

Many wireless software packages include the capability to turn off the communication hardware, so perhaps it would be more convenient to insert two NICs into the host and activate only one at a time. Although many computers have two card slots, it's often difficult to install both together because of the extra height needed for the antenna, and even if they could both be installed, the proximity of the two antennas within a few millimeters of each other would severely affect both antenna radiation patterns. Manual switching is also possible, but still not practical, when Wi-Fi and Bluetooth are both included as built-in features in a host.

Driver-Level Switching For faster switching, the two wireless systems can be controlled electronically at the driver level by the host *operating system* (OS), as shown in Figure 8-16 [Mob01a]. The depicted implementation has two

Figure 8-16
Driver-layer
switching for
multiplexing
operation
between
Bluetooth and
Wi-Fi

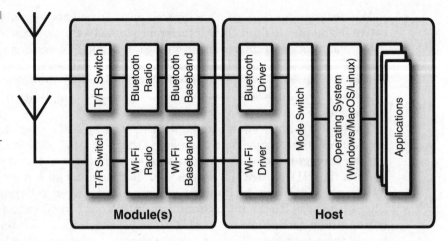

antennas, but with the proper TX and *receive* (RX) switching a single antenna is also feasible. It's likely under many conditions that a user will require access to only one of these wireless systems at any one time; for example, most Web browsing (Wi-Fi) occurs without simultaneously sending a file to a printer (Bluetooth). However, during those times when both systems are operational, the OS attempts to prevent one device from transmitting while the other is actively receiving data.

A link can be suspended either with or without notifying others in the network. Suspension without notification can be done quickly; when the OS determines that a packet is arriving on network A, it simply locks out network B's transmitter. This form of suspension could have an adverse effect on the other participants in network B, however, because any incoming data won't be acknowledged, retransmissions will occur with an associated increased interference potential, and the other users will experience reduced throughput. On the other hand, it may be difficult to notify other users in network B of the suspension because, once a network A packet begins to arrive, it's too late to activate network B's transmitter.

The driver-level switching reaction time is limited by the time between the OS notification of an incoming packet and its response in disabling the other radio's transmitter. This time is highly variable and can routinely exceed the duration of the incoming packet itself, which invalidates this entire process. Long switching delays equate to slow switching times, further reducing potential throughput. The situation is alleviated somewhat when the local host initiates communication and can disable the other node's transmitter first.

MAC-Level Switching The MAC-level switching concept is similar to driver-level switching, but because the process takes place much lower in the respective protocol stacks, the response time is much faster and switching can be matched more closely with dynamic changes in the two networks [Mob01a]. To do this, both Wi-Fi and Bluetooth must be carefully integrated at the lower protocol levels to ensure that proper cross-coordination takes place. This is difficult due to the significant differences among networks such as Wi-Fi and Bluetooth at all protocol levels. Switching at the MAC level is often used as part of an overall integration scheme between Wi-Fi and Bluetooth in advanced collaborative implementations.

Collaborative Adaptive Hopping

We've already shown that noncollaborative AFH is aimed at increasing throughput by placing Wi-Fi and Bluetooth into separate parts of the 2.4 GHz ISM band. Bluetooth AFH is implemented by using the master's AFH_channel_map to determine which channels are to be used in the adapted channel-hopping sequence. The host can inform the Bluetooth node which channel the Wi-Fi node is using, and that information then becomes part of the AFH_channel_map, increasing the chance that Bluetooth operation will occur outside the Wi-Fi bandspread. FCC rules prohibit adjusting the FHSS hop channel set by directly coordinating with other users, however, so the legality of this method must be addressed.

Collaborative Wi-Fi Frame Scheduling

Although the Bluetooth node can avoid the channel occupied by its collocated Wi-Fi node through adaptive hopping, the Wi-Fi node can alternatively adjust its frame transmission times to take place when the collocated Bluetooth node is hopping outside the Wi-Fi channel bandwidth [Chi03]. This method could be used, for example, in situations where high Bluetooth traffic density would be adversely affected by requiring piconet members to use an adapted channel-hopping sequence. Instead, the Wi-Fi node accepts the coexistence burden, and this is manifested by increased initial frame delay.

Deterministic Interference Suppression

The Wi-Fi node can use information from the collocated Bluetooth node to mitigate interference through deterministic interference suppression. Since Bluetooth transmits a narrowband signal on a given hop channel, interference can be suppressed by placing a notch at the same location in the Wi-Fi passband. The notch is necessary only during hop frequencies that coincide with the Wi-Fi passband. Although the 802.15.2 recommended practice advocates deterministic

Figure 8-17
Deterministic
interference
suppression
block diagram

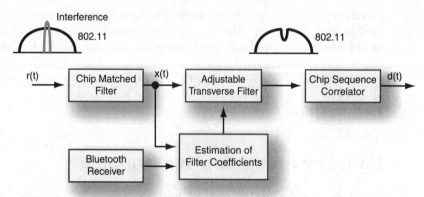

interference suppression for 802.11b, the method may also prove useful during 802.11g operation by noting the three subcarriers affected by the Bluetooth transmission and by using soft decisions or erasures on their data to increase the chance that the FEC can recover the affected bits.

Figure 8-17 shows a block diagram of the deterministic interference suppression system [IEEE802.15.2]. The collocated Bluetooth receiver is used to convey hop and timing information to enable the coefficients to be estimated for an adjustable transverse filter that maintains the proper notch width and depth. The concept is similar to the noncollaborative adaptive interference suppression, but the deterministic system reacts faster because it can obtain accurate hop information. This information may even be available early enough to adjust the transverse filter before the interference arrives.

Figure 7-22 in Chapter 7 shows the performance of 1 Mb/s Wi-Fi in the presence of Bluetooth interference when no interference suppression is employed. The SIR has to be well above 0 dB for the BER to drop below 10^{-3}. The performance of a 1 Mb/s 802.11b receiver using deterministic interference suppression with a seven-stage adaptive filter is shown in Figure 8-18. The improvement is striking; BER is now less than 10^{-3} for all carrier offsets when the SIR is better than about -32 dB. With proper antenna isolation, discussed later in this chapter, a Wi-Fi receiver may operate satisfactorily even when a collocated Bluetooth transmitter is active.

Alternating Wireless Medium Access (AWMA)

Alternating wireless medium access (AWMA) operates at the MAC layer by dedicating a portion of the Wi-Fi beacon interval to Bluetooth operations, so the two networks can operate without a time overlap. The Bluetooth piconet master must be collocated with a Wi-Fi node so it can receive the AWMA timing information, which is coordinated by the common host. If the collocated Wi-Fi node is an STA,

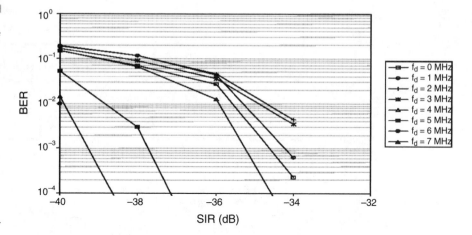

Figure 8-18
Performance of 1 Mb/s 802.11b in the presence of Bluetooth interference when deterministic interference suppression is used [IEEE802.15.2] © IEEE

the timing information must be sent to the AP for AWMA coordination to occur throughout the BSS. If multiple APs operate in an *extended service set* (ESS), they must also be coordinated for an interference-free Bluetooth environment, but this additional complexity may not be necessary if the Bluetooth piconet is confined to an area where coexistence is predominately with a single BSS.

Figure 8-19 depicts the intervals allocated to WLAN and WPAN communications during the beacon interval T_B between the *targeted beacon transmission times* (TBTT) [IEEE802.15.2]. The minimum and maximum WPAN intervals are 0 and 32.768 ms, respectively, but the maximum cannot exceed T_B. An optional guard time, with a maximum value of 10.24 ms, can be appended to the WPAN interval to guarantee that Bluetooth traffic is completed before the next WLAN beacon transmission. The AWMA mechanism generates a *medium-free* signal that is asserted during the WPAN interval. This signal is conveyed to the piconet master at one of the collocated nodes so it can operate the piconet during its allotted interval. Throughput is divided between the two networks in proportion to their respective time intervals.

Wherever a high density of collocated Wi-Fi and Bluetooth nodes exists, AWMA will mitigate interference among all these nodes, and interference among other noncollated nodes will be reduced as well. AWMA works best when access to the channel can be deterministically controlled and is not highly dependent on WLAN or WPAN traffic loads. Bluetooth SCO links usually cannot be supported due to the relatively long intervals dedicated to WLAN operation.

Packet Traffic Arbitration (PTA)

Packet traffic arbitration (PTA) operates at the MAC layer by providing per-packet authorization of all transmissions taking place between collocated WLAN

Figure 8-19
Timing of
WLAN and
WPAN
subintervals
during AWMA

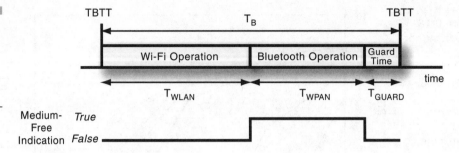

Figure 8-19
Timing of
WLAN and
WPAN
subintervals
during AWMA

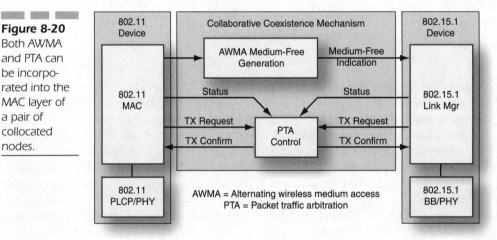

Figure 8-20
Both AWMA
and PTA can
be incorpo-
rated into the
MAC layer of
a pair of
collocated
nodes.

and WPAN nodes. Each node requests TX authorization from the PTA mechanism, which may approve or deny the request depending on the status of the other node. PTA works well when traffic loads are highly variable, because it can dynamically adjust to the needs of each node. Furthermore, PTA can support Bluetooth SCO links. Both AWMA and PTA can be incorporated into the same collocated nodes, as shown in Figure 8-20 [IEEE802.15.2].

When collocated Wi-Fi and Bluetooth nodes are operating, several different collision scenarios can occur at different times. Each radio can be transmitting or receiving, and the Bluetooth radio can be operating on a hop channel that is either within or outside the Wi-Fi radio's passband. Table 8-1 lists the possible combinations and the collision type that may result [IEEE802.15.2]. The four collision types are as follows:

■ **Transmit** Both radios are transmitting in-band. One or both of the packets may be received at their respective remote nodes with errors due to cross-interference.

Table 8-1

Collision scenarios at collocated nodes

Local Wi-Fi activity	Local Bluetooth activity			
	Transmit		Receive	
	In-band	Out-of-band	In-Band	Out-of-band
Transmit	Transmit	None	Transmit-receive	Transmit-receive
Receive	Transmit-receive	Transmit-receive	Receive	None

- **Receive** Both radios are receiving in-band. One or both of the packets may be received with errors due to cross-interference.
- **Transmit-receive** One radio is transmitting while the other is receiving. The locally received packet contains errors due to the high interference level.
- **None** The PER is mostly unaffected by simultaneous operation of the two collocated radios.

At the local collocated nodes, the most serious packet error scenario occurs when one node is transmitting while the other is actively receiving. If one receiver becomes blocked while the other radio is transmitting, the packet is lost even for short TX overlap times. Blocking can be relieved to a large extent through antenna isolation, which we discuss later in this chapter. When the antennas are sufficiently isolated, TX-RX collisions may become rare and can be listed in the "none" category. The PTA mechanism can use the parameters in Table 8-1 to help it decide when to approve TX requests from the WLAN or WPAN node.

According to Figure 8-20, each node provides status information to the PTA mechanism. The Wi-Fi status includes the following:

- **Current state** The current or expected 802.11 MAC activity, whether RX, TX, or idle
- **Channel** The 802.11 operating channel
- **End Time** The time at which the current state ends

The Bluetooth status contains the following:

- **Current state** The current or expected 802.15.1 MAC activity, whether RX, TX, or idle
- **Channel list** The list of hop channels, both current and future
- **Packet type** The type of packet to be sent in current and future slots
- **Duration** The TX duration of the current packet
- **Slot end time** The time at which the current slot ends

Figure 8-21
Wi-Fi through-
put when
Bluetooth is
idle, and
operating both
with and
without PTA
[IEEE802.15.2]
© IEEE

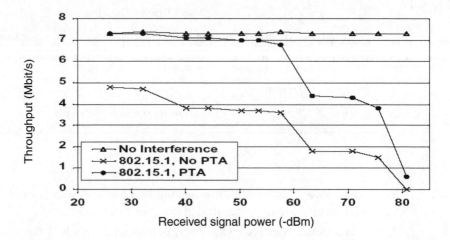

When a TX request arrives from one of the nodes, the PTA unit uses status information to assess the applicable collision scenario, along with any prioritization directives it might have, to decide whether or not to grant the request. For example, an ACK packet from either Wi-Fi or Bluetooth can be given a high priority, because its loss will require the entire exchange to be repeated. Some QoS measure could be used to prioritize Bluetooth SCO packets such that the PER is maintained below a desired threshold.

Wi-Fi throughput for collocated nodes when Bluetooth is idle, operating both with and without PTA, is shown in Figure 8-21. The Bluetooth master and Wi-Fi STA are collocated, and the master is exchanging single-slot packets in every time slot with a slave 1 m away. Both the master and slave operate with +20 dBm of TX power. Although power control is required in an actual operation, these +20 dBm Bluetooth transmissions in every time slot provide worst-case interference to the Wi-Fi receiver. *Transmission control protocol / Internet protocol* (TCP/IP) throughput is measured over the Wi-Fi link using 1,500-byte frames from AP to STA, and 40-byte ACK frames from STA to AP.

When Bluetooth is off, Wi-Fi throughput is above 7 Mb/s for RSSI levels as low as −80 dBm. When Bluetooth is activated without PTA, throughput drops to under 5 Mb/s for high Wi-Fi RSSI levels and rapidly deteriorates to 2 Mb/s when RSSI weakens to about −60 dBm. When PTA is activated, Wi-Fi throughput increases significantly, almost to the original levels for RSSI above −60 dBm.

System-Level Integration Example

System-level integration of collocated Wi-Fi and Bluetooth nodes can facilitate coordinating operations at all protocol layers. One way to do this is to integrate

Figure 8-22
Mobilian's True
Ratio™
[Mob01b]

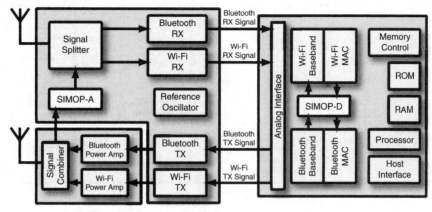

TrueRadio Analog Module   TrueRadio Digital Module

both nodes into a single chipset. One such system-level implementation is shown in Figure 8-22 [Mob01b]. The host can implement some of the remedial collaborative coexistence mechanisms such as driver-level switching. The chipset itself incorporates *simultaneous operation* (SIMOP) at two protocol layers. The SIMOP-A block provides coordination at the PHY layer and has a core functionality to support Wi-Fi operation when a Bluetooth SCO link is active. SIMOP-A also assists in operating low data-rate Bluetooth activities such as polling peripherals to reduce their impact on Wi-Fi communications. MAC layer switching is accomplished through the SIMOP-D block, which can coordinate timing to prevent cross-interference.

Figure 8-23 shows how Wi-Fi throughput improves when SIMOP is active during collocated operation. When Bluetooth is turned on, Wi-Fi throughput drops significantly, especially for Wi-Fi RSSI levels below about −60 dBm, which is also the case in Figure 8-21 when PTA is off. Most of that throughput loss is recovered when SIMOP begins operating. Throughput remains high even for Wi-Fi RSSI values below −60 dBm, when the PTA performance begins to drop (refer to Figure 8-21). This demonstrates that system-level integration can outperform collaborative measures implemented only at the MAC layer.

Antenna Factors

A promising area of research for enhanced coexistence is in the field of antenna engineering. As an impedance-matching device between the transmission line and free space, the antenna at first appears to have a rather mundane function. During collocated node operation, especially if the radios share the same band,

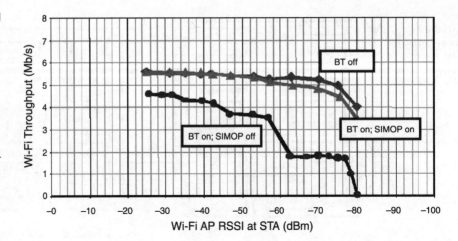

significant cross-coupling of the signals can occur, perhaps rendering one receiver useless while the other transmitter is active. Sufficient isolation between the two antennas will at least partially alleviate this situation.

For devices that are communicating with each other, the omnidirectional antenna works well, especially for portable hosts, because no aiming is required. Power efficiency is poor, though, because most of the transmit energy is wasted by radiating in undesired directions, and this also increases the level of interference on other users.

Antenna Isolation

If two wireless systems are collocated in the same host, the potential for interference is severe. As we've already discovered, interference can be greatly reduced by using collaborative techniques between the systems, but the capability to achieve time orthogonality is limited by the type of packets being exchanged and the desired QoS. When one node is transmitting while the other is receiving, coexistence can be improved through antenna isolation.

Two antennas can be oriented with vertical separation, horizontal separation, or in an echelon configuration, as shown in Figure 8-24. For each of these configurations, we can find the approximate isolation in dB between the two in free space [Tim97]. Vertically separated antennas have isolation given by

$$L_v = 28 + 40 \log\left(\frac{h}{\lambda}\right)$$

(8.14)

whereas antennas that are separated horizontally have isolation

$$L_h = 22 + 20 \log\left(\frac{d}{\lambda}\right) - (G_t + G_r) \tag{8.15}$$

and the echelon configuration has isolation given by

$$L_e = L_h + \frac{2(L_v - L_h)}{\pi} \tan^{-1}\left(\frac{h}{d}\right) \tag{8.16}$$

where λ is the carrier wavelength and G_t and G_r are the TX and RX gains, respectively.

As an example, suppose Bluetooth with a TX power of 0 dBm and Wi-Fi with a TX power of +15 dBm are both installed in a host with antennas separated vertically by $h = 20$ cm, and that free-space propagation is the only coupling mechanism. We want to determine if antenna isolation is sufficient to prevent a TX-RX collision under conditions given in Table 8-1. According to Equation 8.14 the isolation between these two antennas is 37 dB.

The Wi-Fi signal power at the Bluetooth antenna is -22 dBm ($15 - 37$), and with an additional 13 dB rejection in the Bluetooth narrowband RX filter, the interference power is -35 dBm. An incoming Bluetooth signal must be 11 dB above that level to meet co-channel C/I criteria. The required -24 dBm desired signal's received power level is too high to enable a reasonable range to the remote Bluetooth device when the hop channel is within the Wi-Fi transmitted bandwidth.

From the perspective of the Wi-Fi receiver, the entire Bluetooth transmitted signal will enter the Wi-Fi receiver's passband if their frequencies overlap. Therefore, the 37 dB antenna isolation with a 0 dBm Bluetooth TX power requires the incoming Wi-Fi signal to be -27 dBm to achieve a co-channel C/I of 10 dB, which

Figure 8-24
Antenna
configurations
can be vertical,
horizontal, or
echelon.

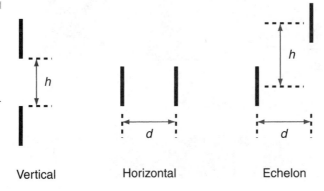

Vertical   Horizontal   Echelon

is still too high to enable Wi-Fi to have a reasonable range. Even when Wi-Fi operates using DSSS at 1 Mb/s, with an additional 10.4 dB of processing gain, the incoming Wi-Fi signal must still exceed about -37 dBm for a co-channel C/I of 10 dB. Isolation between the two antennas must be improved to obtain a reasonable range for collocated nodes in which one receives while the other transmits.

Enhancing Antenna Isolation: A Case Study   When both Wi-Fi and Bluetooth are included in a laptop computer, the antennas are often located in the lid that houses the display. This area is away from the table surface and the decoupling effects of the human body. Antenna isolation can be enhanced by mounting one antenna along the upper edge and the other along the side. Using Equation 8.16 and neglecting cross-polarization effects, this echelon configuration, with $h = 15$ cm, $d = 20$ cm, and 0 dBi antenna gains, provides a free-space isolation of 31 dB. In practice, though, coupling between the two antennas is higher due to electric field propagation around the metal shielding and wiring that is part of the screen and lid assembly, reducing isolation to only about 24 dB in one case study [Rog03].

Isolation can be improved by placing material that impedes the transfer of electromagnetic energy between the two antennas. Three such substances are *magnetic radar-absorbing material* (MAGRAM), *resistive sheet material* (R-card), and *artificial magnetic conductor* (AMC). The first two materials operate by absorbing near-field energy, which has the detrimental affect of reducing antenna radiation efficiency. The AMC attenuates tangential magnetic fields on its surface that can propagate from one end to the other, thus also reducing orthogonal electric fields. Figure 8-25 depicts a laptop computer screen with

Figure 8-25
Wi-Fi and Bluetooth antennas mounted on a laptop computer lid with AMC edge treatments [Rog03]

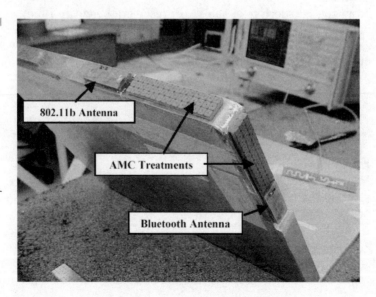

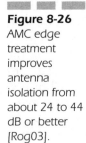

Figure 8-26
AMC edge treatment improves antenna isolation from about 24 to 44 dB or better [Rog03].

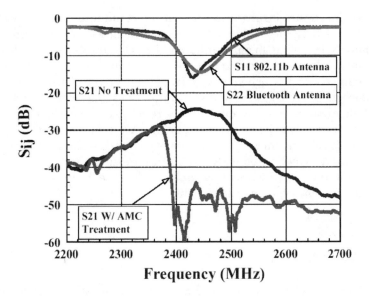

AMC edge treatments between the 802.11 and Bluetooth antennas, and Figure 8-26 shows the resonant bandwidth and S_{21} cross-coupling between the antennas both with and without AMC treatment. Isolation is now improved to about 45 dB. The main disadvantage to this method is that the two antennas become somewhat directional, with maximum radiation occurring in the respective hemispheres away from the AMC strip.

Smart Antennas: The Key to Enhanced Coexistence

As we mentioned in Chapter 2, "Indoor RF Propagation and Diversity Techniques," smart antennas have the potential to greatly enhance the capability of multiple wireless networks to coexist in the same area for three major reasons. First, by automatically steering the primary lobe toward the desired receiver, the destination RSSI is increased by the antenna gain factor. Second, the directional nature of the smart antenna also implies that the interference it causes to others is reduced, provided they are located away from the antenna's main lobe. Finally, by taking the smart antenna a step further and building it into a *multiple-input, multiple-output* (MIMO) array, communication can take place faster over the link, reducing the duty cycle for a given information rate.

By focusing only on the desired receiver's C/I improvements through the use of smart antennas, we can show how coexistence can be improved significantly over using an omnidirectional antenna. Consider an interference-limited channel in which a desired transmitter uses a smart antenna with a modest 5 dB gain in its main lobe, along with a 5 dB rejection in directions away from the main lobe. Compared to using an omni antenna, the desired receiver's C/I is increased by 10 dB, and other receivers experience 5 dB of interference relief merely from the antenna change, which is impressive. However, the desired transmitter can now reduce its power by 10 dB and still obtain the same destination C/I as before, producing another 10 dB of interference relief to other nearby users. Furthermore, if the other transmitters also employ smart antennas, it's easy to visualize a "virtuous spiral" in which TX powers are reduced to the point where the channel becomes AWGN limited instead of interference limited. If the receivers also use smart antennas, which will no doubt be the case for two-way links, their C/I situation becomes even more favorable.

Summary

Although wireless network performance is often satisfactory using passive coexistence, communication reliability can often be greatly improved if a network takes steps to reduce interference to itself and its neighbors. Such active coexistence, either by preventing collisions or reducing the PER during overlapping transmissions, not only increases the desired link's throughput, but enhances the throughput of neighboring links as well.

When two nodes are more than 0.5 m apart, they are assumed to have different hosts and are therefore noncollocated. These nodes have several means by which they can coordinate their channel access activity such that the good neighbor policy is maintained, depending on which protocol they are using. These methods include TX power control, enhanced RX CCA, time and/or frequency orthogonality, adaptive notch filtering, and selecting the proper packet type to match channel conditions.

Two nodes that are collocated have the advantage of being able to communicate through a common host, but their proximity also means that the interference between them can be severe, especially when one node transmits while the other is actively receiving a packet. The IEEE 802.15.2 recommended practice lists several methods to enhance collocated node performance through coexistence. These include AWMA, PTA, and deterministic interference suppression. Because all protocol layers are available for interaction in collocated nodes, active coexistence through system-level integration can achieve throughput values almost as high as if the other links weren't present.

The performance of collocated nodes can also be improved through antenna isolation, but achieving this in the small confines of a single host can be difficult due to the necessary proximity of both antennas. An improvement can be achieved by placing a material between them to reduce their capability to couple signals to each other. Finally, it's possible to reduce interference to insignificant levels if all nodes employ smart antennas.

References

[BT1.2] "Specification of the Bluetooth System," Core Package version 1.2, Bluetooth Special Interest Group, November 5, 2003.

[Chi03] Chiasserini, C., et al., "Coexistence Mechanisms for Interference Mitigation in the 2.4-GHz ISM Band," *IEEE Transactions on Wireless Communications*, September 2003.

[Cor03] Cordeiro, C., et al., "Interference Modeling and Performance of Bluetooth MAC Protocol," *IEEE Transactions on Wireless Communications*, November 2003.

[Gal01] Galli, S., et al., "Bluetooth Technology: Link Performance and Networking Issues," Telcordia white paper, 2001.

[Hay96] Haykin, S., *Adaptive Filter Theory*, Third Edition. Upper Saddle River, N.J.: Prentice Hall, 1996.

[How03] Howitt, I., "Mutual Interference Between Independent Bluetooth Piconets," *IEEE Transactions on Vehicular Technology*, May 2003.

[IEEE802.11] IEEE 802.11-1999, "Wireless LAN Medium Access Control (MAC) and Physical Layer (PHY) Specifications," August 20, 1999.

[IEEE802.11a] IEEE 802.11b-1999, "High-Speed Physical Layer in the 5 GHz Band," September 16, 1999.

[IEEE802.11b] IEEE 802.11b-1999, "Higher-Speed Physical Layer Extension in the 2.4 GHz Band," September 16, 1999.

[IEEE802.11g] IEEE 802.11g-2003, "Amendment 4: Further Higher Data Rate Extension in the 2.4 GHz Band," June 27, 2003.

[IEEE802.11h] IEEE 802.11h-2003, "Amendment 5: Spectrum and Transmit Power Management Extensions in the 5 GHz Band in Europe," October 14, 2003.

[IEEE802.15.1] IEEE 802.15.1-2002, "Wireless Medium Access Control (MAC) and Physical Layer (PHY) Specification for Wireless Personal Area Networks (WPAN)," June 14, 2002.

[IEEE802.15.2] IEEE 802.15.2-2003, "Coexistence of Wireless Personal Area Networks with Other Wireless Devices Operating in Unlicensed Frequency Bands," August 28, 2003.

[IEEE802.15.3] IEEE 802.15.3/D17, "Wireless Medium Access Control (MAC) and Physical Layer (PHY) Specifications for High Rate Wireless Personal Area Networks (WPAN)," draft standard, February 2003.

[IEEE802.15.4] IEEE 802.15.4-2003, "Wireless Medium Access Control (MAC) and Physical Layer (PHY) Specifications for Low-Rate Wireless Personal Area Networks (LR-PANS)," October 1, 2003.

[Mob01a] "Wi-Fi (802.11b) and Bluetooth: An Examination of Coexistence Approaches," Mobilian White Paper, 2001.

[Mob01b] "Sim-Op—Unleashing the Full Potential of Wi-Fi and Bluetooth Coexistence," Mobilian White Paper, 2001.

[Ole01] Oleynik, V., "An Approach to the Problem of Optimizing Channel Parameters," IEEE 802.11 Document IEEE 802.1101/152, March 2001.

[Rog03] Rogers, S., et al., "Artificial Magnetic Conductor (AMC) Technology Enables the Coexistence of 802.11b and Bluetooth," eTenna white paper, rev B, February 14, 2003.

[Tim97] Timiri, S., "RF Interference Analysis for Collocated Systems," *Microwave Journal*, January 1997.

[Zyr03] Zyren, J., et al., "IEEE 802.11g Network Behavior in a Mixed Environment," Intersil white paper, February 2003.

Coexistence with Other Wireless Services

The preceding chapters in this book concentrated on the coexistence capabilities among four different *wireless local area network* (WLAN) and *wireless personal area network* (WPAN) protocols: 802.11 Wi-Fi, 802.15.1 Bluetooth, 802.15.3 WiMedia, and 820.15.4 ZigBee. The issue of coexistence extends far beyond these networking services, which are also able to interfere with, and receive interference from, other users of the *radio frequency* (RF) spectrum. It should be remembered that FCC rules place the burden of interference avoidance on the Part 15 device, which in turn has no legal protection against interference from others. A Part 15 device that causes disruption to a licensed user is legally required to alleviate the situation or cease operations.

The source of interference to a receiver can originate from intentional, unintentional, or incidental radiators. Obviously, it's impossible in this limited space to consider all interference scenarios, so we will concentrate on those that are likely to be of greatest interest. These can be roughly divided into two categories. First, interference can originate at a WLAN or WPAN transmitter and affect the receiver in another service. For example, because *ultra-wideband* (UWB) occupies such a wide bandwidth, much of our discussion will be directed at the potential for Part 15 UWB transmissions to interfere with non-Part 15 receivers. Second, interference can originate within another service and affect a WLAN or WPAN receiver, such as that caused by a microwave oven.

In this chapter, we examine coexistence between WLAN/WPAN and the *global positioning system* (GPS), avionics, cordless telephones in the United States, microwave ovens, and microwave lighting. GPS and avionics are usually recipients of interference from wireless network transmissions, whereas microwave ovens and lighting are sources of interference. Cordless telephones can be both a source and a recipient of interference. We also demonstrate a method for determining interference between a cellular telephone and its collocated Bluetooth node. Because this book is about wireless network coexistence, we won't discuss interference between services outside of this scope such as, for example, microwave lighting and cordless phones.

Although we present some analytical studies, by far most of the work has taken an empirical approach, both on a laboratory test bench and with actual deployed systems. The setup and procedures for obtaining accurate test results for these experiments are extensive. Rather than repeating this information here, we instead concentrate on results. Anyone interested in the complete description of the experiments can refer to the literature.

Global Positioning System (GPS)

Development of the GPS satellite navigation system, also called NAVSTAR, began within the U.S. Department of Defense in 1978. The system was to provide all-weather positioning for armed forces operations by the United States and its allies, but GPS has become increasingly important for the navigational needs in professional and recreational applications. The predominance of GPS as a primary positional and navigational tool for commercial aviation, along with increasing pressure to allow the operation of wireless networking onboard the aircraft, has spawned significant coexistence research.

Operational Overview

The design of the GPS system is based on the following concepts [Riz99]:

- Suitable for all classes of platform in the air, land, sea, and space
- Capable of real-time positioning and velocity to a desired degree of accuracy
- Resistant to intentional jamming and unintentional interference
- Redundant systems for survivability
- Passive positioning that requires only the *receive* (RX) capability at the user level
- Service that is available to an unlimited number of users

The GPS system encompasses three segments. The *space segment* consists of the satellites and their transmitted signals. The *control segment* contains the ground facilities that supervise and direct satellite operations. The *user segment* is the equipment and computational techniques available to the user. Coexistence concerns are directed toward the user segment.

Several launch phases have resulted in a network of 24 GPS satellites, along with a number of spares, orbiting at an altitude of 20,200 km and with an orbital period of 11 h, 58 m. Six different orbital inclinations are each occupied by four satellites. Each satellite makes two revolutions per sidereal day of 23 h, 56 m. Coupled with the earth's rotation, a particular satellite reaches the same position relative to the earth's surface once per sidereal day. At least four satellites are guaranteed to be visible from any position on earth, and this is the minimum number necessary for the three-dimensional positioning of latitude, longitude, and altitude. Sometimes as many as 12 satellites are visible, but the typical number is 6 to 7.

Satellite Signals Figure 9-1 depicts the signals transmitted from each GPS satellite. The *Link-1* (L1) carrier operates at 1,575.42 MHz and L2 operates at 1,227.60 MHz. The L1 carrier contains two ranging codes, the *coarse acquisition* (C/A) code and the *precice* (P) code, along with a military M code being deployed on replacement satellites. The L2 carrier contains only the P code, but it will be enhanced with the M code and eventually include a new civil L2C code. For increased security, the P code is encrypted with a secret W code to form the transmitted Y code that can be decrypted by authorized users. Most

Figure 9-1
GPS satellite
signals are
shown in solid
boxes; the
dashed boxes
are projected
to be fully
operational by
2010.

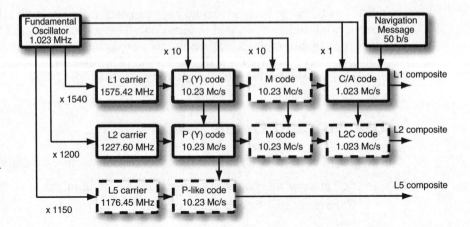

consumer-oriented receivers use only the C/A code and its associated *standard positioning service* (SPS). Service based on the P(Y) code is called the *precise positioning service* (PPS). A new carrier at L5 has been approved at 1,176.45 MHz for safety-of-life applications, with a P-like positioning code for high accuracy. The GPS modernization program includes adding capabilities and increasing the *equivalent isotropic radiated power* (EIRP) of some of the signals.

The ranging codes use *pseudorandom noise* (PRN) generators common to *direct sequence spread spectrum* (DSSS) signaling. Code timing is precisely controlled by onboard atomic clocks. The C/A code is a 1,023-chip sequence that is different and numbered for each GPS satellite. With a chip rate of 1.023 Mchips/s, the C/A code repeats every millisecond. The P code is the same for each satellite, and at a rate of 10.23 Mchips/s the code repeats approximately every 267 days. Each satellite sends a different seven-day portion of the P code, and that particular sequence is restarted each week. Both codes are *binary phase shift keying* (BPSK) modulated onto their carriers. The P code has a chip resolution of 10 times that of the C/A code, but well-designed correlation receivers using the C/A code alone can produce accuracy levels only about a factor of 2 worse than receivers using the P code [Riz99].

The *power spectral density* (PSD) of the L1 C/A, P, and M codes is shown in Figure 9-2 [Sta02]. The bandwidth of the primary lobe is twice the chip rate, so the C/A signal is about 2 MHz wide, and the P (Y) signal is about 20 MHz wide at their respective first zero crossings. However, the signals still contain considerable energy in their sidebands, as shown in Figure 9-3 for a C/A code composed of a sequence called PRN 21.

Authorized users can lock into both the L1 and L2 carriers to compensate for ionospheric propagation delays to obtain positioning information with submeter resolution. A *semicodeless receiver* has been developed for civil use that can

Figure 9-2
PSD of the L1
C/A, P, and M
codes [Sta02]

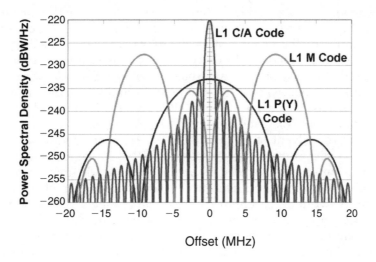

Figure 9-3
Detailed PSD of
L1 C/A code
using PRN 21
[Luo01b]

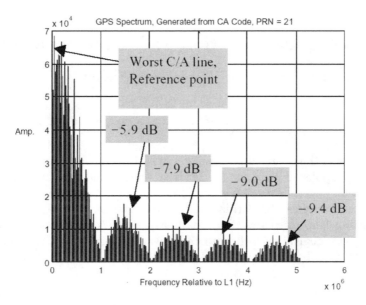

extract enough information from the L2 P(Y) code such that, coupled with acquisition of the L1 C/A signal, accuracy values comparable to PPS can be obtained. Typical signal powers at the earth's surface range between -140 and -123 *decibels relative to 1 milliwatt* (dBm) for L1 C/A, with a nominal value of -130 dBm.

The navigation message contains the satellite ephemeris (orbit) information and clock parameters, among other data. This relatively slow data signal is *exclusive ORed* (XORed) with the C/A code so it can be extracted by ground stations.

By combining ephemerides of the visible satellites with the time required for the signal to travel from each satellite to the GPS receiver, which is determined by correlating on the ranging codes, the user's position can be determined through spherical trigonometry. The control segment can command a satellite to alter its C/A code to reduce accuracy through a policy called *selective availability* (SA). Accuracy is reduced to about 100 m horizontally and 160 m vertically for SPS users by falsifying some of the satellite ephemeris and clock data. SA was activated in March 1990 and discontinued in May 2000.

With the advent of *Enhanced-911* (E-911), considerable attention has been directed at improving the operation of GPS receivers inside buildings so positioning data can be included with any 911 call made from a cell phone. Unfortunately, GPS receiver sensitivities are already near the physical *signal-to-noise ratio* (SNR) limit, so improved signal processing is being designed into GPS chipsets for operation in areas of reduced satellite signal strength. Even so, the capability of these GPS receivers to operate when interference is present will become increasingly important as WLAN and WPAN systems proliferate indoors.

GPS Signal Detection and Interference Effects

The incoming signal to the GPS receiver is given by

$$s(t) = \sqrt{2P}\, a(t - \tau) \cos[2\pi(f + \Delta f) - \phi] \tag{9.1}$$

where P is the received power, $a(t - \tau)$ is the code sequence shifted by time τ, Δf is the Doppler shift of the carrier frequency f, and ϕ is the phase. The receiver performs several processing and filtering operations, but in the end it essentially creates a replica $\hat{s}(t)$ of this signal and performs a correlation with $s(t)$ via its tracking loop until the time, frequency, and phase shifts are duplicated. The receiver's goal is to lock onto the incoming signal and determine its precise arrival time by maximizing the correlation integral, given by

$$R(T) = \frac{1}{2T} \int_{-T}^{T} s(t)\,\hat{s}(t)\, dt \tag{9.2}$$

The integration interval $2T$ is manufacturer-dependent, but is often set to the duration of a GPS navigation message bit at 20 ms [But02].

The faster the tracking loop operates to match the incoming signal, the higher its bandwidth and corresponding noise power. The total noise power N is the product of the noise power density N_0 in W/Hz and the loop bandwidth B_L in Hz.

The receiver adjusts B_L during signal acquisition, so the SNR also changes. Therefore, it is convenient to designate the signal-to-noise density ratio as S/N_0 with units in dB-Hz. (To determine the SNR in dB, simply convert the desired bandwidth [in Hz] to dB and subtract it from S/N_0.) A professional-quality GPS receiver can operate with an S/N_0 of about 32 dB-Hz or higher [Sta02]. The typical noise floor of a GPS receiver is about -111 dBm/MHz, which is the physical kTB noise floor of -114 dBm/MHz adjusted by a receiver noise figure of 3 dB [Sta02].

A GPS unit can estimate the signal power $\hat{S}$ through the average value of the correlation integral in Equation 9.2, and it can estimate the noise power $\hat{N}$ by finding the mean-square deviation of the correlation integral over a long period of time; that is,

$$\hat{N} = \frac{1}{T_N} \int_0^{T_N} [R(t) - \overline{R}(t)]^2 \, dt \qquad (9.3)$$

where $T_N >> T$. Now the post-correlation quantities $\hat{S}/\hat{N}$ or $\hat{S}/\hat{N}_0$ can be found, which are related to the precorrelation quantities C/N or C/N_0.

The precision of the position indicated by processing the GPS satellite signals, sometimes called the *pseudorange* (PSR) accuracy and represented by the standard deviation σ_p, is a function of S/N_0 and the type of tracking loop employed in the GPS unit. Figure 9-4 shows σ_p in meters from a standard correlator (accurate to one code chip) and a narrow correlator (accurate to 0.1 code chip) as a function of postcorrelator S/N_0 [But02]. The PSR accuracy standard deviation is approximately proportional to the square root of the SNR when the noise is *additive white Gaussian noise* (AWGN). Thus, for every 6 dB increase in S/N_0, accuracy improves by roughly a factor of 2. A GPS receiver can nominally process data from satellites that are between 5° and 90° of elevation above the horizon, for which the interference-free S/N_0 varies by about 7.5 dB.

Interference to the GPS Signal   When the incoming GPS satellite signal contains additive *radio frequency interference* (RFI), both the signal and noise are processed by the correlator during despreading at the receiver. The replica code at the receiver spreads narrowband interference, as depicted in Figure 4-8 in Chapter 4, "Advanced Modulation and Coding," with a processing gain that is a function of the spreading sequence length. The C/A code, for example, has a length of 1,023 chips, corresponding to a postcorrelation *processing gain* (PG) of about 30 dB. Wideband RFI that is uncorrelated to the replica code remains spread. In either case, the postcorrelation processing of continuous interference often exhibits noise-like characteristics, so it contributes to the total noise at the

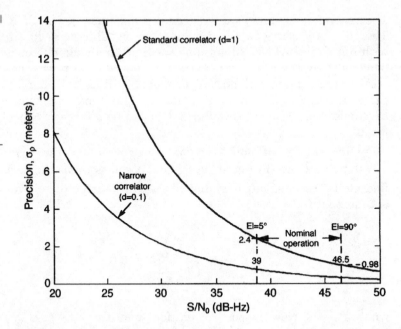

Figure 9-4
GPS location precision as a function of postcorrelator signal-to-noise density ratio [But02]

GPS receiver. RFI and thermal noise are uncorrelated, so their respective powers produce a total noise density of $N_0 + N_0^{(J)}$. (We follow convention by using the term J to refer to both interference and jamming.)

If the interference power is significantly lower than the thermal noise power, the former has little effect on GPS positional accuracy. However, if $N_0^{(J)} = N_0$, then $S/(N_0 + N_0^{(J)})$ is degraded by about 3 dB. Beyond this, the interference predominates and the signal-to-noise density ratio is reduced accordingly. This degradation may be more or less severe than that from an equivalent increase in thermal noise, depending on the nature of the interference.

The GPS L1 *path loss* (PL) equation can be found by substituting the appropriate wavelength value into Equation 2.3 in Chapter 2, "Indoor RF Propagation and Diversity Techniques," which gives

$$PL = 36.4 + 10n \log d \qquad (9.4)$$

This enables us to determine the power at the GPS receiver from an interference source. As an example, suppose a non-UWB FCC Part 15 device radiates a 1 MHz-wide spurious signal in the GPS L1 band at the regulatory limit of −41.3 dBm/MHz. Using free-space propagation and neglecting antenna effects, the device produces −111 dBm of noise at the GPS receiver from a distance of 46 m. This is a fairly long distance, but (fortunately for coexistence) most Part 15

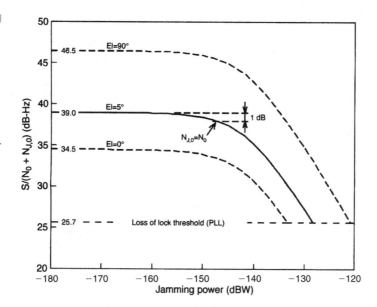

Figure 9-5

Jamming power and its effect on SINR for various satellite elevation angles [But02]

devices emit negligible radiation in the GPS bands. Both indoor and outdoor (handheld) UWB Part 15 devices have a more restrictive FCC spurious radiation limit in the GPS bands of -75.3 dBm/MHz, so a UWB transmitter conforming to these rules must be closer than 1 m to the GPS receiver to produce more than -111 dBm of additional noise.

As $N_0^{(J)}$ increases, PSR accuracy decreases, and the GPS receiver will eventually lose its lock on the satellite signal. This *loss-of-lock threshold* depends on the design of the receiver's *analog-to-digital* (A/D) converters, tracking loop, and interference mitigation methods. Examples of interference (jamming) power and its effect on the *signal-to-interference-and-noise ratio* (SINR) are shown in Figure 9-5 for different satellite elevation angles [But02]. Once the lock is lost, the receiver requires a higher S/N_0 before it can regain the lock,* which is called the *tracking threshold*. Both of these thresholds provide a good indication of a receiver's performance when interference is present.

RFI in the GPS bands can be examined in detail by connecting a GPS antenna to a spectrum analyzer able to plot weak signals. An example of such a plot is given in Figure 9-6 [But02]. The *International Civil Aviation Organization* (ICAO) has standardized a *continuous wave* (CW) interference mask for the GPS bands, which is also shown in Figure 9-6. The GPS L1 signal can be seen at

---

*At this point we begin using the term S/N_0 as a general signal-to-noise density ratio value, including any interference that may be present.

Figure 9-6
Spectrum analyzer plot of signals present in the GPS L1 band, along with the ICAO sinusoidal interference mask and loss-of-lock threshold for a particular receiver [But02]

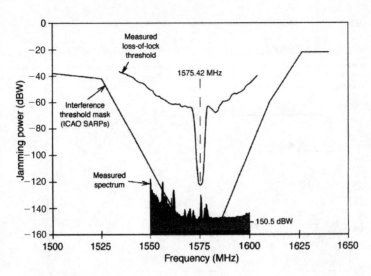

1,575.42 MHz, and it is surrounded by a number of interfering signals. None of these is above the loss-of-lock threshold for the particular GPS receiver being tested, although several exceed the ICAO CW mask.

UWB Interfering with GPS: Analytical Approach

In this subsection, we present a theoretical study of UWB signals based on the Gaussian pulse and the associated interference powers produced in the GPS L1 band [Ham02]. Although some of the interference levels exceed FCC limits without additional filtering, the results are instructive because they show how interference changes as a function of UWB pulse type and duration.

UWB Waveforms and Spectra   Four Gaussian pulse waveforms with duration T_p were included in the study and are given in Figure 9-7, along with their first derivatives from antenna differentiation. The corresponding spectrum of each signal is depicted in Figure 9-8. The three Gaussian-based monopulses have relatively smooth spectra, while the doublet shows spectral nulls with spacing inversely proportional to the time between pulses. The center frequency and -10 dB bandwidths for each of these signaling methods are given in Table 9-1.

Both *time-modulated UWB* (TM-UWB) and *direct-sequence UWB* (DS-UWB) were considered in the study, with basic structures described in Chapter 4 and depicted in Figure 9-9. Spectral lines are reduced in TM-UWB through time-dithering, so we refer to this subclass as *time-hopping UWB* (TH-UWB). A single data bit is encoded with N UWB pulses in both schemes. For TH-UWB, the pulses

Figure 9-7
UWB signals created from the Gaussian monopulse and its second and third derivatives, as well as from the Gaussian doublet. Each is shown with its respective differentiation operation at the antenna [Ham02].

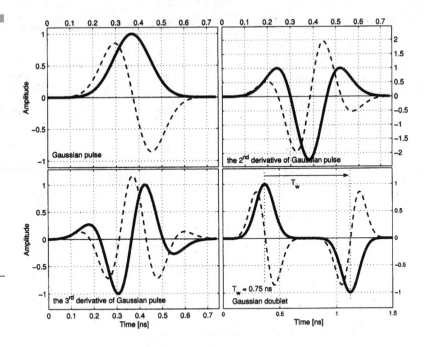

Figure 9-8
UWB signal spectra from the Gaussian pulses in Figure 9-7 [Ham02]

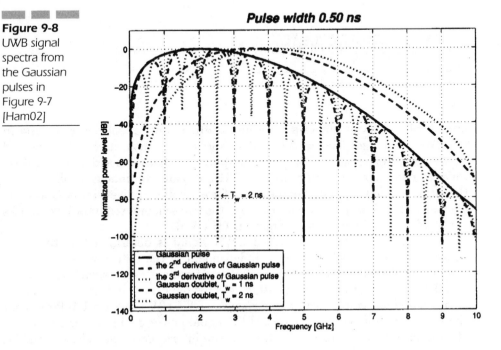

Pulse waveform	Center frequency	−10 dB bandwidth
Gaussian monopulse	$1/T_p$	$2/T_p$
Gaussian monopulse, second derivative	$1.73/T_p$	$2.1/T_p$
Gaussian monopulse, third derivative	$2/T_p$	$2/T_p$
Gaussian doublet	$1/T_p$	$2/T_p$

Figure 9-9

TH-UWB and DS-UWB signal structures [Ham02]

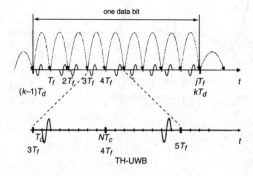

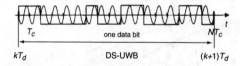

are distributed over windows according to their pulse repetition code, and in DS-UWB the pulses form the equivalent spreading code. Data bits are encoded in the initial excursion of the pulse amplitude, whether positive or negative. Pulse transmissions are intermittent in TH-UWB and continuous in DS-UWB, so the latter method has a higher data rate for equivalent pulse widths. Obtaining equivalent bit rates and energies between the two schemes requires fewer pulses with higher individual pulse energies in TH-UWB.

The spectrum of the transmitted UWB signal is determined by the pulse width and pulse waveform, and is independent of the data rate. This has the advantage of allowing the selection of a waveform with corresponding parameters that minimize interference within selected bands. By maintaining equal bit energies and spreading factors, the spectra of TH-UWB and DS-UWB are kept the same for comparison purposes. UWB signal power during the transmission of each data bit is 0.5 mW (−3 dBm) for the interference analysis, and the bit energy is divided equally among the N pulses sent for each data bit. As N increases, the energy per pulse decreases with a corresponding decrease in interference to other systems. Pulse widths are allowed to vary between 3.5 and 0.2 ns, corresponding to a center frequency between 0.25 and 10 GHz.

Analytical Results Figure 9-10 plots interference in the GPS L1 band generated at the UWB antenna. The GPS receiver is assumed to be tracking the P(Y) code with a bandwidth of 20 MHz. Both TH-UWB and DS-UWB operating with the third derivative Gaussian pulse exhibit lower interference levels for shorter and longer pulse widths, with a peak at a pulse width of about 1.5 ns. The remaining pulse waveforms exhibit gradually increasing interference as the pulse width becomes shorter. The Gaussian doublet with a spacing of 0.63 ns shows a spectral null in the L1 band and thus interferes the least for nearly all pulse widths considered. Otherwise, DS-UWB produces interference levels about 4 dB lower than TH-UWB using the same pulse parameters. These interference levels are reduced by about 10 dB if the GPS receiver is tracking the 2 MHz-wide C/A code.

According to Figure 9-10, the highest UWB interference power over a 20 MHz bandwidth is about −33 dBm. Assuming a flat PSD over the L1 band, this interference power is about −46 dBm/MHz (from −33 − 10log(20)), which is well above the FCC Part 15 indoor/outdoor UWB radiation limit. To meet the FCC limit of −75.3 dBm/MHz, the total radiated power within a 20 MHz bandwidth cannot exceed −62.3 dBm (from −75.3 + 10log(20)), which can be met by each

Figure 9-10
TH-UWB and DS-UWB interference in the GPS L1 band over a bandwidth of 20 MHz [Ham02]

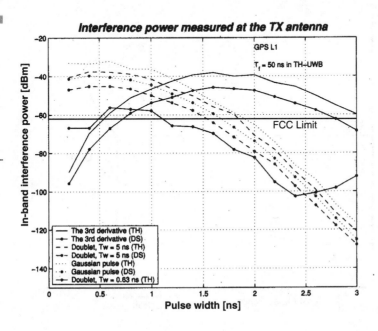

modulation method and pulse waveform somewhere within its particular range of pulse widths.

UWB Interfering with GPS: Empirical Approach

In an effort to verify the theoretical results and to test conditions difficult to model mathematically, much research interest has been directed toward empirical studies of interference between GPS and UWB [And01], [Bru01], [Luo01a], [Sch01]. We will focus on a comprehensive empirical study that determined the interference from a UWB transmission equivalent to a known level of broadband noise that causes the GPS receiver to just meet its performance criteria [Luo00], [Luo01a], [Luo01b].

UWB Noise Equivalency The noise equivalency concept test methodology consists of injecting broadband noise (AWGN) into a GPS receiver and increasing the noise level until a desired ranging error standard deviation is reached. Next, the broadband noise is reduced by a backoff value of, say, 4 dB and the UWB emission level is increased until either σ_p once again reaches the desired value or the GPS receiver loses its lock. The UWB signal level can then be compared to the broadband noise level to determine, among other criteria, which of the two types of RFI is more detrimental to GPS accuracy. Figure 9-11 shows this process in graphical form. UWB parameter set i affects GPS positional errors to

Figure 9-11
Noise equivalency testing of a GPS receiver with UWB interference

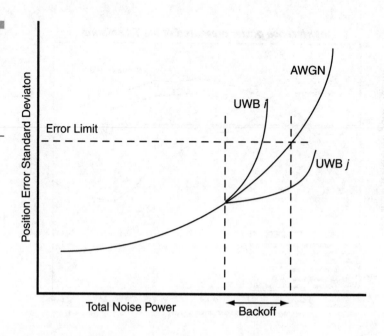

a greater degree than broadband noise, and UWB parameter set *j* affects the GPS receiver to a lesser degree. In this way, the empirical study demonstrates the influence of UWB interference directly.

UWB Test Signal Composition The UWB pulse used in the tests has a Gaussian-like rise and fall with a width of about 0.2 ns, along with some ringing. Figures 9-12 and 9-13 show the time and frequency domain representations of the pulse [Luo01b]. The signal contains a rather high proportion of its energy in the GPS bands, so even modest amounts of *transmit* (TX) power run the risk of violating FCC Part 15 UWB rules. The pulse is spread slightly in time with additional ringing each time it passes through an amplifier or filter in the UWB transmitter.

UWB signals in the tests can be either modulated or unmodulated, with adjustable duty cycles. For unmodulated signals, the *pulse repetition frequency* (PRF) is varied between 100 kHz and 20 MHz. For modulated signals, a *pulse position modulation* (PPM) scheme with either 2 or 10 pulse positions is used, as shown in Figure 9-14. For 2 PPM, the pulse is modulated randomly to be either before or after the nominal pulse time, and for 10 PPM the transmitted pulse is selected randomly to be up to 5*d* before nominal, at the nominal time, or up to 4*d* after nominal.

UWB interference on a GPS receiver can be manifested by one of three predominant characteristics, depending on its structure [And01]. If the PRF is relatively slow compared to the GPS receiver's bandwidth, the interference appears

Figure 9-12
Time domain representation of the UWB pulse used in the experiments [Luo01b]

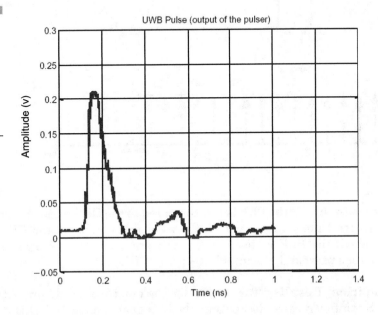

Figure 9-13
Frequency
domain
representation
of the UWB
pulse used
in the
experiments
[Luo01b]

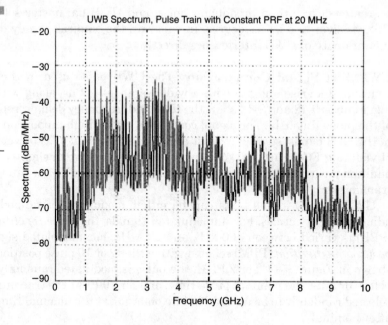

Figure 9-14
Examples of
2 PPM and
10 PPM UWB
signaling

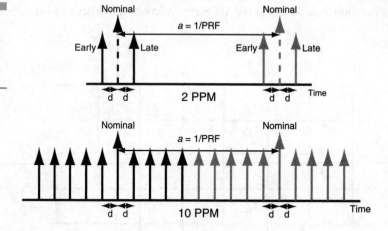

to be *pulse-like*. If the PRF is fast compared to the GPS receiver's bandwidth and
no spectral lines appear within the receiver's passband, the RFI is *noise-like*.
Finally, if the PRF is fast but one or more spectral lines appear in the GPS
receiver's passband, the interference is *CW-like*.

Empirical Results The results for the empirical study are obtained with a
GPS aviation receiver locked onto the L1 signal at a level of −131.3 dBm. Broad-

band noise is added until the accuracy drops to a desired threshold. A backoff of either 2 or 4 dB is set, and the UWB signal is introduced and increased in amplitude until either the generator reaches its maximum level or the GPS receiver loses its lock.

An example of pulse-like RFI occurs when an unmodulated UWB signal with a PRF of 100 kHz is injected into a UWB receiver locked onto the L1 signal with a bandwidth of 2 MHz. Figure 9-15 shows the noise equivalency results for backoff levels of approximately 2 and 4 dB. In both cases, the RFI caused by UWB affects the receiver much less than AWGN; in fact, UWB can increase total GPS in-band noise power by about 20 dB before PSR accuracy is significantly affected, and by about 33 dB before the original PSR accuracy limit is reached. Furthermore, even with the UWB signal at its maximum output level of −57.3 dBm, the GPS receiver holds its lock.

By increasing the UWB PRF to 20 MHz, interference to the GPS receiver becomes more noise-like, as shown in Figure 9-16, again using backoff values of about 2 and 4 dB. In either case, PSR accuracy levels are slightly better and the loss-of-lock performance is slightly worse than for AWGN alone.

Finally, Figure 9-17 depicts noise equivalency results for an unmodulated UWB signal with PRF set at 19.94 MHz. For this PRF, the seventy-ninth harmonic falls on 1,575.260 MHz. This is only 160 kHz away from the L1 center frequency, so the RFI is CW-like. Such RFI has a catastrophic effect on GPS operation, with the receiver losing its lock as soon as the total noise slightly exceeds the backoff value set by AWGN. Note that when the PRF is raised slightly to 20 MHz for the previous example, the seventy-ninth harmonic moves to 1,580 MHz, well outside the GPS receiver's L1 passband. These results

Figure 9-15
GPS receiver response on the L1 band to an unmodulated UWB signal with PRF at 100 kHz [Luo01b]

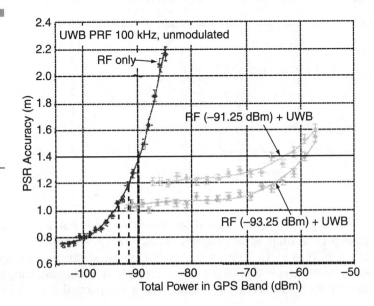

Figure 9-16
GPS receiver
response on
the L1 band
to an
unmodulated
UWB signal
with PRF at 20
MHz [Luo01b]

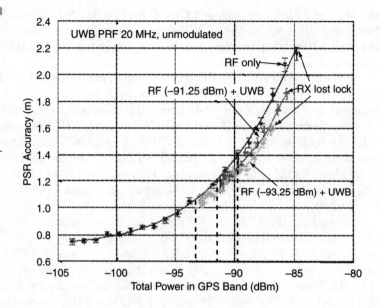

Figure 9-17
GPS receiver
response on
the L1 band
to an
unmodulated
UWB signal
with PRF at
19.94 MHz
[Luo01b]

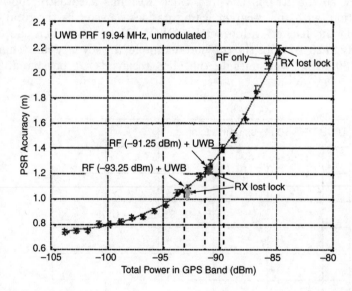

demonstrate the need for time-dithering of the UWB pulse to reduce the possibility of spectral lines appearing within the GPS passband.

Both 2 PPM and 10 PPM with random pulse shifts provide levels of time-dithering to the UWB transmitted waveform. It is important, though, to select a sufficiently high dithering percentage value or spectral lines will remain. For

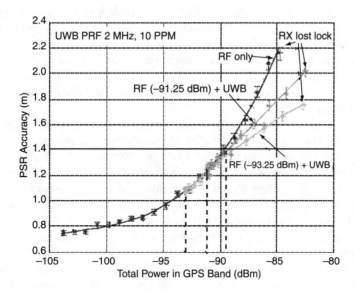

Figure 9-18
GPS receiver response on the L1 band to a 10 PPM UWB signal with nominal PRF at 2 MHz [Luo01b]

example, with the 2 PPM depicted in Figure 9-14, selecting $a = 56$ ns and $d = 2$ ns results in a ratio of position dithering d/a of only 3.6 percent. In this case, a PRF of 15.91 MHz places a spectral line at 1,575.09 MHz, causing the GPS receiver to perform almost identically to the poor results in Figure 9-17. The results are much more promising for the 10 PPM structure shown in Figure 9-14. For $d = 50$ ns (limited by the UWB transmitter capability), the nominal PRF is 2 MHz. The transmitted spectrum looks much like that of an unmodulated 20 MHz PRF, but each spectral line has much lower power due to the 0.1 probability that a particular PPM time slot contains a pulse. (Of course, the average UWB transmitted power is proportionally lower as well.) Figure 9-18 plots the results for 10 PPM, showing that the RFI generated by the UWB signal is more noise-like.

Avionics

Most government entities that regulate commercial aviation, including the *Federal Aviation Administration* (FAA) in the United States, prohibit the use of RF transmission and reception equipment by passengers during certain phases of flight. These rules exist to prevent the possibility of *electromagnetic interference* (EMI)* between these devices and critical aircraft communication and navigation

---

*EMI includes RFI along with interference coupled into sensitive equipment from wiring and other near-field sources.

equipment. However, with the fast proliferation of *portable electronic devices* (PED) among passengers, the potential for interference to navigation and communication avionics has increased, even from unintentional radiators such as a *personal digital assistant* (PDA). Furthermore, increasing interest exists in allowing *industrial, scientific, and medical* (ISM) communications to take place onboard commercial aircraft to give passengers the ability to connect to the Internet while traveling.

Aircraft operations can be roughly divided into three phases. The first, *ground operations*, takes place from engine start to clearance for takeoff and from clearing the runway after landing until engine shutdown. Most avionics receivers in use during ground operations have high desired signal RSSI values, and interference from passenger PEDs is minimal and mostly noncritical. Indeed, some airlines have begun to allow passengers to use their cellular telephones during ground operations at the destination airport.

The second phase of aircraft operations is *takeoff and landing*, which for our purposes includes departure on the outbound segment and approach on the inbound segment. This is the most critical phase from a safety standpoint due to demands on the flight crew and a high level of reliance on avionics. Any interference to aircraft operations during this phase could be life-threatening.

Finally, the *cruise* phase is usually the most relaxed and longest phase of the flight, during which the operation of PEDs is usually allowed but only if they are incidental radiators. Equipment operating as intentional radiators may eventually be allowed as well.

In this subsection we will study the interference potential that exists for avionics from transmissions in the 2.4 GHz ISM band and from UWB transmissions. The research is almost all empirical in nature, with different setup methods and procedures. Naturally, the results differ as well, but nonetheless a general picture of the coexistence capabilities between FCC Part 15 devices and avionics emerges.

EMI Standards and Airborne Wireless Operations

Since the introduction of the first transistor radio broadcast receiver, studies have been performed to assess the interference potential between PEDs and avionics. The *Radio Technical Commission for Aeronautics* (RTCA) is a nonprofit group that develops recommendations in such areas, and these recommendations are often used by the FAA for its regulatory and policy decisions. RTCA report DO-199 in 1988 and DO-233 in 1996 considered the potential for interference from PED spurious emissions real but infrequent [Ely02b].

One of the difficulties with even allowing wireless devices to be onboard an aircraft is the fact that many of them are part of a multifunction PED that can transmit without any user action. Such is the case with cellular telephones, which periodically transmit whenever they are powered on, and such is also the case with many WLAN and WPAN devices that periodically probe for network activity. Fortunately, with the advent of advanced signal processing and spread spectrum techniques, it is possible that these intentional transmissions will be no more disruptive to avionics than unintentional radiators that meet FCC Part 15 spurious emission rules [Ely02b].

Figure 9-19 is a comprehensive plot of spurious radiation limits required by numerous standards bodies. The highest is for cellular and PCS phones with a TX power of up to 1 W, and the most stringent are from RTCA recommendations for operation on aircraft. The RTCA places such electronic equipment into several categories, two of which apply to consumer PEDs:

- **Category M** Equipment and wiring located in the passenger cabin and cockpit, but not directly in view of aircraft radio receiver antennas
- **Category H** Equipment and wiring located directly in view of aircraft radio receiver antennas

As expected, the EMI limits for category H devices are more stringent than those for category M devices. These limits have shaded regions in Figure 9-19 to account for the directivity found in almost all PED radiation patterns. The most obvious characteristic of the RTCA recommendations is that they are sometimes significantly lower than required by the FCC for Part 15 devices.

Within the operational bands of aircraft navigation and communication equipment, the RTCA-recommended EMI limits are lower still. Affected avionics include *VHF omnidirectional ranging* (VOR), *localizer* (LOC) for runway course

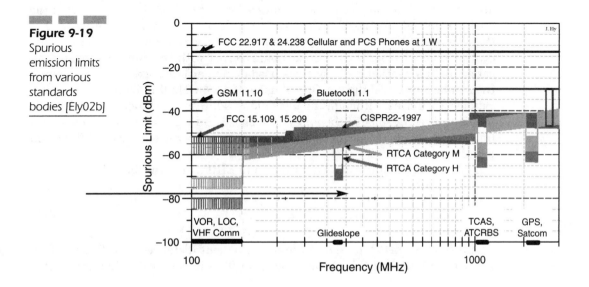

Figure 9-19
Spurious emission limits from various standards bodies [Ely02b]

information, and VHF voice communication using carrier frequencies from 108 to 136 MHz. The avionics also include a *glideslope* (GLS) for runway glidepath information from 328 to 335 MHz, as well as the *traffic alert and collision avoidance system* (TCAS) and the *air traffic control radar beacon system* (ATCRBS). These systems are used for aircraft identification and position information to ground controllers, and both operate in the 1,030 and 1,090 MHz bands. *Distance-measuring equipment* (DME) near 1 GHz and a GPS near 1.5 GHz are included in the affected avionics as well. Localizer and glide path information is combined into the aircraft *instrument landing system* (ILS). Other avionics systems not on the chart include a radar altimeter (4.3 GHz), a microwave landing system (5.03 and 5.09 GHz), and a weather radar (5.4 and 9.3 GHz). These last three systems are all within the legal operating region for FCC Part 15 UWB devices.

Fortunately, almost all commercial PEDs produce EMI values well below even the stringent limits shown in Figure 9-19, but several areas of concern still exist. Customers could make modifications to the equipment to increase EMI, components or shielding could deteriorate with age, and passengers could be operating devices that conform to less stringent rules from other regulatory agencies. Finally, as intentional radiators between 3.1 and 10.6 GHz, UWB transmitters could possibly produce spurious EMI throughout the bands shown in Figure 9-19.

Interference to Avionics from 2.4 GHz ISM Bluetooth-Class Transmissions

An empirical study on the feasibility of using 2.4 GHz "Bluetooth class" devices in aircraft was performed by Intel on wide-body and narrow-body commercial aircraft, as well as on a common business-class jet [Sch00]. The first part of the study examined whether 2.4 GHz was a viable band for communication within the confines of the passenger compartment. The second part of the study concentrated on whether these devices could coexist with the numerous communication and navigation radio systems used during aircraft operation. In general, the results supported the viability of using the 2.4 GHz ISM band for communications on aircraft. Although the tests didn't address non-Bluetooth ISM transmissions directly, some of the analysis is based on FCC Part 15 spurious output limits, so they are applicable to other such devices operating in the 2.4 GHz ISM band.

Field strength values can be converted to EIRP using the formula

$$P = \frac{d^2 E^2 G}{30} \tag{9.5}$$

where P is the EIRP in W, d is the measurement distance in m, E is the field strength in V/m, and G is the antenna gain. This is identical to Equation 1.2 in

Chapter 1, "Introduction," with algebraic simplifications. For all testing in the Intel study, dipole antennas having a gain of 2.14 dBi, or 1.64 in linear terms, were used. In this case, the formula relating EIRP to field strength becomes

$$P = \frac{d^2 E^2 (1.64)}{30} = \frac{d^2 E^2}{49.2} \quad \text{Dipole antenna} \tag{9.6}$$

The EMI plot of the Bluetooth device used for some of the results is shown in Figure 9-20. The TX power is 106 dBμV/m at 1 m, equivalent to 0 dBm EIRP. Interestingly, this particular device produced some spurious emissions at about 50 MHz when operating with its own battery. The highest emissions measured outside the 2.4 GHz band were about 70 dBμV/m, which equates to about 0.2 μW EIRP.

Path Loss (PL) Values   For communications using the 2.4 GHz ISM band, two classes of PL become important. The first is the loss between a passenger's PED and the various avionics antennas mounted on the aircraft exterior. The second is the loss between PEDs communicating with each other in the aircraft cabin. The PL between the PED and avoinics is determined by transmitting a 2.4 GHz signal with either +10 or +30 dBm of output power within the cabin and measuring the power level at the affected avionics receiver. After accounting for cable losses, the PL values are summarized in Table 9-2 for a B747 and in Table 9-3 for a Gulfstream V [Sch00]. It is evident that PL values fell consistently between about 50 and 65 dB.

Figure 9-20
Field strength values at 1 m produced by the Bluetooth device used for the Intel avionics EMI study [Sch00]

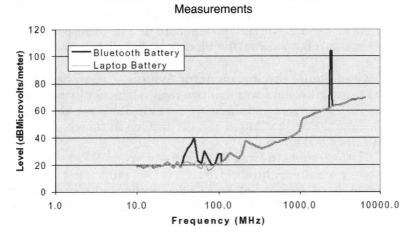

Table 9-2

PL values at
2.4 GHz
between cabin
and avionics
antennas,
B747

Victim antenna	Path loss (dB)
VOR 1	53
VOR 2	53
ATCRBS (upper)	53
ATCRBS (lower)	52

Table 9-3

PL values at
2.4 GHz
between cabin
and avionics
antennas, gulf-
stream V

Victim antenna	Path loss (dB)
VHF comm. (lower)	66
VHF comm. (upper)	66
DME (front)	65
DME (rear)	60
TCAS (lower)	65
TCAS (upper)	66

Within the cabin, the PL values were measured with several different antenna orientations between the two communicating devices. The antenna could be aligned with the fuselage (fus), with the wings (wing) or vertically (vert) at either device, and several combinations were tested to check for cross-polarization differences in PL. These results, measured in a B727 cabin, are shown in Figure 9-21. The values are quite consistent, regardless of how the antennas were oriented. The PL averaged about 63 dB at 7 m, corresponding to a PL exponent of about 2.7. Beyond 7 m, the PL exponent is slightly lower.

Desired Signal Environment for Avionics   ICAO has developed guaranteed field strength values within a coverage volume for avionics systems that operate with a receiver, and these are listed in Table 9-4 for the United States and Canada. Coverage volumes usually include distances in *nautical miles* (NM), equal to 6,000 ft/NM.

The field strength levels in Table 9-4 are typically 20 to 40 dB greater than the receiver sensitivity threshold. Furthermore, within most of the coverage area, the actual signal strength will be considerably higher because the aircraft is closer to the signal source than the maximum operating range. On the other hand, antennas on the aircraft could experience shadowing from the fuselage, wings, tail, and

Figure 9-21

Average PL values in a B727 cabin as a function of distance and antenna polarization between communicating devices [Sch00]

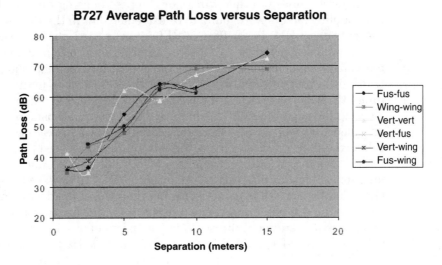

B727 Average Path Loss versus Separation

Legend:
- Fus-fus
- Wing-wing
- Vert-vert
- Vert-fus
- Vert-wing
- Fus-wing

Table 9-4

ICAO-guaranteed field strength values for avionics systems (U.S. and Canada) [Sch00]

System	Guaranteed field strength (dBµV/m)	Coverage volume
VOR	+39	Maximum range 160 NM (aircraft altitude dependent)
DME	+56	Maximum range 160 NM (aircraft altitude dependent)
LOC	+32	±35° to 10 NM; 10° to 18 NM
GLS	+46	±8° to 10 NM
VHF comm	+37	Maximum range 150 NM (aircraft altitude dependent)
GPS	+33	Global
ATCRBS	+65	N/A

engines, especially for antennas mounted on the upper fuselage. The tests allowed for up to 10 dB of shadowing when calculating SIR values from interfering transmissions.

Interference Analysis Results Interference from EMI sources can enter the avionics equipment through its antenna, the connecting cables (either antenna transmission lines or other wires), or the case housing the receiver. The latter two coupling factors were found to be extremely low in the Intel study, so we will concentrate on EMI in the form of RFI entering through the receiver's antenna.

A worst-case analysis for determining interference from 2.4 GHz ISM transmissions to a particular avionics system is done by assuming that the interfering transmitter produces spurious signals at the FCC Part 15 limit, finding the minimum PL from the interfering transmitter to the affected antenna, and then calculating the maximum EMI. Cumulative EMI was not considered because it was assumed that worst-case interference from the worst-case location within the cabin would predominate. Next, the minimum guaranteed field-strength levels for the desired avionics signals given in Table 9-4 are adjusted for aircraft antenna shadowing of 10 dB (upper fuselage) or 3 dB (lower fuselage), except for GPS. Finally, the SIR is found by taking the ratio of the desired and interfering signals. The safety margin is the difference (in dB) between the calculated SIR and the SIR operational threshold for the particular avionics system being considered. These safety margins, in most cases rounded to the nearest dB, are given in Table 9-5 for the aircraft tested. (The Gulfstream V is not listed because the only system tested on this aircraft was the GPS, which showed a safety margin of 17 dB.)

It's obvious from Table 9-5 that 2.4 GHz ISM transmissions will not affect tested DME, ATCRBS, and TCAS signals, because the safety margin is at least 20 dB for all of these. All three *very high frequency* (VHF) communications radios exhibit a safety margin greater than 10 dB, so operation here is not affected either. For VOR, LOC, and GLS, the typical average safety margin for

Table 9-5

Worst-case SIR safety margin (dB) for avionics affected by 2.4 GHz ISM transmissions [Sch00]

System	B757	B737	L1011	MD80	A320	B747
VOR	1	27	22	17	43	36
DME	N/A	35	N/A	28	27	36
LOC	3	24	12	N/A	0.05	16
GLS	2	14	9	8	29	−0.6
VHF comm. 1	20	23	36	27	22	11
VHF comm. 2	15	36	N/A	42	39	40
VHF comm. 3	30	30	39	32	33	49
GPS	5	N/A	N/A	N/A	N/A	N/A
ATCRBS	36	42	N/A	N/A	N/A	42
TCAS	44	N/A	N/A	N/A	30	N/A

all aircraft tested is 11 to 43 dB, but for specific aircraft the safety margin is closer to 0 dB. However, for EMI to actually affect performance, the 2.4 GHz ISM transmitter must radiate spurious signals at the FCC limit within the relatively narrow operational bandwidths of these systems, which is extremely unlikely. The test was inconclusive for GPS, but we covered that topic in detail earlier in this chapter.

The authors of the Intel study concluded that operating 2.4 GHz ISM wireless devices within the passenger cabin of the tested aircraft had no significant effect on the reliability of either communication or navigation avionic equipment. The study therefore recommends that FAA rules be amended to allow the operation of Bluetooth-equipped devices during noncritical phases of flight. Europe's Airbus Industries has certified the A340-600 aircraft for in-flight Wi-Fi operation, and Boeing has conducted its own Wi-Fi tests with encouraging results.

Interference to Avionics from UWB Transmissions

We've already discussed the effect of UWB on GPS, but it would be interesting to discover whether UWB transmissions will interfere with other avionics systems. An analytical study performed by the *National Telecommunications and Information Administration* (NTIA) [Bru01] examined coexistence between UWB and selected federal systems, most of which were associated with aviation. Measurements were used to corroborate the analysis. An empirical study by the *National Aeronautics and Space Administration* (NASA) [Ely02a] was directed at interference caused by UWB to aircraft radios.

Analytical Approach   The NTIA study was accomplished under an early assumption that the existing FCC Part 15 spurious emission levels would be sufficient to prevent unwarranted UWB interference to other wireless services operating with narrower bandwidths. The FCC adopted rules that allowed indoor and handheld UWB transmitters to operate between 3.1 and 10.6 GHz, with emissions in this region conforming to existing Part 15 spurious limits of −41.3 dBm/MHz, along with more restrictive limits outside the band of operation. As such, results in the NTIA study are most applicable to avionics operating in the 3.1 to 10.6 GHz region, with more pessimistic results otherwise. Furthermore, the study considered that peak power levels may be up to 20 dB above the average spurious limit above 1 GHz, as allowed by the FCC in Part 15.35(b). However, this provision does not apply to UWB transmissions.

The NTIA study concentrated on the possibility that ground-based UWB systems would interfere with non-UWB operations, although some of the separation criteria used could be applied to in-cabin use. The avionics systems included in the test were DME, ATCRBS, radar altimeters at 4.2 to 4.4 GHz, and *microwave landing systems* (MLS) between 5.030 and 5.091 GHz. Of these, the last two fall within the UWB intentional emission band, and the others are subject to the low UWB spurious emissions below 3.1 GHz. The general results are given here:

- **DME** Safety margins for the DME ground-based interrogator are exceeded for EMI of −41.3 dBm/MHz closer than 90 m. However, because FCC UWB spurious limits in the 960 to 1,200 MHz operating bands of DME are actually −75.3 dBm/MHz, UWB interference becomes significant only for separation distances of less than 1 m. The safety margins for the airborne DME transponder are exceeded for EMI of −41.3 dBm/MHz closer than 300 m. With the actual FCC limits, UWB interference is significant for separation distances of less than 10 m.

- **ATCRBS** The safety margins for the ATCRBS ground-based interrogator receiver are exceeded for EMI of −41.3 dBm/MHz closer than 270 m. However, because FCC UWB spurious limits in the 1,030 and 1,090 MHz operating bands of ATCRBS are actually −75.3 dBm/MHz, UWB interference is significant for separation distances of less than 10 m. UWB EMI was not a factor for the airborne ATCRBS transponder.

- **Radar altimeter** Ground-based UWB transmissions at −41.3 dBm/MHz in the 4.2 to 4.4 GHz band are too weak to interfere with a proper operation of the radar altimeter, with typical safety margins of more than 60 dB. This applies to CW or pulsed radar altimeters.

- **MLS** The safety margins for airborne MLS operating in the 5.030 to 5.091 GHz band are exceeded for EMI of −41.3 dBm/MHz closer than about 160 m horizontal separation, measured from the centerline of the aircraft at 30 m altitude. This would suggest placing a restriction on the operation of UWB devices in this area, which is typically within the first few thousand feet of the approach end of a runway.

The study concludes that some UWB devices may require minimum separation distances from avionics to prevent detrimental EMI. For measuring the effect of such EMI, the RMS detector is better than the average logarithmic detector. If the affected receiver uses a high-gain antenna, EMI from a UWB emitter may be significantly greater. The study also discovered that UWB EMI could affect the performance of some satellite ground station receivers and some airport radar systems. Finally, the interference caused by multiple UWB transmitters may be more severe than from a single transmitter at the same equivalent power level.

Empirical Approach The NASA empirical study examined the limited functional testing of UWB EMI on six B737 aircraft at window and door exits for VOR, LOC, GLS, VHF-1 comm, TCAS, and GPS systems. The testers also examined four B747 aircraft for LOC, GLS, VHF-1 comm, TCAS, GPS, and Satcom radio systems. The B747 shares antenna and RF pathway with the B737 for VOR and LOC, so these tests weren't repeated on the B747.

The UWB signal pulse was 6.7 *volts peak-to-peak* (Vp-p) with a rise time of 259 *picoseconds* (ps), a fall time of 116 ps, and a width of 239 ps. If we use a Gaussian pulse approximation, its center frequency is about 4.2 GHz and the −10 dB bandwidth spans about 8.4 GHz. Additional filtering would be needed for this pulse to conform to FCC UWB rules given nominal TX power levels. Many of the tests performed by NASA were severe, in which the UWB source was placed within a few meters' *line of sight* (LOS) from the victim receiver. Unless noted otherwise, the UWB source was pulsed at 10 MHz and fed to a 60–140 MHz adjustable dipole antenna tuned to the carrier frequency used by the tested system.

A summary of the B737 test results is given here:

- **VHF voice communications**   The UWB antenna was placed 1 m LOS from the aircraft's VHF-1 upper antenna. The UWB TX power was −23 dBm at 119.9 MHz and less than −80 dBm at 118.0 MHz. The disparity was due to the UWB pulse repetition rate placing a harmonic at 120 MHz. No discernable affect on the audio quality was noted.

- **VOR**   The local VOR signal was acquired by the test aircraft. The UWB antenna was placed 1 m LOS from the aircraft VOR/LOC antenna in its tail. No UWB effects were observed.

- **LOC**   The localizer could not be acquired from the test location.

- **GLS**   The glideslope was marginally acquired at the test location. The UWB antenna was placed 1 m from the aircraft GLS antenna in the nose. No UWB effects were observed.

- **ATCRBS and TCAS**   The local ATCRBS interrogator and TCAS transponders on the aircraft in the local airspace were acquired by the test aircraft. The "ATC Fail" indicator illuminated and targets disappeared from the TCAS display when the UWB transmitter was activated in the following locations: 1.5 m LOS from the TCAS antenna, inside the aircraft from all first-class window seats on the port side, and at the third window location in coach on the port side.

For testing on the B747, the UWB signal was given *on-off keying* (OOK) and dithering capabilities, and an ILS test set was available to provide calibrated signals to the aircraft's ILS receivers. The UWB signal in dithered form caused the LOC and GLS to fail under the test conditions, but ILS operated properly with continuous UWB pulses or with OOK modulation. ATCRBS and TCAS on the B747 could also be made to fail under UWB EMI. However, no adverse effect was noted on the aircraft's GPS system, even when the pulse rate was selected to place a harmonic at the 1,575.42 MHz L1 carrier, with the UWB antenna placed 3 m from the aircraft GPS antenna. Tests were also performed on the B737 with the new OOK and dithering capabilities with similar results.

In conclusion, under severe test criteria UWB EMI can affect some aircraft avionics, particularly the ATCRBS/TCAS and ILS systems. The former system is used during all phases of flight from takeoff to landing, and the latter is used during the landing phase. The study recommends further testing, both in the laboratory and in the field, followed by the development of appropriate regulatory policies.

Integrating Bluetooth into a Cell Phone

One of the largest markets for Bluetooth applications is in the portable cellular telephone. The most common use is a Bluetooth link between the phone and a wireless headset, so the phone can remain on the belt or in the briefcase while in use. Also, automobile manufacturers have begun offering Bluetooth links built into the car to enable hands-free use of the phone as required in many parts of the world. In this section, we will study the requirements for integrating the RF sections of a Bluetooth module and a *global system for mobile communications* (GSM) cellular telephone for interference-free operation [Bro01], [Bug02].

The challenge of placing a Bluetooth node into a GSM cellular telephone at first appears to be extremely difficult due to the requirement that one must be transmitting while the other is receiving for normal operation to take place. Fortunately, the two systems operate in different bands of the RF spectrum, with Bluetooth at 2,402 to 2,480 MHz while traditional GSM is at 890 to 915 MHz (TX) and at 935 to 960 MHz (RX). Other cellular systems also operate below the 2.4 GHz ISM band, with the closest being at 2.1 GHz for some *third-generation* (3G) cellular applications.

Out-of-Band Spurious Transmissions

Generally, out-of-band spurious signals from a non-UWB transmitter can be either CW-like or noise-like. Signals that are used within the transmitter's internal circuitry can leak to the antenna either through the primary signal path or through component or trace cross-coupling. These signals usually appear as CW-like spurs on a spectrum analyzer and can be devastating should they appear within the passband of a collocated receiver. Fortunately, the designer has significant control over these responses and can ensure that the design carries no undesired spurs in the affected band.

Noise-like spurious signals originate from thermal effects within the early stages of the TX circuitry that are amplified and transmitted along with the desired modulated carrier. These products have the potential to raise the RX noise floor of the collocated receiver and consequently to reduce its sensitivity accordingly. In the next two subsections, we will show examples of how to determine the maximum broadband noise from a transmitter that can be tolerated at the other receiver without significantly affecting its performance. Calculations will be normalized to dBm/Hz to compensate for bandwidth differences.

Protecting the GSM Receiver

Figure 9-22 depicts the signal levels required for adequate operation of the GSM receiver while the collocated Bluetooth transmitter is active. The GSM channel is 160 kHz (52 dB-Hz) wide, and the receiver must operate with a minimum sensitivity of −102 dBm. Most receivers can successfully demodulate a signal with an RSSI as low as −108 dBm, which equates to −160 dBm/Hz. The required *carrier-to-interference ratio* (C/I) for GSM operation is +5 dB [Bug02], and an additional 10 dB of safety margin is included to ensure that the resulting Bluetooth TX noise raises the GSM RX noise floor by less than 1 dB. Finally, if 15 dB of antenna isolation exists, the Bluetooth transmitter can create as much as −160 dBm/Hz of noise in the GSM band.

Protecting the Bluetooth Receiver

The situation for the Bluetooth receiver under GSM TX noise is shown in Figure 9-23. The Bluetooth hop channel is 1 MHz (60 dB-Hz) wide, and the receiver must operate with a minimum sensitivity of −70 dBm. Most commercial receivers work with an RSSI of −85 dBm, or −145 dB/Hz. The required C/I is +11 dB, and an additional 10 dB of safety margin maintains the Bluetooth receiver noise floor within 1 dB of interference-free operation. The same 15 dB of antenna isolation enables the GSM transmitter to generate as much as −151 dBm/Hz of noise in the 2.4 GHz ISM band. Although the allowable noise density is higher for the GSM transmitter than for the Bluetooth transmitter, GSM TX power can be up to +33 dBm, presenting a greater design challenge.

Figure 9-22
Method of determining the maximum allowable Bluetooth TX broadband noise into a GSM receiver

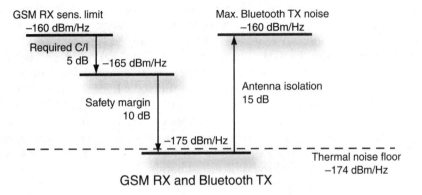

GSM RX sens. limit
−160 dBm/Hz

Max. Bluetooth TX noise
−160 dBm/Hz

Required C/I
5 dB   −165 dBm/Hz

Antenna isolation
15 dB

Safety margin
10 dB

−175 dBm/Hz

Thermal noise floor
−174 dBm/Hz

GSM RX and Bluetooth TX

Figure 9-23
Method of
determining
the maximum
allowable GSM
TX broadband
noise into a
Bluetooth
receiver

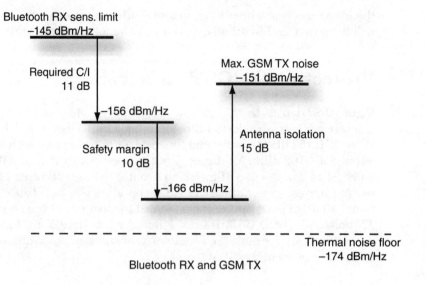

Bluetooth RX sens. limit
−145 dBm/Hz

Required C/I
11 dB

−156 dBm/Hz

Safety margin
10 dB

−166 dBm/Hz

Max. GSM TX noise
−151 dBm/Hz

Antenna isolation
15 dB

Thermal noise floor
−174 dBm/Hz

Bluetooth RX and GSM TX

Cordless Telephones

Cordless telephones in the United States are generally of proprietary design and are required to meet the FCC Part 15 limitations. These phones use a wide variety of carrier frequencies from about 1.8 MHz up to and including the 2.4 and 5.x GHz ISM bands. Both digital and analog narrowband modulation are used, along with digital DSSS or *frequency hopping spread spectrum* (FHSS) if the phone operates in the ISM bands. Sometimes one band is used for the uplink (handset to base) and the other is used for the downlink (base to handset). As a result, cordless telephone interference can take many different forms, depending on the phone's make and model.

Early digital spread spectrum cordless phones used DSSS and were marketed for their improved range, voice clarity, and privacy. However, manufacturers have shifted almost entirely to FHSS for their spread spectrum cordless phones. Interference from 2.4 GHz phones to WLAN and WPAN depends primarily on the manufacturer and model of the cordless telephone. Some phones will disrupt a 802.11b network throughout a large area, whereas others have no effect even when they are less than 1 m from a Wi-Fi *access point* (AP) or STA antenna.

Generally, cordless phones that operate in the 900 MHz or 5.x GHz ISM bands have no effect on 2.4 GHz wireless operations. Phones that operate under the FCC Part 15.249 "low power" rules have little effect on WLAN or WPAN operations within any of the unlicensed bands.

Microwave Ovens

Several hundred million microwave ovens exist in the world, and many of them are physically located in the same areas where WLAN and WPAN devices function. These ovens transmit several hundred watts at a frequency of 2.45 GHz (near the upper end of Wi-Fi channel 6), and this energy is ideally confined to the interior of the oven itself. In reality, some of the energy escapes and thus the oven can be classified as an unintentional radiator of RF. Therefore, it's important to discover what effect these ovens will have on communications within the 2.4 GHz ISM band.

Regulation on microwave ovens in the United States falls under FCC Part 18 rules, which place limits on radiated field strength *outside* the ISM bands. The limit is $25\sqrt{P/500}$ μV/m at 300 m, where P is the oven power in W. According to Equation 9.5, a 1,000 W oven can legally radiate power outside the ISM bands up to 3.8 μW EIRP. In any case, the field strength cannot exceed 10 μV/m at 1,600 m, equivalent to about 8.5 μW EIRP. However, Part 18 places no limits on external radiation *within* the ISM bands. A real possibility exists, then, for the microwave oven to significantly disrupt ISM communications taking place in the 2.4 GHz band.

Quantifying Microwave Oven Emissions   One of the most detailed studies on microwave oven emissions was performed by NTIA in 1994 [Gaw94]. This study was mainly confined to new consumer ovens from different manufacturers. A calibrated horn was placed 3 m from the front face of the oven, and measurements were taken in the time and frequency domains. Figure 9-24 shows the

Figure 9-24
Typical microwave oven emissions as a function of elevation and azimuth [Gaw94]

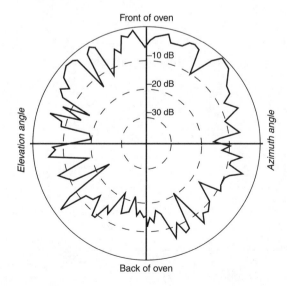

radiation pattern from a typical oven relative to the highest emission level. The right side of the plot is the radiation as the azimuth angle changes from the front to the back of the oven, and the left side of the plot shows the emissions as the elevation angle changes from front to back. It is apparent that the highest levels of emissions emanate from the front, probably around the door seal.

Figure 9-25 shows the plots of typical external emissions in both time and frequency domains taken from the front of two different ovens. The frequency domain resolution is 3 MHz. The left scale shows the measured field strength at

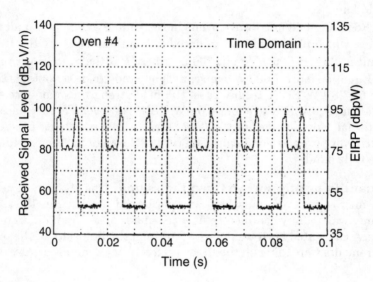

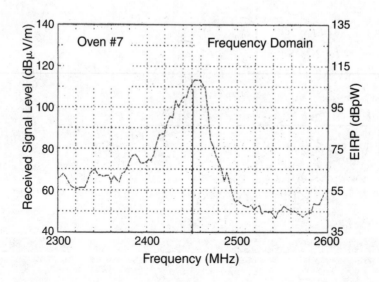

Figure 9-25
Typical microwave oven emissions in the time and frequency domains [Gaw94]

1 m, and the right scale converts these measurements to EIRP in *dB relative to 1 picowatt* (dBpW). (To convert dBpW to dBm, subtract 90 dB.) These measurements were taken using peak detection methods in an effort to capture useful broadband data.

When an oven is operating, its magnetron is energized by a high-voltage power supply that uses a half-wave rectifier for cost savings. This causes the characteristic pulsed RF output with the period tied to the AC line frequency. In the United States, where the line frequency is 60 Hz, the magnetron is on for 7 to 8 ms and off for 8 to 9 ms while the oven is operating at 100 percent power. For lower power settings, the magnetron operates normally for a few seconds and then is turned off for a few seconds, with the on/off ratio depending on the power level selected.

The −20 dB bandwidth for oven #7 is about 50 MHz, covering slightly more than half the ISM band. Some ovens use magnetrons that rapidly change their nominal operating frequency, and these often produce −20 bandwidths of about 80 MHz. Other ovens have composite bandwidths closer to 20 MHz [Buf00]. Loading the magnetron from contents in the oven plays a relatively small role in the overall level of RF escaping from the oven, but some variation exists in the shape of the PSD. Antenna polarization has little effect on the measured field strength.

The transient spikes, called rabbit ears, which are evident in the time domain plot of Figure 9-25, are common to pulsed magnetron operations. Each spike is 500 μs to 2 ms in width and carries most of the broadband emission power. The relatively stable operating regions during the time between the rabbit ears produce essentially CW emissions, with a frequency that may sweep over a few MHz between 2,450 and 2,460 MHz [Kam97]. Negligible emissions occur between magnetron pulses.

Research at the University of Michigan investigated a method to azimuthally vary the axial magnetic field in a magnetron by attaching exterior permanent magnets [Nec03]. This resulted in a significant reduction of on-off transients and a narrow CW-like PSD while the magnetron is energized. The physical mechanism behind the transient noise reduction is not completely understood, so additional research is expected to occur in this area.

The maximum, average, and minimum envelopes of external PSD from the 13 different ovens tested by NTIA are plotted in Figure 9-26. The minimum envelope slightly exceeds +85 dBpW, or −5 dBm, corresponding to 0.3 mW EIRP in a 3 MHz bandwidth. For the same 3 MHz resolution, the mean envelope is about +120 dBpW (+30 dBm) over a 20 MHz-wide region centered at 2,455 MHz. This area falls predominately at the lower end of Wi-Fi channel 11. These tests were performed on new ovens, so as the door seal ages or becomes soiled, EIRP levels could easily increase further.

Unlike consumer ovens, commercial microwave ovens often contain two magnetrons to speed the heating process, and these are each operated from different AC-line half-cycles. These ovens can cause greater interference to communications because no "off" time exists while the magnetrons are energized [Kam97].

Figure 9-26
Maximum,
average, and
minimum
external PSD
plots for 13
commercial
ovens [Gaw94]

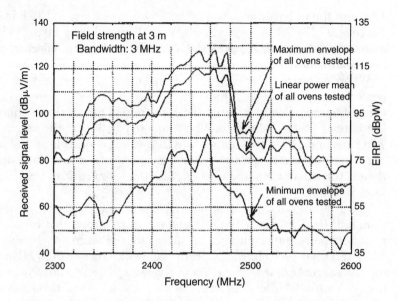

Figure 9-26
Maximum, average, and minimum external PSD plots for 13 commercial ovens [Gaw94]

Commercial ovens are more powerful than consumer ovens and their RF emissions cover a wider bandwidth, with swept CW-like products occurring after the magnetron turn-on transient. Furthermore, they tend to be in operation for longer cooking times and over longer hours. Commercial ovens are usually better shielded with heavy stainless steel enclosures, and high cleanliness standards result in lower leakage around door seals.

Wireless Network Coexistence with Microwave Ovens   Disruption to wireless networks from consumer microwave ovens has been lower than expected for two main reasons. First, the oven's magnetron is essentially a CW device, so between on-off transients the transmitted bandwidth may be as low as 2 MHz [Buf00]. Second, the 8 to 9 ms off period during each power cycle is long enough that devices can still maintain reasonable average throughput. For example, a Bluetooth piconet can exchange 6 pairs of single-slot packets, or 2 pairs of 5-slot packets and their single-slot *acknowledgment* (ACK) packets, in 7.5 ms. Even if the oven completely jams communications when its magnetron is energized, aggregate ACL throughput will drop only about 50 percent. Actual disruption will probably be less because Bluetooth will often use unaffected hop frequencies during magnetron operation. SCO performance degradation, however, may be noticeable due to the potential for burst error events, especially during magnetron on-off transients. For Wi-Fi operating in the 2.4 GHz band, the longest possible *PLCP protocol data unit* (PPDU) sent at 5.5 Mb/s or faster, along with the associated ACK, can be accommodated between magnetron pulses. Data rates of 1 or 2 Mb/s can be supported using shorter PPDU lengths or through packet fragmentation.

An empirical study examined the throughput reduction from microwave oven interference on a Bluetooth point-to-point link [Seg01]. With 0.1 m master-slave separation, ACL throughput dropped 45 percent with the oven 0.1 m from one of the nodes. When the oven was moved 5 m or farther away, throughput was unaffected. For master-slave separations of 7.5 m, the oven at 0.1 m from one of the nodes caused a 65 percent throughput reduction, but moving the oven 10 m away removed its influence on network performance. Placing plastic or steel partitions between the oven and piconet nodes improved performance between 10 and 20 percent. We can conclude that wireless networks operating in different rooms than the ovens will in most cases experience negligible oven-produced interference.

To summarize, interference from microwave ovens can be reduced by the following methods:

- Deploying wireless networks away from areas where ovens are located
- Using adaptive frequency hopping (Bluetooth)
- Operating in the low end of the 2.4 GHz band (Wi-Fi, WiMedia, ZigBee)
- Improving oven shielding
- Redesigning the oven magnetron for reduced pulse operation transients

Microwave Lighting

A lighting system called the *sulfur lamp* has been developed that provides high efficiency and a spectrum closely matching that of sunlight. The bulb consists of a spherical quartz envelope containing a few milligrams of sulfur and argon as an inert gas. A magnetron operating at 2.45 GHz irradiates the argon, which in turn heats the sulfur into a gaseous state, forming diatomic sulfur molecules. These emit a broad spectrum of light as they drop to a lower energy level through a process called *molecular emission*. The visible spectrum of the sulfur lamp is almost identical to sunlight but without the undesirable infrared or ultraviolet radiation. Other types of light built around *atomic emission*, such as mercury and neon lamps, produce an artificial-looking light that emphasizes one predominant color. Sulfur lamps also maintain efficiency and lumen level for long periods of time. The magnetron is the only part that requires replacement every 15,000 to 20,000 hours.

If the lamp isn't properly shielded, the potential exists for these devices to interfere with communications in the 2.4 GHz ISM band. Operating under FCC Part 18 rules, the lamps must conform to maximum emission levels outside the ISM bands, but radiation within the ISM bands is not limited by these rules. The microwave-powered sulfur lamps tend to be high-power devices (1 kW) suitable for stadium lighting. Lamps have been developed by Lawrence Berkeley Laboratory to operate with lower power (100 W) for indoor lighting [Rub95]. These lamps are already in place at the U.S. Department of Energy and the Smithsonian Institution. The *Institute of Electrical and Electronics Engineers* (IEEE) is working with the FCC to insure that microwave lighting and communication systems can coexist in the 2.4 GHz band [Zyr99].

Summary

GPS receivers operate with extremely low RSSI levels. Under the FCC Part 15 limits for non-UWB devices, the potential to interfere with GPS receivers is high. In reality, these devices emit very little energy within the relatively narrow bandwidth occupied by the GPS satellite signals. Interference to a GPS receiver is manifested by reduced ranging accuracy, a loss of lock, and an inability to reacquire the signal. UWB interference to a GPS receiver can be pulse-like, noise-like, or CW-like. The latter is the most serious. Time dithering can give CW-like interference more noise-like characteristics. The FCC has placed much lower spurious emission restrictions on UWB transmitters such that one of these needs to be closer than about 1 m from the GPS receiver to affect its operation. These restrictions can be met with UWB pulse shaping and/or filtering.

Interference to avionics is significant because it often involves safety of flight. The RTCA has determined the maximum interference levels within the various avionics receiving bands, and these often are below those required by the FCC for Part 15 devices. Empirical tests show that Bluetooth and Wi-Fi transmissions in the 2.4 GHz band produce little interference to avionics. Some aircraft have been certified for Wi-Fi operations while in flight. UWB transmissions can affect the operation of avionics, but in most cases such interference is from a source placed physically close to the affected system's antenna.

Cordless telephones in the United States operate using proprietary technology from the different manufacturers. A particular phone's capability to interfere with WLAN and WPAN operations is a function of its operating frequency, TX power, and modulation method. Most phones with higher TX power use FHSS; some of these operating on 2.4 GHz interfere severely with Wi-Fi, whereas others have little effect. When integrating Bluetooth into a cellular telephone, separate bands of operation coupled with good engineering practice allow both to operate without a significant loss of performance.

RF leakage in the 2.4 GHz ISM band from microwave ovens is highly variable, with some ovens producing over a watt while others transmit less than 1 mW. Wireless communications can still take place near consumer microwave ovens because the magnetron is powered by a half-wave rectifier, and interference is CW-like between on-off transients. Interference from most ovens becomes tolerable for wireless nodes operating a few meters away or in other rooms. Finally, the sulfur lamp operates with high-power microwave radiation produced by a magnetron, and more research is necessary to determine how these devices might affect wireless communications in the 2.4 GHz band.

References

[And01] Anderson, D., et al., "Assessment of Compatibility between Ultrawideband (UWB) Systems and Global Positioning System (GPS) Receivers," NTIA Special Publication 01-45, February 2001.

[Bro01] Brown, S., et al., "Integrating Bluetooth in the GSM Cell Phone Infrastructure," *RF Design*, September 2001.

[Bru01] Brunson, L., et al., "Assessment of Compatibility between Ultrawideband Devices and Selected Federal Systems," NTIA Special Publication 01-43, January 2001.

[Buf00] Buffler, C., and Risman, P., "Compatibility Issues Between Bluetooth and High Power Systems in the ISM Bands," *Microwave Journal*, July 2000.

[Bug02] Buggs, P. and Murray, B., "Piecing Together the Bluetooth/Mobile Phone Puzzle," *EE Times*, March 28, 2002.

[But02] Butsch, F., "Radiofrequency Interference and GPS," *GPS World*, October 2002.

[Ely02a] Ely, J., et al., "Ultrawideband Electromagnetic Interference to Aircraft Radios," NASA TM-2002-211949, October 2002.

[Ely02b] Ely, J., and Nguyen, T., "EMI Standards for Wireless Voice and Data On Board Aircraft," National Aeronautics and Space Administration white paper, 2002.

[FAA93] FAA Reference Advisory Circular AC-91.21-1-1A, "Use of Portable Electronic Devices Aboard Aircraft," August 20, 1993.

[Gaw94] Gawthrop, P., et al., "Radio Spectrum Measurements of Individual Microwave Ovens," NTIA Report 94-303-1, March 1994.

[Gre02] Greczyn, M., "FCC Staff Report Underscores Conservative Assumptions for UWB," *Communications Daily*, October 23, 2002.

[Ham02] Hämäläinen, M., et al., "On the UWB System Coexistence with GSM900, UMTS/WCDMA, and GPS," *IEEE Journal on Selected Areas in Communications*, December 2002.

[Hor97] Horne, J., and Vasudevan, S., "Modeling and Mitigation of Interference in the 2.4 GHz ISM Band," *Applied Microwave & Wireless*, March/April 1997.

[Jah03] Jahn, A., et al., "Evolution of Aeronautical Communications for Personal and Multimedia Services," *IEEE Communications Magazine*, July 2003.

[Kam97] Kamerman, A., and Erkoçevic, N., "Microwave Oven Interference on Wireless LANs Operating in the 2.4 GHz ISM Band," Eighth IEEE International Symposium on Personal, Indoor, and Mobile Radio Communications, Helsinki, Finland, September 1997.

[Luo00] Luo, M., et al., "Potential Interference to GPS from UWB Transmitters: Phase I," Stanford University white paper, May 1, 2000.

[Luo01a] Luo, M., et al., "Potential Interference to GPS from UWB Transmitters: Phase II Test Results," Stanford University white paper, March 16, 2001.

[Luo01b] Luo, M., et al., "Testing and Research on Interference to GPS from UWB Transmitters," Stanford University white paper, 2001.

[Nec03] Neculaes, V., et al., "Low-Noise Microwave Magnetrons by Azimuthally Varying Axial Magnetic Field," *Applied Physics Letters*, September 8, 2003.

[Pro02] Prophet, G., "Zero In—GPS Options Expand with Applications," *EDN*, September 19, 2002.

[Riz99] Rizos, C., "Introduction to GPS," University of New South Wales, 1999.

[Rub95] Rubenstein, F., "Sulfur Lamps—The Next Generation of Efficient Light?" *Center for Building Science Newsletter*, Spring 1995.

[Sch00] Schiffer, J., and Waltho, A., "Safety Evaluation of Bluetooth Class ISM Band Transmissions on Board Commercial Aircraft," Revision 2, Intel white paper, December 28, 2000.

[Sch01] Schnaufer, B., "NETEX Program: Networking in Extreme Environments," Rockwell Collins presentation, 2002.

[Seg01] Sega, K., "Measuring Interference Among Bluetooth, Microwaves, and IEEE 802.11b," *Nikkei Electronics Asia*, October 2001.

[Sta02] Stansell, T., "UWB Coexistence with GPS," Ultra-Wideband Technology Workshop, October 4, 2002.

[Zyr99] Zyren, J., et al., "Letter to Secretary of FCC on Alternative Emission Limit for Microwave Lighting Devices," IEEE Document IEEE 802.1199/083, March 10, 1999.

Acronyms and Abbreviations

3G	third generation
ACK	acknowledgment
ACL	asynchronous connectionless
A/D	analog-to-digital
AF	attenuation factor
AFH	adaptive frequency hopping
AGC	automatic gain control
AM	amplitude modulation
AMC	artificial magnetic conductor
AP	access point
ARQ	automatic repeat request
ASCII	American Standard Code for Information Interchange
ASK	amplitude shift keying
ATCRBS	air traffic control radar beacon system
ATIM	ad hoc traffic indication message
ATM	automated teller machine
AWGN	additive white Gaussian noise
AWMA	alternating wireless medium access
BB	baseband
BCH	Bose-Chadhuri-Hocquenghem
BD_ADDR	Bluetooth device address
BER	bit error rate (same as PBE)
BFSK	binary frequency shift keying
BPF	bandpass filter
BPSK	binary phase shift keying
BS	base station
b/s	bits per second (also bps)
BSIG	Bluetooth Special Interest Group
BSS	basic service set
BW	bandwidth
CA	collision avoidance
CAD	computer-aided design
CAP	contention access period
CCA	clear channel assessment
CCI	co-channel interference
CCITT	International Telegraph and Telephone Consultative Committee
CCK	complimentary code keying
CDMA	code division multiple access
CFP	contention-free period
C/I	carrier-to-interference ratio (also CIR)
CID	channel identifier
CL	connectionless
CLH	cluster head
CMOS	complementary metal oxide semiconductor
CO	connection-oriented

CP	circularly polarized
CP	cyclic prefix
CPU	central processing unit
CRC	cyclic redundancy check
CSI	channel state information
CSMA	carrier sense multiple access
CSMA/CA	CSMA with collision avoidance
CSMA/CD	CSMA with collision detection
CTA	channel time allocation
CTAP	channel time allocation period
CTS	clear to send
CW	contention window
CW	continuous wave
dB	decibels
dBd	decibels relative to a dipole antenna
dBi	decibels relative to an isotropic source
dBm	decibels relative to 1 milliwatt
DBPSK	differential binary phase shift keying
DCF	distributed coordination function
DEV	device
DFIR	diffuse infrared
DFS	dynamic frequency selection
DH	data high speed
DIFS	DCF interframe space
DM	data medium speed
DME	distance-measuring equipment
DPSK	differential phase shift keying
DQPSK	differential QPSK
DS/CDMA	direct sequence code division multiple access
DSP	digital signal processor
DSSS	direct sequence spread spectrum
DS-UWB	direct-sequence ultra-wideband
E-911	enhanced 911
EBSS	extended basic service set
ECC	error correction code
ED	energy detect
EIA	Electronic Industries Association
EIFS	extended interframe space
EIRP	equivalent isotropic radiated power
EMI	electromagnetic interference
EMP	electromagnetic pulse
ERP	extended-rate PHY
eSCO	extended synchronous connection-oriented
EU	European Union
FAA	Federal Aviation Administration

FAF	floor attenuation factor
FCC	Federal Communications Commission
FCS	frame check sequence
FDD	frequency division duplexing
FDMA	frequency division multiple access
FEC	forward error correction
FER	frame error rate
FFD	full function device
FFT	fast Fourier transform
FH/CDMA	frequency hop code division multiple access
FHSS	frequency hopping spread spectrum
FIR	finite impulse response
FM	frequency modulation
FSK	frequency shift keying
FTP	file transfer protocol
FUT	frequency usage table
GaAs	gallium-arsenide
GFSK	Gaussian-filtered frequency shift keying
GHz	gigahertz
GLS	glideslope
GPS	global positioning system
GSM	global system for mobile communications
GTS	guaranteed time slot
HCI	host controller interface
HCS	header check sequence
HDTV	high-definition television
HEC	header error check
HID	human interface device
HR	high-rate
HTTP	hypertext transfer protocol
HV	high-quality voice
IAPP	interaccess point protocol
IBSS	independent basic service set (802.11)
IC	integrated circuit
ICAO	International Civil Aviation Organization
IEE	Institution of Electrical Engineers (UK)
IEEE	Institute of Electrical and Electronics Engineers
IF	intermediate frequency
IFFT	inverse fast Fourier transform
IIR	infinite impulse response
ILS	instrument landing system
Im	imaginary
IMD	intermodulation distortion
I/O	input/output
IP	intellectual property

IP	Internet protocol
IR	infrared
ISI	intersymbol interference
ISM	industrial, scientific, and medical
ISO	International Standards Organization
ISP	Internet service provider
ITU	International Telecommunications Union
ITU-R	ITU Radiocommunications
kb/s	kilobits per second (also kbps)
kHz	kilohertz
L2CAP	logical link control and adaptation protocol
LAN	local area network
LC	link controller
LFSR	linear feedback shift register
LIFS	long interframe space
LLC	logical link control
LM	link manager
LMP	link manager protocol
LNA	low-noise amplifier
LO	local oscillator
LOC	localizer
LOS	line of sight
LP	linearly polarized
LQI	link quality indication
LR	low-rate
LSB	least significant bit
MAC	medium access control layer in the OSI model
MAGRAM	magnetic radar-absorbing material
MAI	multiple access interference
MBOA	Multiband OFDM Alliance
Mb/s	megabits per second (also Mbps)
MCT	multicarrier transmission
MCTA	management channel time allocation
MC-UWB	multicarrier ultra-wideband
MHz	megahertz
MIMO	multiple-input multiple-output
MLS	microwave landing system
MMI	man-machine interface
MPDU	MAC protocol data unit
MRA	mutual recognition agreement
MRC	maximal ratio combining
MS	mobile station
MSB	most significant bit
MSDU	MAC service data unit
MSK	minimum shift keying

Ms/s	megasymbols per second
MSS	mobile satellite service
MTU	maximum transmission unit
mW	milliwatt
NAK	negative acknowledgment
NASA	National Aeronautics and Space Administration
NAV	network allocation vector
NDA	nondisclosure agreement
NIC	network interface card
NLOS	non-line-of-sight
NM	nautical miles
ns	nanosecond
NTIA	National Telecommunications and Information Administration
NZIF	near-zero intermediate frequency
OET	Office of Engineering and Technology
OFDM	orthogonal frequency division multiplexing
OLOS	obstructed line of sight
OOK	on-off keying
OS	operating system
OSI	Open Systems Interconnection
PA	power amplifier
PAN	personal area network
PB	packet boundary
PBCC	packet binary convolutional coding
PBE	probability of bit error (same as BER)
PC	point cocrdinator
PCB	printed circuit board
PCF	point coordination function
PCM	pulse code modulation
PCS	personal communications service
PDA	personal digital assistant
pdf	probability density function
PDU	protocol data unit
PED	portable electronic device
PER	packet error rate
PG	processing gain
PHY	physical layer in the OSI model
PIFA	planar inverted "F" antenna
PIFS	PCF interframe space
PIN	personal identification number
PL	path loss
PLCP	physical layer convergence protocol
PLL	phase-locked loop
PM	phase modulation
PMD	physical medium dependent

PN	pseudonoise
PNC	piconet controller
POS	personal operating space
PPDU	PLCP protocol data unit
ppm	parts per million
PPM	pulse position modulation
PPS	precise positioning service
PRBS	pseudorandom binary sequence
PRF	pulse repetition frequency
PRI	pulse repetition interval
PRN	pseudorandom noise
PSD	power spectral density
PSDU	PLCP service data unit
PSK	phase shift keying
PSR	pseudorange
PTA	packet traffic arbitration
QAM	quadrature amplitude modulation
OBEX	object exchange
QoS	quality of service
QPSK	quadrature phase shift keying
Re	real
RF	radio frequency
RFCOMM	radio frequency communication (Bluetooth serial port)
RFD	reduced function device
RFI	radio frequency interference
RFID	radio frequency identification
RLSL	recursive least-squares lattice
RSSI	received signal strength indication
R/T	receive/transmit
RTCA	Radio Technical Commission for Aeronautics
RTOS	real-time operating system
RTS	request to send
RX	receive
RXD	receive data
SA	selective availability
SAR	specific absorption rate
SAW	surface acoustic wave
SB	stuff bits
SCM	single-carrier modulation
SCO	synchronous connection-oriented
SDM	space-division multiplexing
SDP	service discovery protocol
SER	symbol error rate
SF	spectrum factor
SFD	start of frame delimiter

SIFS	short interframe space
SIMOP	simultaneous operation
SINR	signal-to-interference-and-noise ratio
SIR	signal-to-interference ratio
SLC	square-law combining
S/N	signal-to-noise
SNR	signal-to-noise ratio
SPS	standard positioning service
SSID	service set identifier
STA	station
SWAP	shared wireless access protocol
TAG	Technical Advisory Group
TBTT	targeted beacon transmission times
TCAS	traffic alert and collision avoidance system
TCB	Telecommunications Certification Body
TCM	trellis coded modulation
TCP	transport control protocol
TCS	telephony control protocol specification
TDD	time division duplexing
TDMA	time division multiple access
TH-UWB	time-hopping ultra-wideband
TM-UWB	time-modulated ultra-wideband
TPC	transmit power control
T/R	transmit/receive
TS	tail symbols
TX	transmit
TXD	transmit data
U-NII	Unlicensed National Information Infrastructure
UART	universal asynchronous receiver/transmitter
USB	universal serial bus
UWB	ultra-wideband
VCO	voltage-controlled oscillator
VHF	very high frequency
VoIP	voice over Internet protocol
VOR	VHF omnidirectional ranging
W	watt
WAP	wireless application protocol
WECA	Wireless Ethernet Compatibility Alliance
WEP	wired equivalent privacy
Wi-Fi	Wireless Fidelity
WLAN	wireless local area network
WPAN	wireless personal area network
WRC	World Radiocommunications Conference
XOR	exclusive-OR
ZIF	zero intermediate frequency

INDEX